Nanocarbons for Electroanalysis

Nanocarbon Chemistry and Interfaces

Series Editor

Nianjun Yang, Institute of Materials Engineering, University of Siegen, Germany

Titles in the Series

Nanocarbons for Electroanalysis

*Sabine Szunerits, Rabah Boukherroub, Alison Downard,
Jun-Jie Zhu*

Forthcoming Titles

Carbon Nanomaterials for Bioimaging, Bioanalysis and Therapy

Huan-Cheng Chang, Yuen Yung Hui, Haifeng Dong, Xueji Zhang

**Novel Carbon Materials and Composites: Synthesis,
Properties and Applications**

Xin Jiang, Zhenhui Kang, Xiaoning Guo, Hao Zhuang

Nanocarbon Electrochemistry

Nianjun Yang, Guohua Zhao, John S. Foord

Nanocarbons and their Hybrids

Jean-Charles Arnault, Dominik Eder

Nanocarbons for Electroanalysis

Edited by

Sabine Szunerits
Institute of Electronics
Microelectronics and Nanotechnology
(IEMN)
Villeneuve d'Ascq
France

Rabah Boukherroub
Institute of Electronics
Microelectronics and Nanotechnology
(IEMN)
Villeneuve d'Ascq
France

Alison Downard
Department of Chemistry
University of Canterbury
Christchurch, New Zealand

Jun-Jie Zhu
School of Chemistry and Chemical Engineering
Nanjing University
Nanjing, China

Registered Office(s)
John Wiley & Sons, Inc., 111 River Street, Hoboken, NJ 07030, USA
John Wiley & Sons Ltd, The Atrium, Southern Gate, Chichester, West Sussex, PO19 8SQ, UK

Editorial Office
9600 Garsington Road, Oxford, OX4 2DQ, UK

For details of our global editorial offices, customer services, and more information about Wiley products visit us at www.wiley.com.

Wiley also publishes its books in a variety of electronic formats and by print-on-demand. Some content that appears in standard print versions of this book may not be available in other formats.

Library of Congress Cataloging-in-Publication Data applied for
ISBN: 9781119243908

Cover Design: Wiley
Cover Image: © adventtr/Gettyimages

Set in 10/12pt WarnockPro by SPi Global, Chennai, India
Printed and bound in Malaysia by Vivar Printing Sdn Bhd
10 9 8 7 6 5 4 3 2 1

Contents

List of Contributors   *ix*
Series Preface   *xiii*
Preface   *xv*

1   **Electroanalysis with Carbon Film-based Electrodes**   *1*
Shunsuke Shiba, Tomoyuki Kamata, Dai Kato and Osamu Niwa
1.1   Introduction   *1*
1.2   Fabrication of Carbon Film Electrodes   *2*
1.3   Electrochemical Performance and Application of Carbon Film Electrodes   *4*
1.3.1   Pure and Oxygen Containing Groups Terminated Carbon Film Electrodes   *5*
1.3.2   Nitrogen Containing or Nitrogen Terminated Carbon Film Electrodes   *8*
1.3.3   Fluorine Terminated Carbon Film Electrode   *11*
1.3.4   Metal Nanoparticles Containing Carbon Film Electrode   *13*
      References   *19*

2   **Carbon Nanofibers for Electroanalysis**   *27*
Tianyan You, Dong Liu and Libo Li
2.1   Introduction   *27*
2.2   Techniques for the Preparation of CNFs   *28*
2.3   CNFs Composites   *30*
2.3.1   NCNFs   *30*
2.3.2   Metal nanoparticles-loaded CNFs   *32*
2.4   Applications of CNFs for electroanalysis   *32*
2.4.1   Technologies for electroanalysis   *32*
2.4.2   Non-enzymatic biosensors   *33*
2.4.3   Enzyme-based biosensors   *40*
2.4.4   CNFs-based immunosensors   *44*
2.5   Conclusions   *47*
      References   *47*

3   **Carbon Nanomaterials for Neuroanalytical Chemistry**   *55*
Cheng Yang and B. Jill Venton
3.1   Introduction   *55*
3.2   Carbon Nanomaterial-based Microelectrodes and Nanoelectrodes for Neurotransmitter Detection   *57*

3.2.1   Carbon Nanomaterial-based Electrodes Using Dip Coating/Drop Casting Methods *57*

3.2.2   Direct Growth of Carbon Nanomaterials on Electrode Substrates *59*

3.2.3   Carbon Nanotube Fiber Microelectrodes *61*

3.2.4   Carbon Nanoelectrodes and Carbon Nanomaterial-based Electrode Array *62*

3.2.5   Conclusions *64*

3.3   Challenges and Future Directions *65*

3.3.1   Correlation Between Electrochemical Performance and Carbon Nanomaterial Surface Properties *65*

3.3.2   Carbon Nanomaterial-based Anti-fouling Strategies for *in vivo* Measurements of Neurotransmitters *67*

3.3.3   Reusable Carbon Nanomaterial-based Electrodes *70*

3.4   Conclusions *73*

    References *74*

4   **Carbon and Graphene Dots for Electrochemical Sensing** *85*

    Ying Chen, Lingling Li and Jun-Jie Zhu

4.1   Introduction *85*

4.2   CDs and GDs for Electrochemical Sensors *86*

4.2.1   Substrate Materials in Electrochemical Sensing *86*

4.2.1.1   Immobilization and Modification Function *86*

4.2.1.2   Electrocatalysis Function *87*

4.2.2   Carriers for Probe Fabrication *93*

4.2.3   Signal Probes for Electrochemical Performance *95*

4.2.4   Metal Ions Sensing *96*

4.2.5   Small Molecule Sensing *97*

4.2.6   Protein Sensing *100*

4.2.7   DNA/RNA Sensing *101*

4.3   Electrochemiluminescence Sensors *101*

4.4   Photoelectrochemical Sensing *107*

4.5   Conclusions *110*

    References *110*

5   **Electroanalytical Applications of Graphene** *119*

    Edward P. Randviir and Craig E. Banks

5.1   Introduction *119*

5.2   The Birth of Graphene *120*

5.3   Types of Graphene *122*

5.4   Electroanalytical Properties of Graphene *124*

5.4.1   Free-standing 3D Graphene Foam *124*

5.4.2   Chemical Vapour Deposition and Pristine Graphene *125*

5.4.3   Graphene Screen-printed Electrodes *127*

5.4.4   Solution-based Graphene *129*

5.5   Future Outlook for Graphene Electroanalysis *132*

    References *133*

6   **Graphene/gold Nanoparticles for Electrochemical Sensing**   *139*
          Sabine Szunerits, Qian Wang, Alina Vasilescu, Musen Li and Rabah Boukherroub
6.1   Introduction   *139*
6.2   Interfacing Gold Nanoparticles with Graphene   *141*
6.2.1   *Ex-situ* Au NPs Decoration of Graphene   *142*
6.2.2   *In-situ* Au NPs Decoration of Graphene   *143*
6.2.3   Electrochemical Reduction   *145*
6.3   Electrochemical Sensors Based on Graphene/Au NPs Hybrids   *146*
6.3.1   Detection of Neurotransmitters: Dopamine, Serotonin   *146*
6.3.2   Ractopamine   *151*
6.3.3   Glucose   *152*
6.3.4   Detection of Steroids: Cholesterol, Estradiol   *153*
6.3.5   Detection of Antibacterial Agents   *154*
6.3.6   Detection of Explosives Such as 2, 4, 6-trinitrotoluene (TNT)   *154*
6.3.7   Detection of NADH   *154*
6.3.8   Detection of Hydrogen Peroxide   *155*
6.3.9   Heavy Metal Ions   *156*
6.3.10   Amino Acid and DNA Sensing   *156*
6.3.11   Detection of Model Protein Biomarkers   *157*
6.4   Conclusion   *161*
          Acknowledgement   *162*
          References   *162*

7   **Recent Advances in Electrochemical Biosensors Based on Fullerene-C60
          Nano-structured Platforms**   *173*
          Sanaz Pilehvar and Karolien De Wael
7.1   Introduction   *173*
7.1.1   Basics and History of Fullerene (C60)   *174*
7.1.2   Synthesis of Fullerene   *175*
7.1.3   Functionalization of Fullerene   *175*
7.2   Modification of Electrodes with Fullerenes   *176*
7.2.1   Fullerene (C60)-DNA Hybrid   *177*
7.2.1.1   Interaction of DNA with Fullerene   *178*
7.2.1.2   Fullerene for DNA Biosensing   *179*
7.2.1.3   Fullerene as an Immobilization Platform   *179*
7.2.2   Fullerene(C60)-Antibody Hybrid   *183*
7.2.3   Fullerene(C60)-Protein Hybrid   *185*
7.2.3.1   Enzymes   *185*
7.2.3.2   Redox Active Proteins   *188*
7.3   Conclusions and Future Prospects   *190*
          References   *191*

8   **Micro- and Nano-structured Diamond in Electrochemistry: Fabrication
          and Application**   *197*
          Fang Gao and Christoph E. Nebel
8.1   Introduction   *197*
8.2   Fabrication Method of Diamond Nanostructures   *198*

8.2.1   Reactive Ion Etching   *198*
8.2.2   Templated Growth   *200*
8.2.3   Surface Anisotropic Etching by Metal Catalyst   *204*
8.2.4   High Temperature Surface Etching   *204*
8.2.5   Selective Material Removal   *206*
8.2.6   sp^2-Carbon Assisted Growth of Diamond Nanostructures   *207*
8.2.7   High Pressure High Temperature (HPHT) Methods   *209*
8.3   Application of Diamond Nanostructures in Electrochemistry   *209*
8.3.1   Biosensors Based on Nanostructured Diamond   *209*
8.3.2   Energy Storage Based on Nanostructured Diamond   *211*
8.3.3   Catalyst Based on Nanostructured Diamond   *214*
8.3.4   Diamond Porous Membranes for Chemical/Electrochemical Separation
         Processes   *216*
8.4   Summary and Outlook   *218*
         Acronyms   *219*
         References   *219*

9   **Electroanalysis with C$_3$N$_4$ and SiC Nanostructures**   *227*
         Mandana Amiri
9.1   Introduction to g-C$_3$N$_4$ *227*
9.2   Synthesis of g-C$_3$N$_4$ *229*
9.3   Electrocatalytic Behavior of g-C$_3$N$_4$ *231*
9.4   Electroanalysis with g-C$_3$N$_4$ Nanostructures   *233*
9.4.1   Electrochemiluminescent Sensors   *233*
9.4.2   Photo-electrochemical Detection Schemes   *236*
9.4.3   Voltammetric Determinations   *239*
9.5   Introduction to SiC   *241*
9.6   Synthesis of SiC Nanostructures   *243*
9.7   Electrochemical Behavior of SiC   *244*
9.8   SiC Nanostructures in Electroanalysis   *246*
9.9   Conclusion   *250*
         Acknowledgements   *250*
         References   *250*

         Index   *259*

List of Contributors

Mandana Amiri
University of Mohaghegh Ardabili
Iran

Craig E. Banks
Manchester Metropolitan University
Manchester
UK

Rabah Boukherroub
Institute of Electronics
Microelectronics and Nanotechnology
(IEMN)
Villeneuve d'Ascq
France

Ying Chen
School of Chemistry and Chemical
Engineering
Nanjing University
China

Karolien De Waelt
AXES Research Group
Department of Chemistry
University of Antwerp
Belgium

Fang Gao
Fraunhofer Institute
Freiburg
Germany

Tomoyuki Kamata
National Institute of Advanced
Industrial Science and Technology
Tsukuba
Ibaraki
Japan

and

Chiba Institute of Technology
Japan

Dai Kato
National Institute of Advanced
Industrial Science and Technology
Tsukuba
Ibaraki
Japan

Libo Li
School of Agricultural Equipment
Engineering
Institute of Agricultural Engineering
Jiangsu University
China

Lingling Li
School of Chemistry and Chemical
Engineering
Nanjing University
China

Musen Li
Key Laboratory for Liquid–solid
Structural Evolution and Processing of
Materials
Shandong University
Jinan
China

Dong Liu
School of Agricultural Equipment
Engineering
Institute of Agricultural Engineering
Jiangsu University
China

Christoph Nebel
Fraunhofer Institute
Freiburg
Germany

Osamu Niwa
Advanced Science and Research
Laboratory
Saitama Institute of Technology
Japan

and

National Institute of Advanced
Industrial Science and Technology
Tsukuba
Ibaraki
Japan

Sanaz Pilehvar
AXES Research Group
Department of Chemistry
University of Antwerp
Belgium

Edward Randviir
Manchester Metropolitan University
Manchester
UK

Shunsuke Shiba
Advanced Science and Research
Laboratory
Saitama Institute of Technology
Japan

and

National Institute of Advanced
Industrial Science and Technology
Tsukuba
Ibaraki
Japan

and

Chiba Institute of Technology
Japan

Sabine Szunerits
Institute of Electronics, Microelectronics
and Nanotechnology (IEMN)
University of Lille
Villeneuve d'Ascq
France

Alina Vasilescu
International Center of Biodynamics
Bucharest
Romania

B. Jill Venton
Department of Chemistry
University of Virginia
Charlottesville
Virginia
USA

Qian Wang
Key Laboratory for Liquid–solid
Structural Evolution and Processing of
Materials
Shandong University
Jinan
China

Cheng Yang
Department of Chemistry
University of Virginia
USA

Tianyan You
School of Agricultural Equipment
Engineering
Institute of Agricultural Engineering
Jiangsu University
China

Jun-Jie Zhu
School of Chemistry and Chemical
Engineering
Nanjing University
China

Series Preface

Carbon, the 6th element in the periodic table, is extraordinary. It forms a variety of materials because of its ability to covalently bond with different orbital hybridizations. For millennia, there were only two known substances of pure carbon atoms: graphite and diamond. In the mid-1980s, a soccer-ball shaped buckminsterfullerene, namely a new carbon allotrope C_{60}, was discovered. Together with later found fullerene-structures (C_{70}, C_{84}), the nanocarbon researcher was spawned. In the early 1990s, carbon nanotubes were discovered. They are direct descendants of fullerenes and capped structures composed of 5- and 6-membered rings. This was the next major advance in nanocarbon research. Due to their groundbreaking work on these fullerene materials, Curl, Kroto and Smalley were awarded the 1996 Nobel Prize in Chemistry. In the beginning of the 2000s, graphene was prepared using Scotch tape. It is a single sheet of carbon atoms packed into a hexagonal lattice with a bond distance of 0.142 nm. For their seminal work with this new nanocarbon material, Geim and Novoselov were awarded the 2010 Nobel Prize in Physics. As new members, carbon nanoparticles, such as diamond nanoparticles, carbon dots, and graphene (quantum) dots, have emerged in the family of nanocarbon materials. Although all these materials only consist of the same carbon atoms, their physical, chemical, and engineering features are different, which are fully dependent on their structures.

The purpose of this series is to bring together up-to-date accounts of recent developments and new findings in the field of nanocarbon chemistry and interfaces, one of the most important aspects of nanocarbon research. The carbon materials covered in this series include diamond, diamond nanoparticles, graphene, graphene-oxide, graphene (quantum) dots, carbon nanotubes, carbon fibers, fullerenes, carbon dots, carbon composites, and their hybrids. The formation, structure, properties, and applications of these carbon materials are summarized. Their relevant applications in the fields of electroanalysis, biosensing, catalysis, electrosynthesis, energy storage and conversion, environment sensing and protection, biology and medicine are highlighted in different books.

I certainly want to express my sincere thanks to Miss Sarah Higginbotham from Wiley's Oxford office. Without her efficient help or her valuable suggestions during this book project, the publication of this book series would not be possible.

Last, but not least, I want to thank my family, especially my wife, Dr. Xiaoxia Wang and my children Zimo and Chuqian, for their constant and strong support as well as for their patience in letting me finalize such a book series.

February 2017

Nianjun Yang
Siegen,
Germany

Preface

Recent developments in materials science and nanotechnology have propelled the development of a plethora of materials with unique chemical and physical properties. Carbon-based nanomaterials such as carbon nanotubes, carbon dots, carbon nanofibers, fullerenes and, more recently graphene, reduced graphene oxide and graphene quantum dots have gained a great deal of interest for different applications including electroanalytical applications. Diamond nanostructures as well as silicon carbide and carbon nitride nanostructures have to be added to the spectrum of carbon-based nanomaterials widely used nowadays for electrochemical sensing.

It is the objective of this book to present the most widely employed carbon-based electrode materials and the numerous electroanalytical applications associated with them. It seems that several elements underlie research in electroanalysis today. Advances made in nanotechnology and nanosciences have made the fabrication of novel carbon-based materials and their deposition onto electrical interfaces in the form of thin and 3D films possible. The different nanostructures of electrodes have led to a wealth of electrical interfaces with improvements in terms of sensitivity, selectivity, long-term stability and reproducibility together with the possibility for mass construction in good quantities at low cost. Besides the exceptional physico-chemical features of these materials, the presence of abundant functional groups on their surface and good biocompatibility make them highly suitable for electroanalysis. This has motivated a number of researchers over the last decade to explore different chemical and physical routes to obtain nanomaterials with superior electrochemical properties.

The first part of the book deals with the value of carbon nanomaterials in the form of fibres, particles and thin films for electroanalysis. Chapter 1 (by Osama Niwa) explores the properties of nanocarbon films for electroanalysis. Chapter 2 (by Tianyan You, Dong Liu and Libo Li) reviews electroanalytical application of carbon nanofibers and related composites. The state of the art of the fabrication of carbon nanofibers will be provided followed by an overview their applications for the construction of non-enzymatic and enzyme-based biosensors as well as immunosensors. The value of carbon nanomaterials for neuroanalytical chemistry is presented in Chapter 3 (by Chen Yang and Jill Venton). The high electrocatalytic activity of neurotransmitters such as dopamine on carbon surfaces allows for the development of highly sensitive direct neurotransmitter detection. The challenges towards implementing the electrodes routinely *in vivo* will be discussed furthermore. This first part will be concluded by Chapter 4 (by Junjie Zhu, Lingling Li and Ying Chen) on the use of carbon and graphene dots for electrochemical analysis.

The second part of the book considers the value of graphene for electroanalytical applications. Chapter 5 (by Edward Randviir and Craig Banks) gives an excellent insight into the use of graphene for electoanalysis. This chapter discusses the origins of graphene, the types of graphene available and their potential electroanalytical properties of the many types of graphene available to the researcher today. Chapter 6 (by Sabine Szunerits and Rabah Boukherroub) demonstrates that loading of graphene nanosheets with gold nanoparticles generates a new class of functional materials with improved properties and thus provides new opportunities of such hybrid materials for catalytic biosensing.

The use of the most recent applications of fullerene-C60 based electrochemical biosensors is presented in Chapter 7 (by Sanaz Pilehavar and Karolien De Wael) Taking into account the biocompatibility of fullerene-C60, different kind of biomolecules such as microoganisms, organelle, and cells can be easily integrated in biosensor fabrication making the interfaces of wide interest.

The third part of the book describes the value of diamond and other carbon-based nanomaterials such as carbon nitride (C_3N_4) and silicon carbide (SiC). Chapter 8 (by Christophe Nebel) is focused on the different aspects of diamond nanostructures for electrochemical sensing. Chapter 9 (by Mandana Amiri) is focused on the interest of carbon nitrides and silicon carbide nanoparticles for the fabrication of new electroanalytical sensing platforms.

It is hoped that this collection of papers provides an overview of a rapidly advancing field and are resources for those whose research and interests enter into this field either from sensing or material scientific perspectives. While many topics are presented here, there are many that were not able to be included but are also of current interest or are emerging. All of the contributors are thanked for their brilliant and valuable contributions.

June 2017

Sabine Szunerits
Villeneuve d'Ascq
France

Rabah Boukherroub
Villeneuve d'Ascq
France

Alison Downard
Christchurch
New Zealand

Jun-Jie Zhu
Nanjing
China

1

Electroanalysis with Carbon Film-based Electrodes

Shunsuke Shiba[1,2,3], Tomoyuki Kamata[2,4], Dai Kato[2] and Osamu Niwa[1,2]

[1] *Advanced Science and Research Laboratory, Saitama Institute of Technology, Japan*
[2] *National Institute of Advanced Industrial Science and Technology, Ibaraki, Japan*
[3] *Graduate School of Pure and Applied Sciences, University of Tsukuba, Ibaraki, Japan*
[4] *Chiba Institute of Technology, Japan*

1.1 Introduction

As electrode materials for analytical applications, carbon-based electrodes have been widely employed as detectors for high performance liquid chromatography (HPLC), capillary electrophoresis (CE) and various biosensors. Carbon materials usually shows wider potential window compared with those of novel metals such as platinum and gold electrode. These electrodes are chemically stable, highly conductive and low cost. A recent review article has well described the electrochemistry of certain carbon-based electrodes [1]. Glassy carbon (GC) and highly oriented pyrolytic graphite (HOPG) have been traditionally utilized for various electroanalytical methods. Later, carbon paste electrodes have been used mainly to develop enzymatic biosensors because carbon paste is low cost and the electrode can be fabricated only by printing and various bio-molecules can be modified only by mixing with carbon ink.

In the last 20 years, electrochemical measurements using boron-doped diamond (BDD) electrodes have become more intensively studied by many groups [2–4]. A BDD electrode shows extremely wider potential window due to its chemical stability and lower background noise level than other electrode materials. Due to such unique performances, BDD electrodes are advantageous in terms of detecting various species including heavy metal ions (Pb^{2+}, Cd^{2+}) [5], chlorinated phenols [6], histamine and serotonin [7, 8], and even nonmetal proteins [9]. The BDD electrodes have also been employed to fabricate modified electrodes including As^{3+} detection with iridium-implanted BDD [10], DNA modified BDD [11] and cytochrome c modified BDD [12]. In spite of excellent performance of BDD electrodes, high temperature between 400–700°C is needed for BDD fabrication, which limits the substrates only to inorganic materials such as silicon wafer, metals and glass plate.

More recently, nanocarbon materials including carbon nanotubes (CNTs), carbon nanofibers (CNFs) and graphene nanosheet have been more intensively studied with a view to using them as electrode materials for fuel and biofuel cells [13–15]. For electroanalytical application CNT and graphene have been employed to fabricate various

Nanocarbons for Electroanalysis, First Edition.
Edited by Sabine Szunerits, Rabah Boukherroub, Alison Downard and Jun-Jie Zhu.
© 2017 John Wiley & Sons Ltd. Published 2017 by John Wiley & Sons Ltd.

biosensors because nanocarbon electrodes have large surface area suitable to immobilize large amount of enzymes and antibodies [16–20]. The surface area of such nanocarbon film with immobilizing large amount of biomolecules can achieve sufficient sensitivity and longer stability. More recently, the graphene was modified onto interdigitated array electrode and applied for electrochemical immunoassay [21].

In spite of some works using nanocarbons as film electrode, the nanocarbon materials have been mainly used by modifying them on the solid electrode and larger surface area of nanocarbons also show large capacitive and background currents and reduce signal to noise (S/N) ratio when detecting trace amount of analytes.

In contrast, carbon film electrodes have been used for direct measurement of electroactive molecules such as neurotransmitters and nucleic acids. Various kinds of carbon film materials have been developed using various fabrication processes including pyrolysis of organic films, sputter deposition, chemical vapor deposition. However, carbon film electrodes are needed to improve the electron transfer rate of analytes in order to retain diffusion-limited electrochemical reactions because their smooth surface has fewer active sites than the surfaces of nanocarbon materials. Therefore, it is required to fabricate carbon films with better electroactivity. Another important advantage is that carbon film can be patterned to any shape and size with high reproducibility for use as platforms for chemical or biochemical sensors by utilizing conventional photolithographic process [22]. In this chapter, the fabrication processes of carbon film electrodes are introduced. Then, we described structure and electrochemical properties of various carbon film electrodes. Finally, we describe the application of carbon film electrodes for electroanalysis of mainly biomolecules.

1.2 Fabrication of Carbon Film Electrodes

In order to fabricate carbon film electrodes, the pyrolysis of organic films including various polymers and deposited aromatic compounds have been employed by many groups as summarized in Table 1.1.

Kaplan *et al.* deposited 3, 4, 9, 10-perylenetetracarboxylic dianhydride (PTDA) films on the substrate, pyrolyzed them above 700° C and obtained conducting carbon film [23]. The conductivity was comparable to that of a GC electrode. Rojo *et al.* obtained carbon film using a similar method to Kaplan *et al.* and employed it for electrochemical measurements of catechol and catecholamines [24]. Tabei and Niwa *et al.* employed this process to microfabricate interdigitated array electrodes by lithographic technique [25].

The conducting polymers are also suitable to make highly conducting carbon film because the film already has π–conjugated structure. Tabei *et al.* used poly(p-Phenylene Vinylene):PPV coated on the substrate and prepared carbon film electrode by the pyrolysis at 1100° C, then fabricated to microdisk array electrode [26]. The carbon films have been fabricated by pyrolyzing conventional polymers. Positive photoresist, which mainly consist of phenol resin was used as precursor polymer and pyrolyzed the film at high temperature because positive photoresist can be easily spin-coated into uniform films [27]. The resistivity was between 2×10^{-2} to 2×10^{-3} Ω cm depending on the pyrolysis temperature. The electrochemical performance of pyrolyzed photoresist films (PPF) has been intensively studied by McCreery and Madou's groups [28, 29]. PPF film

Table 1.1 Fabrication of carbon film electrodes by pyrolysis process.

Carbon film	Procedures and properties	References
Pyrolysis of PTDA[1]	PTDA is deposited in quartz tube and pyrolyzed at 850°C at 0.01 torr Conductivity :250 S cm^{-1} (Kaplan et al.)	Kaplan *et al.* 1980 [23], Rojo *et al.* 1986 [24]
Pyrolyzed poly-(phenylene vinylene) film	Microdisk electrode from pyrolyzed PPV films around 1100°C	Tabei et al., 1993 [26]
Pyrolysis of phenol-formaldehyde resin around 1000°C	Spin-coat with phenolic resin solution on the substrate and pyrolysis at 800 or 1050°C. Conductivity: from 2×10^{-2} to $2 \times 10^{-3} \Omega$ cm	Lyons *et al.* 1983 [27]
	Pyrolysis of photoresist AZ4330 from 600 to 1100°C. Near atomic flatness <0.5 nm	Kim et al., 1998 [28] Ranganathan *et al.* 2001 [29]
	Pyrolysis of photoresist AZ4620 at 1100°C. Conductivity comparable to GC	Brooksby *et al.* 2004 [30]
	Pyrolysis of photoresist AZ4562 by rapid Thermal process (140°C min^{-1} to 1000°C.	Campo *et al.* 2011 [31]
Pyrolyzed polyimide film	IDA electrode fabricated by pyrolysis of thick polyimide films and photolithography on quartz substrates.	Morita *et al.* 2015 [32]

1) 3, 4, 9, 10-perylenetetracarboxylic dianhydride.

has a lower O/C ratio than a GC electrode and relatively larger peak separations were observed from the voltammograms of $Fe^{3+/2+}$ and DA. The carbon film obtained by photoresist has very smooth surface. In fact, Ranganathan *et al.* observed that the average roughness is less than 0.5 nm by the atomic force microscopy (AFM) measurement of PPF carbon film. The modification of PPF film by diazonium reduction was performed by Brooksby *et al.* [30]. The modification of such carbon films is very important to use them as platforms of various electrochemical biosensors. More recently, the relationship between fabrication processes of PPF such as types of resists, and heating programs, and their resistivity and surface roughness, were well summarized by Compton's group [31]. Morita *et al.* carbonized polyimide (PI) film and fabricated IDA electrode [32]. The height of the electrode is ranging from 0.1 to 4.5 µm since PI is suitable to obtain thicker film.

On the other hand, carbon film electrodes have been developed by using various vacuum deposition techniques including magnetron or radio frequency (RF) or electron cyclotron resonance sputtering deposition, electron beam evaporation, plasma-assisted chemical vapor deposition (PACVD), radio-frequency plasma enhanced chemical vapor deposition (RF-PECVD). Most well known carbon film is diamond like carbon(DLC), which is very widely used for coating of drills and cutting tools because DLC is extremely hard. Smooth and inert surface of DLC is also suitable to improve biocompatibility and applied for the coating of medical devices. A Ternary phase diagram of amorphous carbons including DLC was reported by Ferrari *et al.* [33] .

As an electrode materials, DLC shows high S/N ratio and low capacitance [34]. Blackstock *et al.* reported ultraflat carbon film (~0.1 nm) whose electrochemical response is similar to that of GC [35]. Swains' group has been studied nitrogen-containing amorphous carbon films and their electrochemical performance as discussed in the later section [36]. Hirono *et al.* developed a very smooth and hard carbon film using electron cyclotron resonance (ECR) sputtering [37]. The film consists of sp^2 and sp^3 hybrid bonds with a nanocrystalline structure and the sp^2 and sp^3 ratio can be easily controlled by changing ion acceleration voltage from 20 to 85 V. Figure 1.1 shows surface image and line scan data of ECR sputtered carbon film obtained by AFM. The average roughness (Ra) is 0.07 nm, indicating atomic level flatness [38]. The film contains nanocrystalline graphite like structure different from amorphous carbon film, which contributes to improve electrochemical performance as described later. In fact, a parallel layered structure identified as a nano-order graphite crystalline structure can be observed at a low ion acceleration voltage, but a curved and closed nanostructure is dominant at a high ion acceleration voltage. More recently, Kamata *et al.* fabricated the

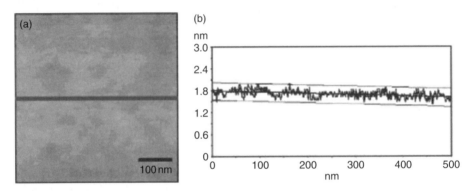

Figure 1.1 AFM image of ECR sputtered carbon surface (a) and line profile (b). Reprinted with permission from [38]. Copyright 2006 American Chemical Society.

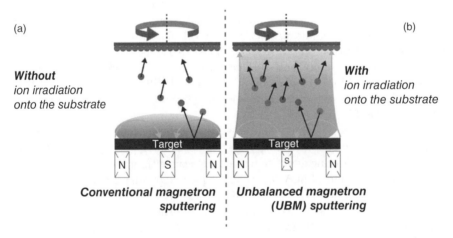

Figure 1.2 Comparison of schematic diagram of UBM sputtering equipment (b) compared with conventional magnetron sputtering (a).

carbon film with similar structure and electrochemical properties to those of ECR nanocarbon film by using unbalanced magnetron (UBM) sputtering [39]. Figure 1.1 shows schematic diagram of UBM sputtering equipment (Figure 1.2a) compared with conventional magnetron sputtering (Figure 1.2b).

The plasma is only distributed near the target in case of conventional magnetron sputtering. In contrast, the plasma is distributed near the substrate and the ion irradiation occurs onto the substrate, which can widely control the structure of carbon film including sp^3 and sp^2 ratio.

1.3 Electrochemical Performance and Application of Carbon Film Electrodes

When fabricating carbon film based electrode, other atoms such as nitrogen and oxygen or even metal nanoparticles can be contained. For example, nitrogen doping can be performed in the presence of small amount of N_2 during vacuum process. Surface termination with other atoms such as hydrogen and nitrogen can be easily performed because the conducting carbon film contains certain amount of sp^2 bonds, which is chemically reactive. Metal nanoparticles which usually show better electrocatalytic performance for analytes have been developed by pyrolysis and vacuum technique. In this section, the electrochemical performance and applications of pure, surface terminated and hybrid carbon films are summarized.

1.3.1 Pure and Oxygen Containing Groups Terminated Carbon Film Electrodes

The carbon films prepared by pyrolyzing organic and polymer films usually contains graphite layers. Figure 1.3a is Raman spectra of the carbon film prepared by Niwa *et al.* [25] on the basis of the process reported by Rojo *et al.* [24].

The two relatively broad peaks were observed at 1590 and 1340 cm^{-1}, and assigned to disordered graphite structure. As an electrode material, Rojo *et al.* reported that the electrochemical response of catechol is irreversible, but became ideal after electrochemical treatment at 1.8 V. Figure 1.3b compared voltammograms of 100 µM dopamine (DA) at the carbon-based IDA electrode before (1) and after (2) electrochemical treatment. Carbon film-based IDA was fabricated by photolithographic technique. After electrochemical treatment, the current increases more rapidly compared with that before treatment. The electrochemical pretreatment increases surface area caused by etching the surface and introduces oxygen containing groups.

In contrast, carbon films prepared by vacuum process have wide variety of the structure as described above. Figure 1.4 shows relationship between potential window and sp^3 [$sp^3/(sp^2+sp^3)$] concentration of the UBM sputtered nanocarbon film. The width of potential window increases with increasing sp^3 ratio [39]. However, the peak separations of $Fe(CN)_6^{4-}$ and DA becomes larger when sp^3 concentration is around 50%. The wide potential window of UBM sputtered nanocarbon film electrode is advantageous to measure biomolecules with high oxidation potential.

The flat surface of nanocarbon film also contributes to suppress the fouling of electrode surface. With a conventional electrode such as a GC electrode, the relatively rougher surface adsorbs the molecules. In contrast, the molecules easily desorbed from

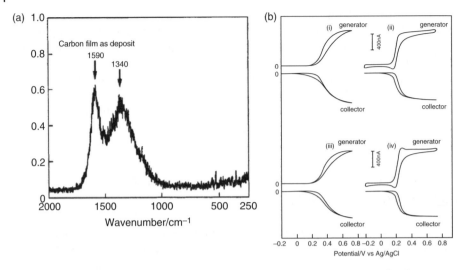

Figure 1.3 (a) Raman spectrum of carbon film deposited on an oxldlzed silicon wafer. (b) Generation-collection voltammograms of 100 µM dopamine in pH 6 phosphate buffer at carbon-based IDA electrodes with different pretreatment conditions: (i) neither electrode pretreated; (ii) generator electrode pretreated; (iii) collector electrode pretreated; (iv) both electrodes pretreated. The collector potential was held at –0.2 V, and the generator potential was cycled at a scan rate of 50 mV s^{-1}. The IDA bandwidth and gap are 3 and 2 µm, respectively. Adapted with permission from [25]. Copyright 1994 American Chemical Society.

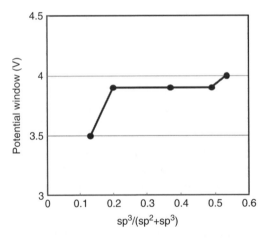

Figure 1.4 Relationship between potential window and sp^3 [sp^3/(sp^2+sp^3)] concentration of the UBM sputtered nanocarbon film.

the nanocarbon film electrode surface after electrochemical reaction because of its flat and chemically stable surface. For example, we achieved much better reproducibility and detection limit compared with GC when measuring 8-OHdG which is known as oxidative stress marker [40]. The suppression of fouling can be enhanced at hydrophilic surface. The electrochemical treatment simply introduces oxygen containing groups, which can be confirmed by reduction of contact angle and XPS measurements [41].

The electrochemical treatment of the carbon electrodes such as GC often make the surface very rough, but nanocarbon film still maintain smooth surface after electrochemical treatment. The electrochemical response of serotonin and thiol was greatly improved after electrochemical treatment at ECR nanocarbon film electrode. This performance is particularly advantageous when measuring biomolecules with large molecular weight since large biomolecules often strongly adsorb on the electrode surface and interfere with the electron transfer between the analytes and electrode. Simple electrochemical DNA analysis techniques such as DNA methylation [42] and single nucleotide polymorphism (SNP) [43] detection have been reported based on the quantitative measurement of all the bases by direct electrochemical oxidation. Figure 1.5 shows background-subtracted differential pulse voltammograms (DPVs) of 3 μM of oligonucleotides (**1**: 5′-CAG-CAG-CAG-3′, **2**: 5′- CAG-CA<u>A</u>-CAG -3′, **3**: 5′- CA<u>A</u>-CA<u>A</u>-CAG -3′, **4**: 5′- CA<u>A</u>-CA<u>A</u>-CA<u>A</u>-3′, the underline base represents a mismatch base) at the nanocarbon film electrode.

The peaks assigned by G oxidation decreases and A oxidation increases with increasing number of A in the oligonucleotide. However, the oxidation of C cannot be observed at GC electrode due to narrower potential window compared with those at ECR

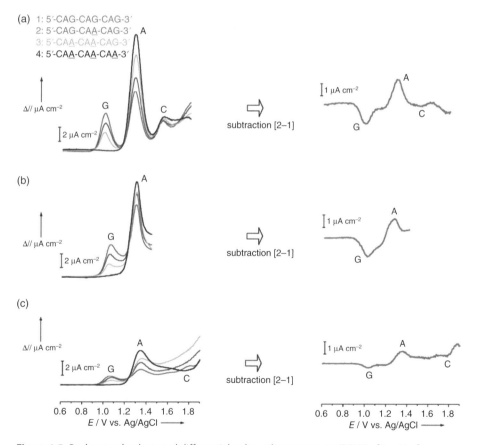

Figure 1.5 Background-subtracted differential pulse voltammograms (DPVs) of 3 mM of oligonucleotides (**1**: 5′-CAG-CAG-CAG-3′, **2**: 5′- CAG-CA<u>A</u>-CAG -3′, **3**: 5′- CA<u>A</u>-CA<u>A</u>-CAG -3′, **4**: 5′- CA<u>A</u>-CA<u>A</u>-CA<u>A</u>-3′) at the (a) ECR nanocarbon film, GC, and BDD electrodes in 50 mM pH5.0 acetate buffer.

nanocarbon film and BDD electrodes. We also observed that oxidation current of oligonucleotide reduced rapidly by continuous measurement at GC, but not at ECR nanocarbon films due to their flat and hydrophilic surface. The response of each base is sharper at ECR nanocarbon film compared with BDD, indicating relatively rapid electron transfer. Furthermore, we also measured each base content of longer oligonucleotides (60mers) that constitute a non-methylated and a methylated CpG dinucleotide with some different methylation ratios [44].

1.3.2 Nitrogen Containing or Nitrogen Terminated Carbon Film Electrodes

It has been reported nitrogen containing carbon materials shows interesting electrocatalytic performances, particularly oxygen reduction reaction(ORR). Ozakis' group developed carbon alloy which enhances oxygen reduction activity by simultaneous doping of boron and nitrogen into carbon materials [45]. In particular, nitrogen doped carbon materials have been studied by many groups to apply as electrodes for fuel cell. In 2009, Dai *et al.* reported nitrogen-doped carbon nanotube arrays which show high ORR activity and long time stability [46]. Their group also developed nitrogen-doped graphene by thermally annealed with ammonia and realized n-type field-effect transistor at room temperature [47]. More recently, Uchiyama *et al.* observed hydrogen oxidation wave using glassy carbon electrode fabricated by stepwise electrolysis in ammonium carbamate aqueous solution and hydrochloric acid [48]. At holding the electrode at 0 V (vs Ag/AgCl), the oxidation current increases by bubbling hydrogen gas and decreases after stopping hydrogen gas supply.

Beside such bulk carbons and nanocarbon materials, nitrogen containing carbon film electrodes have been studied by many groups because the films have a wide variety of structure such as sp^2/sp^3 ratio and show improved electrocatalytic activity. Yoo *et al.* reported that nitrogen-incorporated tetrahedral amorphous carbon electrode shows more active charge transfer properties on a variety of systems relative to the H-terminated BDD and excellent stability [49]. Swain's group reported the nitrogen-doped nanocrystalline diamond thin-film deposited by Gruen and co-workers using microwave-assisted chemical vapor deposition (CVD) from C_{60}/argon and methane/nitrogen gas mixtures consisted of hemispherical features about 150 nm in diameter with a height of 20 nm [50]. The film is active for redox species such as $Fe(CN)_6^{3-/4-}$ and $Ru(NH_3)_6^{2+/3+}$ without any conventional pretreatment and shows semimetallic electronic properties between 0.5 and −1.5 V (vs. Ag/AgCl). The same group also fabricated similar film electrode by plasma-enhanced CVD, which also shows high electrochemical activity [51]. Tanaka *et al.* fabricated nitrogen-doped hydrogenated carbon films also by plasma-enhanced CVD and studied their structure by XPS and basic electrochemical properties [52]. In contrast, Lagrini *et al.* used radio-frequency (RF) magnetron sputtering to fabricate amorphous carbon nitride electrode and studied their microstructure and electronic properties such as conductivity [53]. They also studied about correlation between the local microstructure and the electrochemical behavior by using XPS, FTIR, Raman spectroscopy, and electrochemical measurements [54]. The potential windows and voltammograms of $Fe(CN)_6^{3-/4-}$ were changed by changing nitrogen partial pressure during deposition. Hydrogen and oxygen evolution at nitrogen doped amorphous carbon film electrodes formed with a filtered cathodic vacuum arc in a N_2 atmosphere were also studied by Zeng *et al.* [55].

Recently, Yang *et al.* [36] reported electrochemical responses of $Ru(NH_3)_6^{2+/3+}$ and $Fe(CN)_6^{3-/4-}$ at nitrogen-containing tetrahedral amorphous carbon thin-film electrodes by changing N_2 flow rate during deposition. The peak separation of former species was almost unchanged, but the latter shows lower peak separation when N_2 flow rate increases. The resistivity also decreases with increasing incorporation of nitrogen. Kamata *et al.* studied electrochemical properties of nitrogen-containing carbon film electrodes by widely changing nitrogen concentration. The carbon films were fabricated on boron doped silicon wafer by using ECR sputtering or UBM sputtering equipment [56, 57]. The nitrogen concentration was changed from 0 to 30.4% characterized with XPS, and the surface image was obtained with AFM. The surface average roughness was almost unchanged when the nitrogen concentration was widely changed. The sp^3 concentration was 20% for pure nanocarbon film and nanocrystalline layered structures can be observed by TEM as shown in Figure 1.6a. However, sp^3 concentration increases with increasing nitrogen concentration and became 53.8% when nitrogen concentration was 30.4%.

Circle and closed structures containing sp^3 bonds also increases with increasing nitrogen concentration as shown in Figure 1.6b. The potential window of the film becomes wider but the electrochemical activity for $Fe(CN)_6^{3-/4-}$ decreases with increasing nitrogen concentration from 9.0 to 30.4%, although the film containing 9.0% nitrogen shows smaller peak separation than that of pure nanocarbon film despite lower sp^2 concentration. The ORR peak of nitrogen containing nanocarbon film (9.0%) is more positive than that of pure nanocarbon film, suggesting improved electrocatalytic activity (Figure 1.7).

Nitrogen-containing carbon films have been applied for electroanalysis including heavy metal ions and biomolecules. Table 1.2 summarizes examples of electroanalytical applications with nitrogen-containing carbon film electrodes.

Zeng *et al.* applied for the analysis of heavy metal ions including Pb^{2+}, Cd^{2+}, Cu^{2+} by differential pulse anodic stripping voltammetry [58]. A linear dependence of lead concentration between 5×10^{-7} to 2×10^{-6} M was obtained. Swains' group mainly applied their nitrogen-incorporated tetrahedral amorphous carbon thin film electrodes for

Circle and closed structure
sp^2 layer structure

Circle and closed structure

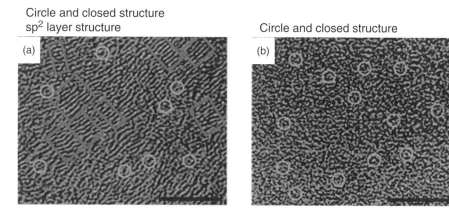

Figure 1.6 Plain views of (a) pure-ECR and (b) N-ECR (N = 9.0 at. %) (b) observed by TEM. N_2 gas contents during deposition are 0 for (a) and 2.5% for (b). Scale bar = 5 nm. Reprinted with permission from [56]. Copyright 1994 American Chemical Society.

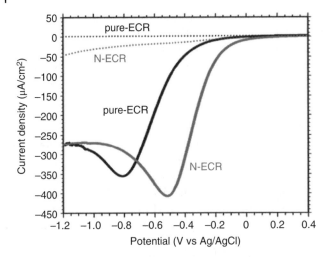

Figure 1.7 Voltammograms of pure-ECR and N-ECR (N = 9.0 at. %) electrodes for oxygen reduction reaction in O_2 saturated 0.5 M H_2SO_4. Dotted lines are background scans. Adapted with permission from [56]. Copyright 2013 American Chemical Society.

Table 1.2 Electroanalytical application of nitrogen containing carbon film electrodes.

Carbon film	Analytes and procedure	References
Nitrogen-doped Diamond-like carbon film by DC magnetron sputter	Differential pulse anodic stripping voltammetry (DPASV)	Zeng *et al.* 2002 [58]
Nitrogen containing nanocarbon film by ECR sputtering	Square wave voltammetry of guanosine and adenosine	Kamata *et al.* 2013 [56]
Nitrogen-incorporated tetrahedral amorphous carbon thin film	Flow injection analysis of norepinephrine	D'N. Hamblin *et al.* 2015 [59]
	Flow injection analysis of tryptophan and tyrosine	Jarosova *et al.* 2016 [60]
	Propranole and hydrochlorothiazide oxidation in standard and synthetic biological fluid.	Lourencao *et al.* 2014 [61]
Nitrogen containing nanocarbon film by ECR and UBMsputtering	Reductive Detection of hydrogen peroxide	Kamata *et al.* 2013 [56] 2015 [57]

detecting small biomolecules. Norepinephrine [59] and tryptophan and tyrosine [60] were detected with their film electrode using flow injection analysis. They also applied to detect pharmaceuticals, propranolol (PROP) and hydrochlorothiazide (HTZ) by square wave voltammetry in standard and synthetic biological fluids [61]. PROP is a non-selective β–adrenergic antagonist drug (blocker) and HTZ is a diuretic drug belonging to the thiazide class. Low detection limits of ~194 ng/ml for PROP and

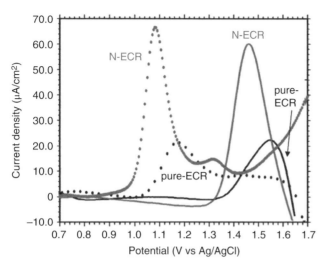

Figure 1.8 Background-subtracted SWVs of 100 μM guanosine (dotted) and adenosine (solid) at pure-ECR and N-ECR (N = 9.0 at. %) electrodes measured in 50 mM acetate buffer (pH 5.0). Adapted with permission from [56]. Copyright 2013 American Chemical Society.

~744 ng/ml for HTZ were obtained. The oxidation peak potentials for guanosine and adenosine were compared at pure ECR nanocarbon and nitrogen containing nanocarbon films by using square wave voltammetry [56]. Much sharper and larger oxidation peaks of both analytes were observed at more negative potential with nitrogen containing nanocarbon film compared with those with pure nanocarbon films as shown in Figure 1.8. In case of hydrogen peroxide detection, the larger reduction peaks can be obtained at more positive potential by containing nitrogen due to improved electrocatalytic activity.

1.3.3 Fluorine Terminated Carbon Film Electrode

Fluorination, one of the most attractive surface terminations, has been reported for various carbon-based electrodes including graphite, GC, carbon nanotube, graphene, and diamond [62–67]. These fluorinated carbon electrodes provide unique characteristics, such as improved hydrophobicity and a different electron transfer rate compared with the original carbon electrodes. However, some fluorinated carbon electrodes have serious problems, including lower stability due to loss of fluorine atoms and/or damage due to oxidation [41, 62, 65]. In contrast, fluorinated BDD electrodes exhibit better long-term stability [64, 67], suggesting that a fluorinated surface containing sp^3 carbon experiences less oxidization and damage during anodic polarization than GC.

To fabricate electrochemically stable fluorine-terminated nanocarbon (F-nanocarbon) film electrodes, the surface of the nanocarbon films was shortly treated with CF_4 plasma [65, 68–70]. The fluorinated surface is easily prepared without losing the surface conductivity and surface flatness of the nanocarbon film electrode as summarized in Table 1.3. After fluorination, the sp^2 content decreased from 58.1 to 45.0 %. At the same time, the F/C ratio was 0.2 [65]. These results clearly indicate that the sp^2 bond is selectively fluorinated by the CF_4 plasma. The contact angle of the film surface increased after surface fluorination (Table 1.3 and Figure 1.9c).

Table 1.3 Surface properties of the O-nanocarbon and F-nanocarbon films.

		Original	O-terminated	F-terminated
C 1s %[a]	sp^2 content	58.1	45.0	45.0
	sp^3 content	41.9	55.0	55.0
F/C [a]		—	—	0.2
O/C [a]		0.06	0.10	0.02
R_a nm^{-1}		0.067	0.21	0.075
Contact angle°		72–75	14	93
C^0 µF cm^{-2} [b]		11.1–12.3	20.1	3.32

a) The chemical components of C, F, and O were obtained and analyzed using XPS analysis.
b) The average roughness (Ra) were obtained by AFM measurements.

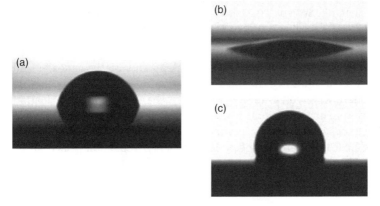

Figure 1.9 Contact angles of (a) the original, (b) O-nanocarbon, and (c) F-nanocarbon films.

The fluorinated electrodes exhibited a C^0 value of 3.32 µF cm^{-2}, which was only one-quarter that of the original nanocarbon film electrode (11.1–12.3 µF·cm^{-2}) [65]. However, the GC electrode was treated with fluorination under the same conditions, and exhibited a larger C^0 value of 9.79 µF cm^{-2}. Extremely low electrochemical double layer capacity is highly advantageous in terms of the S/N ratio to rule measurement limits.

The electron transfer rates at the F-nanocarbon film electrodes was investigated by using some typical redox species such as $Ru(NH_3)_6^{3+/2+}$ and $Fe^{3+/2+}$. The responses of outer-sphere $Ru(NH_3)_6^{3+/2+}$ was almost unchanged after fluorination owing to the high electron transfer rate and its surface insensitivity [65]. On the other hand, the electron transfer rate of $Fe^{3+/2+}$ was greatly affected by surface fluorination. Figure 1.10a shows a CV of $Fe^{3+/2+}$ at the F-nanocarbon film electrode.

It is well known that inner-sphere $Fe^{2+/3+}$ is generally very sensitive to the presence of oxygen-containing functionalities on an sp^2 carbon electrode surface [65, 71]. The F-nanocarbon film electrode provided much larger ΔE_p at the original nanocarbon film electrode. The O/C ratio decreased after fluorination. These results clearly

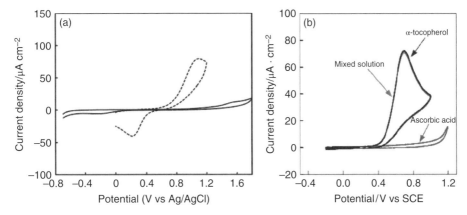

Figure 1.10 (a) Typical CVs of 1.0 mM $Fe^{2+/3+}$ at the original nanocarbon film (dotted lines) and the F-nanocarbon film electrode (solid lines) electrodes, measured in 0.1 M $HClO_4$. at 0.1 V s^{-1}. (b) CVs of a mixed solution of 1 mM ascorbic acid and 1 mM α-tocopherol using the F-terminated nanocarbon film electrodes at 0.1 V s^{-1}. *Source:* Adapted from Terashima 2002 and Tabei 1993.

demonstrated that surface fluorination contributed to give very slow electron transfer of $Fe^{3+/2+}$ unlike that of surface oxygen functionalities.

The F-nanocarbon film electrode can be also used to selectively detect lipophilic antioxidant α-tocopherol (vitamin E), which is one of typical major antioxidant components in liquid foods such as olive oil, in combination with use of bicontinuous microemulsion (BME) [69, 70]. Figure 1.10b shows typical CVs of ascorbic acid (vitamin C), and α-tocopherol, and their mix solution using the F-nanocarbon film electrodes in the BME solution. Irreversible oxidation peaks for lipophilic α-tocopherol was observed. In contrast, electrochemical response for hydrophilic ascorbic acid was effectively suppressed. The oxidation peak observed with the F-nanocarbon film electrodes was proportional to the square root of the scan rate, but not to the scan rate, indicating the simple diffusion control of these species in the BME [69]. Moreover, even in the mixed solution, the peak current at the F-nanocarbon film electrode was in good agreement with that for the α-tocopherol alone. These results indicate that there is no interference. This is highly advantageous in terms of constructing a simple assay of antioxidant because an extraction process is usually required prior to the conventional assay [70]. Indeed, the F-nanocarbon film electrode could provide direct qualitative and quantitative electrochemical analysis of lipophilic antioxidants in some olive oil samples including extra virgin olive oils (EVOOs) as a novel methodology without an extraction process [71].

1.3.4 Metal Nanoparticles Containing Carbon Film Electrode

Although various kind of carbon films have been employed for electroanalysis, it is difficult to realize high electrocatalytic activity similar to novel metal electrodes such as Pt. Carbon-based electrodes (including carbon films) containing small amount of nitrogen atom have been reported to improve ORR or hydrogen peroxide reduction, but the activity is still not as high as that of Pt. It is well known that metal nanoparticles (NPs) such as Pt, Pd and Au have high electrocatalytic activity compared with metal bulk electrodes. Since carbon films usually have low background noise current compared with metal electrodes, the metal NPs modified carbon electrodes realize not only high electrocatalytic activity but also relatively low noise level compared with metal film electrodes.

NPs are in general modified onto the carbon electrode by drop casting after their chemical synthesis, or electrodeposition as summarized in Table 1.4. For example, Compton's group reported the GC electrode with AuNP for As(III) [72] and H_2O_2 [73] detection. They also reported more sensitive As(III) [74] and tinidazole [75] detection by co-modificatin of AuNP and CNT. Such sp^2 bond carbon materials has a problem of low stability and reproducibility due to the passivation of the electrode surface [76], which lead to nanoparticle desorption. BDD electrode has also been preferentially utilized as a nanoparticle scaffold because of its excellent morphological and microstructural stability in addition to low and stable background current [77–81]. BDD thin film electrode modified electrodeposited Au nanoparticles (AuNPs) have been reported by Swain *et al.* [82] In detecting arsenic ion As(III), they co-electrodeposited not only Au ion but also As(III) to form intermetallic Au-As nanoparticles, and then oxidized As(III) component to obtain the reproducible and quantitative stripping peak currents. Similar procedure was also reported by Compton *et al.* for Cd and Pd analysis [78, 79]. Einaga *et al.* also reported AuNP modified BDD electrode. They modified AuNPs during preconcentration step of As(III) to improve preconcentration efficiency [83]. However, since BDD surface is chemically inert, the adsorptive force between nanoparticle and BDD thin film surface is weak. This could result in desorption of the NPs from the electrode surface or their aggregation, in particular, in case of a hydrodynamic system.

In contrast, NPs embedded carbon film electrodes can be fabricated with much simpler processes, also summarized in Table 1.4. In the 1990s, McCreery *et al.* reported the small Pt cluster(~1 nm) dispersed glassy carbon film electrodes prepared by pyrolysis of both carbon and metal precursors at 600° C, and demonstrating high electrocatalytic activity toward the hydrogen evolution reaction and ORR [84]. This method was applied to fabricate carbon nanofiber electrodes containing various metal NPs and nanoalloys in order to apply them for electroanalysis of sugars and ORR. Ion implantation of metals into BDD was reported to fabricate iridium NPs containing BDD electrode [10]. This method can suppress the detachment of NPs from BDD because the most of the NPs were embedded and only part of NPs appeared on the electrode surface. However, ion implantation system is very expensive as equipment of electrode fabrication.

You and Niwa *et al.* reported the metal NPs embedded carbon film electrodes by using radio-frequency co-sputtering by placing metal pellets on the carbon target. Due to poor immiscibility of metals and carbon, various metals such as Pt, Ni, Cu can be employed to fabricate such hybrid film electrodes[85–87]. These electrodes have atomically flat surface, which was characterized by AFM. The carbon matrix composed of disordered graphite-like carbon with partially exposing top body of the NPs, observed by TEM. The NPs of average diameter of 2.5 nm was embedded in the carbon film when Pt was used and that of 4 nm was obtained by using palladium (Pd) as indicated in Figure 1.11 [85, 88].

The former electrode shows high electrocatalytic performance for hydrogen peroxide detection compared with bulk Pt electrode. Figure 1.12 shows amperometric response of glucose at glucose oxidase (GOD) modified Pt-NPs embedded carbon film electrode compared with that at GOD modified Pt bulk electrode. Due to low Pt amount (1–7 at. %), the absolute current value at the GOD modified bulk Pt electrode is larger

Table 1.4 Summarization of the metal nanoparticle modified carbon electrode for electroanalysis.

Metal NP	Carbon Supports	NP size	Electrode fabrication	Analyte	References
NP modification onto the electrode					
Au-NP	GC	9–11 nm	Electrodeposition	As(III)	Dai *et al.* 2004 [72]
	GC-CNT	30–90 nm	Electrodeposition	tinidazole	Shahrokhi *et al.* 2012 [75]
	GC-CNT	~10 nm	Chemical reduction	As(III)	Xiao *et al.* 2008 [74]
	BDD thin film	~22 nm	Electrodeposition	As(III)	Song *et al.* 2007 [77]
Ag-NP	GC	~100 nm	Electrodeposition and stripping	H_2O_2	Welch *et al.* 2005 [73]
Ni-NP	BDD	38.5–321 nm	Electrodeposition	adenine and DNA	Harfield *et al.* 2011 [80]
Sb-NP	BDD	~100 nm	Electrodeposition	Pd(II) and Cd(II)	Toghill *et al.* 2009 [79]
Bi-NP	BDD	~45 nm	Electrodeposition	Pd(II) and Cd(II)	Toghill *et al.* 2008 [78]
NP-embedded carbon film					
Ir-NP	BDD	Non-detectable size by SEM	Ion implantation	As(III)	Ivandini *et al.* 2006 [10]
Pt-NP	GC	~1.5 nm	Pyrolysis	Hydrogen reduction and ORR (not analytical application)	Hutton *et al.* 1993 [84]
Pt-NP	Graphite-like carbon film (sp^2 bond only)	~2.5 nm	RF co-sputtering	H_2O_2	You *et al.* 2003 [85]
Pd-NP		~4 nm		H_2O_2	Niwa *et al.* 2007 [88]
Ni-NP		~3 nm		Sugar mixture	You *et al.* 2003 [86]
Cu-NP		4–5 nm		Glucose	You *et al.* 2002 [87]
Au-NP	sp^2/sp^3 hybrid nanocarbon film	3–5 nm	UBM co-sputtering	As(III)	Kato *et al.* 2016 [90]
Ni-Cu nanoalloy		~3 nm		D-mannitol	Shiba *et al.* 2016 [91]

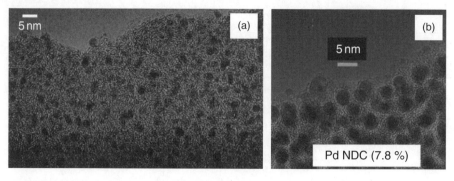

Figure 1.11 Plain view of the co-sputtered (a) Pt and (b) Pd nanoparticle embedded carbon film electrode obtained by TEM. Adapted in part with permission from [88] and [87]. Copyright 2002 American Chemical Society and 2007 MYU KK, respectively.

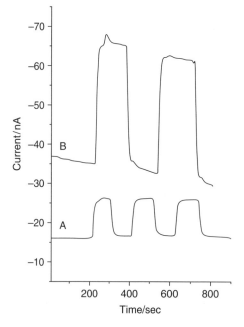

Figure 1.12 Amperometric response of glucose at GOD modified (A) Pt-NPs embedded carbon film and (B) Pt bulk electrodes. Detection potential: 0.6 V vs. Ag/AgCl; run buffer: 0.1 M PBS, pH 7; flow rate: 10 μL/min. Reprinted with permission from [85]. Copyright 2003 American Chemical Society.

than that at GOD modified Pt-NPs embedded carbon film electrode, but latter electrode shows much flatter baseline current compared with that at bulk electrode, indicating Pt-NPs embedded carbon film is extremely stable. The high electrocatalytic activity for oxygen reduction and hydrogen evolution was also observed at Pt-NPs embedded carbon film electrodes [89]. In contrast, non-noble Ni-NP and Cu-NP embedded graphite-like carbon film are applied to sugar detection [86]. Figure 1.13 shows the hydrodynamic voltammograms of glucose at Ni-NP embedded graphite-like carbon film compared with Ni bulk electrode. Electrocatalytic oxidation of glucose

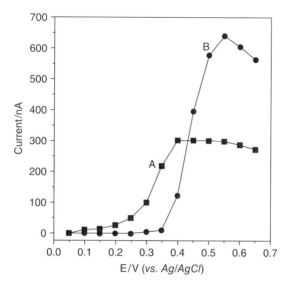

Figure 1.13 Hydrodynamic voltammograms of glucose at (A) 0.8 at.% NiNP embedded carbon film and (B) Ni bulk electrodes. Reprinted with permission from [86]. Copyright 2003 American Chemical Society.

begin at slightly negative potential (~0.15 V) than that of Ni bulk electrode, and exhibiting analytical performance such as high stability with relative standard deviation (RSD) of 1.75 % (n = 40) and low limit of detection (LOD) of 20~50 nM for glucose, fructose, sucrose and lactose.

Recently, UBM co-sputtering with multiple targets, which can independently control each target power of metal and carbon have been developed to widely control the metal NPs concentration and carbon film structure. In fact, sp^2 and sp^3 ratio in the carbon film can be widely controlled by changing acceleration voltage between target and substrate [39]. Kato *et al.* recently reported Au-NP embedded carbon film electrode with a variety of Au-NP concentrations (from 13 to 21 at. %) fabricated by UBM co-sputtering., [90] This electrode was applied to As(III) detection, which demonstrates that Au-NP embedded electrode at the optimized Au concentration (17 at. %) exhibit not only lower LOD of 1 ppb than Au bulk electrode, but also sufficient long-term stability with RSD of 11.7 % (n = 15 / 5 days) as shown in Figure 1.14. This higher stability is ascribed to tightly embedded Au-NPs in carbon matrix. In fact, As stripping current with the Au-NPs electrodeposited on the sputtered carbon film is sharply decreased because of weak interactions between Au-NPs and carbon surface, resulting in desorption of Au-NPs.

More recently, bimetallic nanoparticles (nanoalloy) embedded carbon film electrode composed of small Ni-Cu nanoalloy (~3 m) are also fabricated by Shiba *et al.* (Figure 1.15) the composition of which can be dynamically changed without changing nanoalloy size [91]. The electrocatalytic activity for D-mannitol is significantly affected by Ni/Cu ratio (including monometallic Ni-NP and Cu-NP), resulting in getting maximum with Ni/Cu ratio of 64/36. Figure 1.16a and 1.16b show CVs comparing the electrocatalytic activities of the $Ni_{64}Cu_{36}$ nanoalloy embedded carbon film and the $Ni_{70}Cu_{30}$ alloy (without a carbon film matrix) for D-mannitol, which is one of the intestinal permeability indicators.

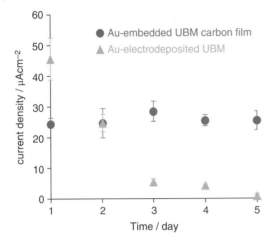

Figure 1.14 Stability of As measurement at the Au-embedded UBM carbon film (circle, Au = 17 at %) and the Au-electrodeposited UBM carbon film (triangle) electrodes. The average peak current density was obtained from repetitive As measurements (3 measurements per day). ASV parameters: deposition at −0.8 V vs Ag/AgCl for 60 s, potential scan rate of 1.5 V s^{-1}. Adapted in part with permission from [90]. Copyright 2003 American Chemical Society.

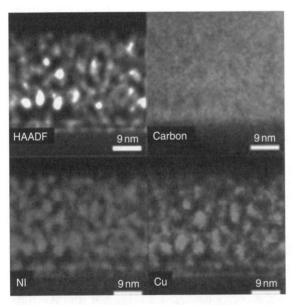

Figure 1.15 Nanostructural observation and elemental mapping obtained by HAADF-STEM and STEM-EDS measurements. (*See color plate section for the color representation of this figure.*)

At the Ni$_{64}$Cu$_{36}$ nanoalloy embedded carbon film, a smaller D-mannitol oxidation current (red solid line) and background current (black solid line) were obtained because of the smaller metal surface area. After correction with the surface metal concentration(15.6 at%, dotted line), a background-subtracted electrocatalytic current density (S–B current) of 1529 μA cm^{-2} at 0.6 V for the Ni$_{64}$Cu$_{36}$ nanoalloy embedded carbon

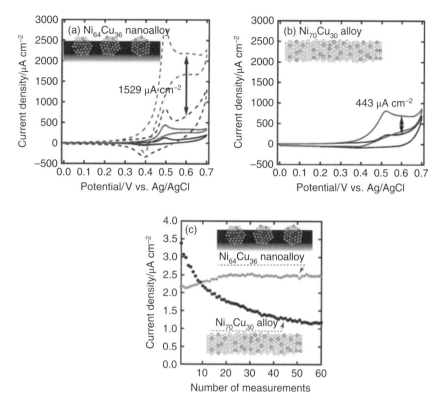

Figure 1.16 CVs obtained with (red line) or without (black line) 300 μM D-mannitol in 0.1 M NaOH solution using (a) the Ni$_{64}$Cu$_{36}$ nanoalloy embedded carbon film and (b) the Ni$_{70}$Cu$_{30}$ alloy film before (solid line) and after (dotted line) correction by metal concentration. (c) Electrocatalytic oxidation currents after background subtraction with various compositions of nanoalloy embedded carbon films for 500 μM D-mannitol at 0.6 V during an anodic scan (n = 3). All the potential sweep measurements are conducted at a scan rate of 0.1 V s^{-1} after a background scan at the same scan rate. Adapted in part with permission from [91]. Copyright 2016 Royal Society of Chemistry.

film was obtained, which is 3.4 times larger than that of the Ni$_{70}$Cu$_{30}$ alloy film (443 μA cm^{-2}). These results clearly demonstrate that the nanoalloys exhibit higher electrocatalytic activity due to the small size of the nanoalloys separated from each other by the carbon matrix. Similar to the electrode mentioned above, most of the nanoalloy body are embedded in carbon film demonstrated by AFM measurements, where convection size of the surface is less than nanoalloy radius (<1.5 nm, Ra of 0.21 nm). Therefore, the electrocatalytic activity and stability toward D-mannitol oxidation are extremely high as shown in Figure 1.16c (RSD: 4.6 %, n = 60), which is superior to Ni-Cu alloy film electrode with a similar Ni/Cu composition (32.2 %, n = 60) [91].

References

1 R.L. McCreery. Advanced carbon electrode materials for molecular electrochemistry. Chemical Reviews 2008, 108, 2646–2687.

2 R.G. Compton, J.S. Foord, F. Marken. Electroanalysis at diamond-like and doped-diamond electrodes. Electroanalysis 2003, 15, 1349–1363.

3 G.M. Swain, R. Ramesham. The electrochemical activity of boron-doped polycrystalline diamond thin-film electrodes. Analytical Chemistry 1993, 65, 345–351.

4 R. Tenne, K. Patel, K. Hashimoto, A. Fujishima. Efficient electrochemical reduction of nitrate to ammonia using conductive diamond film electrodes. Journal of Electroanalytical Chemistry 1993, 347, 409–415.

5 A. Manivannan, R. Kawasaki, D.A. Tryk, A. Fujishima. Interaction of Pb and Cd during anodic stripping voltammetric analysis at boron-doped diamond electrodes. Electrochimica Acta 2004, 49, 3313–3318.

6 C. Terashima, T. N. Rao, B. V. Sarada, *et al.* Electrochemical oxidation of chlorophenols at a boron-doped diamond electrode and their determination by high-performance liquid chromatography with amperometric detection. Analytical Chemistry 2002, 74, 895–902.

7 B.V. Sarada, T.N. Rao, D.A. Tryk, A. Fujishima. Electrochemical oxidation of histamine and serotonin at highly boron doped diamond electrodes. Analytical Chemistry 2000, 72, 1632–1638.

8 Y.S. Singh, L.E. Sawarynski, H.M. Michael, *et al.* Boron-doped diamond microelectrodes reveal reduced serotonin uptake rates in lymphocytes from adult rhesus monkeys carrying the short allele of the 5-HTTLPR. ACS Chemical Neuroscience 2010, 1, 49–64.

9 M. Chiku, J. Nakamura, A. Fujishima, Y. Einaga. Conformational change detection in nonmetal proteins by direct electrochemical oxidation using diamond electrodes. Analytical Chemistry 2008, 80, 5783–5787.

10 T.A. Ivandini, R. Sato, Y. Makide, *et al.* Electrochemical detection of arsenic(III) using iridium-implanted boron-doped diamond electrodes. Analytical Chemistry 2006, 78, 6291–6298.

11 N. Yang, H. Uetsuka, E. Osawa, C.E. Nebel. Vertically aligned diamond nanowires for DNA sensing. Angewandte Chemie-International Edition 2008, 47, 5183–5185.

12 S. Haymond, G.T. Babcock, G.M. Swain. Direct electrochemistry of cytochrome c at nanocrystalline boron-doped diamond. Journal of the American Chemical Society 2002, 124, 10634–10635.

13 L.T. Qu, Y. Liu, J.B. Baek, L.M. Dai. Nitrogen-doped graphene as efficient metal-free electrocatalyst for oxygen reduction in fuel cells. ACS Nano 2010, 4, 1321–1326.

14 Z.Q. Tian, S.P. Jiang, Y.M. Liang, P.K. Shen. Synthesis and characterization of platinum catalysts on muldwalled carbon nanotubes by intermittent microwave irradiation for fuel cell applications. Journal of Physical Chemistry B 2006, 110, 5343–5350.

15 T. Miyake, S. Yoshino, T. Yamada, *et al.* Self-regulating enzyme-nanotube ensemble films and their application as flexible electrodes for biofuel cells. Journal of the American Chemical Society 2011, 133, 5129–5134.

16 M. Pumera, A. Ambrosi, A. Bonanni, *et al.* Graphene for electrochemical sensing and biosensing. Trac-Trends in Analytical Chemistry 2010, 29, 954–965.

17 A. Qureshi, W.P. Kang, J.L. Davidson, Y. Gurbuz. Review on carbon-derived, solid-state, micro and nano sensors for electrochemical sensing applications. Diamond and Related Materials 2009, 18, 1401–1420.

18 S.K. Vashist, D. Zheng, K. Al-Rubeaan, *et al.* Advances in carbon nanotube based electrochemical sensors for bioanalytical applications. Biotechnology Advances 2011, 29, 169–188.

19 M. Zhou, Y.M. Zhai, S.J. Dong. Electrochemical sensing and biosensing platform based on chemically reduced graphene oxide. Analytical Chemistry 2009, 81, 5603–5613.

20 Q. Xue, D. Kato, T. Kamata, *et al*. Human cytochrome P450 3A4 and a carbon nanofiber modified film electrode as a platform for the simple evaluation of drug metabolism and inhibition reactions. Analyst 2013, 138, 6463–6468.

21 Y. Ueno, K. Furukawa, K. Hayashi, *et al*. Graphene-modified interdigitated array electrode: fabrication, characterization, and electrochemical immunoassay application. Analytical Sciences 2013, 29, 55–60.

22 O. Niwa. Electroanalytical chemistry with carbon film electrodes and micro and nano-structured carbon film-based electrodes. Bulletin of the Chemical Society of Japan 2005, 78, 555–571.

23 M.L. Kaplan, P.H. Schmidt, C.H. Chen, W.M. Walsh. Carbon-films with relatively high conductivity. Applied Physics Letters 1980, 36, 867–869.

24 A. Rojo, A. Rosenstratten, D. Anjo. Characterization of a conductive carbon-film electrode for voltammetry. Analytical Chemistry 1986, 58, 2988–2991.

25 O. Niwa, H. Tabei. Voltammetric measurements of reversible and quasi-reversible redox species using carbon-film based interdigitated array microelectrodes. Analytical Chemistry 1994, 66, 285–289.

26 H. Tabei, O. Niwa, T. Horiuchi, M. Morita. Microdisk array electrode based on carbonized poly (p-phenylene vinylene) film. Denki Kagaku 1993, 61, 820–822.

27 A.M. Lyons, C.W. Wilkins, M. Robbins. Thin pinhole-free carbon-films. Thin Solid Films 1983, 103, 333–341.

28 J. Kim, X. Song, K. Kinoshita, *et al*. Electrochemical studies of carbon films from pyrolyzed photoresist. Journal of the Electrochemical Society 1998, 145, 2314–2319.

29 S. Ranganathan, R.L. McCreery. Electroanalytical performance of carbon films with near-atomic flatness. Analytical Chemistry 2001, 73, 893–900.

30 P.A. Brooksby, A.J. Downard. Electrochemical and atomic force microscopy study of carbon surface modification via diazonium reduction in aqueous and acetonitrile solutions. Langmuir 2004, 20, 5038–5045.

31 F.J. del Campo, P. Godignon, L. Aldous, *et al*. Fabrication of PPF electrodes by a rapid thermal process. Journal of the Electrochemical Society 2011, 158, H63–H68.

32 M. Morita, K. Hayashi, T. Horiuchi, S. Shibano, K. Yamamoto, K.J. Aoki. Enhancement of redox cycling currents at interdigitated electrodes with elevated fingers. Journal of the Electrochemical Society 2014, 161, H178–H182.

33 A. C. Ferrari, J. Robertson. Interpretation of Raman spectra of disordered and amorphous carbon. Physical Review B 2000, 61, 14095–14107.

34 R. Schnupp, R. Kuhnhold, G. Temmel, *et al*. Thin carbon films as electrodes for bioelectronic applications. Biosensors and Bioelectronics 1998, 13, 889–894.

35 J.J. Blackstock, A.A. Rostami, A.M. Nowak, *et al*. Ultraflat carbon film electrodes prepared by electron beam evaporation. Analytical Chemistry 2004, 76, 2544–2552.

36 X.Y. Yang, L. Haubold, G. DeVivo, G.M. Swain. Electroanalytical performance of nitrogen-containing tetrahedral amorphous carbon thin-film electrodes. Analytical Chemistry 2012, 84, 6240–6248.

37 S. Hirono, S. Umemura, M. Tomita, R. Kaneko. Superhard conductive carbon nanocrystallite films. Applied Physics Letters 2002, 80, 425–427.

38 O. Niwa, J. Jia, Y. Sato, *et al.* Electrochemical performance of Angstrom level flat sputtered carbon film consisting of sp(2) and sp(3) mixed bonds. Journal of the American Chemical Society 2006, 128, 7144–7145.

39 T. Kamata, D. Kato, H. Ida, O. Niwa. Structure and electrochemical characterization of carbon films formed by unbalanced magnetron (UBM) sputtering method. Diamond and Related Materials 2014, 49, 25–32.

40 D. Kato, M. Komoriya, K. Nakamoto, *et al.* Electrochemical determination of oxidative damaged DNA with high sensitivity and stability using a nanocarbon film. Analytical Sciences 2011, 27, 703–707.

41 N. Sekioka, D. Kato, A. Ueda, *et al.* Controllable electrode activities of nano-carbon films while maintaining surface flatness by electrochemical pretreatment. Carbon 2008, 46, 1918–1926.

42 D. Kato, N. Sekioka, A. Ueda, *et al.* A nanocarbon film electrode as a platform for exploring DNA methylation. Journal of the American Chemical Society 2008, 130, 3716.

43 D. Kato, N. Sekioka, A. Ueda, *et al.* Nanohybrid carbon film for electrochemical detection of SNPs without hybridization or labeling. Angewandte Chemie-International Edition 2008, 47, 6681–6684.

44 K. Goto, D. Kato, N. Sekioka, *et al.* Direct electrochemical detection of DNA methylation for retinoblastoma and CpG fragments using a nanocarbon film. Analytical Biochemistry 2010, 405, 59–66.

45 J. Ozaki, T. Anahara, N. Kimura, A. Oya. Simultaneous doping of boron and nitrogen into a carbon to enhance its oxygen reduction activity in proton exchange membrane fuel cells. Carbon 2006, 44, 3358–3361.

46 K.P. Gong, F. Du, Z.H. Xia, *et al.* Nitrogen-doped carbon nanotube arrays with high electrocatalytic activity for oxygen reduction. Science 2009, 323, 760–764.

47 X.R. Wang, X.L. Li, L. Zhang, Y*et al.* N-doping of graphene through electrothermal reactions with ammonia. Science 2009, 324, 768–771.

48 S. Uchiyama, H. Matsuura, Y. Yamawaki. Observation of hydrogen oxidation wave using glassy carbon electrode fabricated by stepwise electrolyses in ammonium carbamate aqueous solution and hydrochloric acid. Electrochimica Acta 2013, 88, 251–255.

49 K.S. Yoo, B. Miller, R. Kalish, X. Shi. Electrodes of nitrogen-incorporated tetrahedral amorphous carbon: A novel thin-film electrocatalytic material with diamond-like stability. Electrochemical and Solid State Letters 1999, 2, 233–235.

50 B. Fausett, M.C. Granger, M.L. Hupert, *et al.* The electrochemical properties of nanocrystalline diamond thin-films deposited from C_{60}/argon and methane/nitrogen gas mixtures. Electroanalysis 2000, 12, 7–15.

51 Q. Chen, D.M. Gruen, A.R. Krauss, *et al.* The structure and electrochemical behavior of nitrogen-containing nanocrystalline diamond films deposited from $CH_4/N_2/Ar$ mixtures. Journal of the Electrochemical Society 2001, 148, E44–E51.

52 Y. Tanaka, M. Furuta, K. Kuriyama, R*et al.* Electrochemical properties of N-doped hydrogenated amorphous carbon films fabricated by plasma-enhanced chemical vapor deposition methods. Electrochimica Acta 2011, 56, 1172–1181.

53 A. Lagrini, S. Charvet, M. Benlahsen, *et al.* Microstructure and electronic investigations of carbon nitride films deposited by RF magnetron sputtering. Thin Solid Films 2005, 482, 41–44.

54 A. Lagrini, S. Charvet, M. Benlahsen, *et al*. On the relation between microstructure and electrochemical reactivity of sputtered amorphous carbon nitride electrodes. Diamond and Related Materials 2007, 16, 1378–1382.

55 A.P. Zeng, M.M. Bilek, D.R. McKenzie, *et al*. Correlation between film structures and potential limits for hydrogen and oxygen evolutions at a-C:N film electrochemical electrodes. Carbon 2008, 46, 663–670.

56 T. Kamata, D. Kato, S. Hirono, O. Niwa. Structure and electrochemical performance of nitrogen-doped carbon film formed by electron cyclotron resonance sputtering. Analytical Chemistry 2013, 85, 9845–9851.

57 T. Kamata, D. Kato, S. Umemura, O. Niwa. Structure and electroanalytical application of nitrogen-doped carbon thin film electrode with lower nitrogen concentration. Analytical Sciences 2015, 31, 651–656.

58 A. Zeng, E. Liu, S.N. Tan, *et al*. Stripping voltammetric analysis of heavy metals at nitrogen doped diamond-like carbon film electrodes. Electroanalysis 2002, 14, 1294–1298.

59 D. Hamblin, J. Qiu, L. Haubold, G.M. Swain. The performance of a nitrogen-containing tetrahedral amorphous carbon electrode in flow injection analysis with amperometric detection. Analytical Methods 2015, 7, 4481–4485.

60 R. Jarosova, J. Rutherford, G.M. Swain. Evaluation of a nitrogen-incorporated tetrahedral amorphous carbon thin film for the detection of tryptophan and tyrosine using flow injection analysis with amperometric detection. Analyst 2016, 141, 6031–6041.

61 B.C. Lourencao, T.A. Silva, O. Fatibello, G.M. Swain. Voltammetric Studies of propranolol and hydrochlorothiazide oxidation in standard and synthetic biological fluids using a nitrogen-containing tetrahedral amorphous carbon (ta-C:N) electrode. Electrochimica Acta 2014, 143, 398–406.

62 T. Nakajima, M. Koh, R.N. Singh, M. Shimada. Electrochemical behavior of surface-fluorinated graphite. Electrochimica Acta 1999, 44, 2879–2888.

63 V.N. Khabashesku, W.E. Billups, J.L. Margrave. Fluorination of single-wall carbon nanotubes and subsequent derivatization reactions. Accounts of Chemical Research 2002, 35, 1087–1095.

64 S. Ferro, A. De Battisti. Physicochemical properties of fluorinated diamond electrodes. Journal of Physical Chemistry B 2003, 107, 7567–7573.

65 A. Ueda, D. Kato, N. Sekioka, *et al*. Fabrication of electrochemically stable fluorinated nano-carbon film compared with other fluorinated carbon materials. Carbon 2009, 47, 1943–1952.

66 S. Boopathi, T.N. Narayanan, S.S. Kumar. Improved heterogeneous electron transfer kinetics of fluorinated graphene derivatives. Nanoscale 2014, 6, 10140–10146.

67 E. Silva, A.C. Bastos, M. Neto, *et al*. New fluorinated diamond microelectrodes for localized detection of dissolved oxygen. Sensors and Actuators B-Chemical 2014, 204, 544–551.

68 A. Oda, D. Kato, K. Yoshioka, *et al*. Fluorinated nanocarbon film electrode capable of signal amplification for lipopolysaccharide detection. Electrochimica Acta 2016, 197, 152–158.

69 E. Kuraya, S. Nagatomo, K. Sakata, *et al*. Simultaneous Electrochemical analysis of hydrophilic and lipophilic antioxidants in bicontinuous microemulsion. Analytical Chemistry 2015, 87, 1489–1493.

70 H.V. Patten, S.C. Lai, J.V. Macpherson, P.R. Unwin. Active sites for outer-sphere, inner-sphere, and complex multistage electrochemical reactions at polycrystalline boron-doped diamond electrodes (pBDD) revealed with scanning electrochemical cell microscopy (SECCM). Analytical Chemistry 2012, 84, 5427–5432.

71 E. Kuraya, S. Nagatomo, K. Sakata, *et al.* Direct analysis of lipophilic antioxidants of olive oils using bicontinuous microemulsions. Analytical Chemistry 2016, 88, 1202–1209.

72 X. Dai, O. Nekrassova, M.E. Hyde, R.G. Compton. Anodic stripping voltammetry of arsenic(III) using gold nanoparticle-modified electrodes. Analytical Chemistry 2004, 76, 5924–5929.

73 C.M. Welch, C.E. Banks, A.O. Simm, R.G. Compton. Silver nanoparticle assemblies supported on glassy-carbon electrodes for the electro-analytical detection of hydrogen peroxide. Analytical and Bioanalytical Chemistry 2005, 382, 12–21.

74 L. Xiao, G.G. Wildgoose, R.G. Compton. Sensitive electrochemical detection of arsenic (III) using gold nanoparticle modified carbon nanotubes via anodic stripping voltammetry. Analytica Chimica Acta 2008, 620, 44–49.

75 S. Shahrokhian, S. Rastgar. Electrochemical deposition of gold nanoparticles on carbon nanotube coated glassy carbon electrode for the improved sensing of tinidazole. Electrochimica Acta 2012, 78, 422–429.

76 E.A. Viltchinskaia, L.L. Zeigman, D.M. Garcia, P.F. Santos. Simultaneous determination of mercury and arsenic by anodic stripping voltammetry. Electroanalysis 1997, 9, 633–640.

77 Y. Song, G.M. Swain. Total inorganic arsenic detection in real water samples using anodic stripping voltammetry and a gold-coated diamond thin-film electrode. Analytica Chimica Acta 2007, 593, 7–12.

78 K.E. Toghill, G.G. Wildgoose, A. Moshar, *et al.* The fabrication and characterization of a bismuth nanoparticle modified boron doped diamond electrode and its application to the simultaneous determination of cadmium(II) and lead(II). Electroanalysis 2008, 20, 1731–1737.

79 K.E. Toghill, L. Xiao, G.G. Wildgoose, R.G. Compton. Electroanalytical Determination of cadmium(II) and lead(II) using an antimony nanoparticle modified boron-doped diamond electrode. Electroanalysis 2009, 21, 1113–1118.

80 J.C. Harfield, K.E. Toghill, C. Batchelor-McAuley, *et al.* Nickel nanoparticle modified BDD electrode shows an electrocatalytic response to adenine and DNA in aqueous alkaline media. Electroanalysis 2011, 23, 931–938.

81 W.T. Wahyuni, T.A. Ivandini, E. Saepudin, Y. Einaga. Development of neuraminidase detection using gold nanoparticles boron-doped diamond electrodes. Analytical Biochemistry 2016, 497, 68–75.

82 Y. Song, G.M. Swain. Development of a method for total inorganic arsenic analysis using anodic stripping voltammetry and a Au-coated, diamond thin-film electrode. Analytical Chemistry 2007, 79, 2412–2420.

83 Y. Nagaoka, T.A. Ivandini, D. Yamada, *et al.* Selective detection of As(V) with high sensitivity by As-deposited boron-doped diamond electrodes. Chemistry Letters 2010, 39, 1055–1057.

84 H.D. Hutton, N.L. Pocard, D.C. Alsmeyer, *et al.* Preparation of nanoscale platinum(0) clusters in glassy-carbon and their catalytic activity. Chemistry of Materials 1993, 5, 1727–1738.

85 T.Y. You, O. Niwa, M. Tomita, S. Hirono. Characterization of platinum nanoparticle-embedded carbon film electrode and its detection of hydrogen peroxide. Analytical Chemistry 2003, 75, 2080–2085.

86 T.Y. You, O. Niwa, Z.L. Chen, *et al*. An amperometric detector formed of highly dispersed Ni nanoparticles embedded in a graphite-like carbon film electrode for sugar determination. Analytical Chemistry 2003, 75, 5191–5196.

87 T.Y. You, O. Niwa, M. Tomita, H. *et al*. Characterization and electrochemical properties of highly dispersed copper oxide/hydroxide nanoparticles in graphite-like carbon films prepared by RF sputtering method. Electrochemistry Communications 2002, 4, 468–471.

88 O. Niwa, D. Kato, R. Kurita, *et al*. Electro catalytic detection of hydrogen peroxide using palladium-nanoparticle dispersed carbon film electrodes. Sensors and Materials 2007, 19, 225–233.

89 T. You, O. Niwa, T. Horiuchi, *et al*. Co-sputtered thin film consisting of platinum nanoparticles embedded in graphite-like carbon and its high electrocatalytic properties for electroanalysis. Chemistry of Materials 2002, 14, 4796–4799.

90 D. Kato, T. Kamata, D. Kato, *et al*. Au nanoparticle-embedded carbon films for electrochemical as^{3+} detection with high sensitivity and stability. Analytical Chemistry 2016, 88, 2944–2951.

91 S. Shiba, D. Kato, T. Kamata, O. Niwa. Co-sputter deposited nickel-copper bimetallic nanoalloy embedded carbon films for electrocatalytic biomarker detection. Nanoscale 2016, 8, 12887–12891.

2

Carbon Nanofibers for Electroanalysis

Tianyan You, Dong Liu and Libo Li

School of Agricultural Equipment Engineering, Institute of Agricultural Engineering, Jiangsu University, China

2.1 Introduction

Modern electrochemical techniques have significantly promoted the development of electroanalysis, particularly electrochemical biosensors. It is of great importance to achieve the higher sensitivity, selectivity and long-term stability in the field of electrochemical biosensors. For these objectives, numbers of researchers have paid extensively attention to the synthesis of nanomaterials with superior electrochemical properties to fabricate high-performance electrochemical biosensors. Generally, the discovery of novel materials with excellent properties could result in the great advances in many fields, and a famous example would be carbon nanotubes (CNTs) discovered in 1991 [1]. Since then, many carbon nanomaterials have been developed and investigated, such as graphene [2] and carbon nanohorn [3], and most of them are found to be suitable electrode materials for the electrochemical biosensors due to the high conductivity, large surface-to-volume ratio and excellent biocompatibility. Up to date, lots of investigations have demonstrated that carbon nanomaterials-based biosensors could display enhanced analysis properties compared with that constructed by conventional materials, such as higher sensitivities and better selectivity.

Carbon nanofibers (CNFs) exhibit many similarities with CNTs, for example, they both have hollow structure and high electrical conductivity. Briefly, CNFs are composed of a graphene layer, and exhibit a cylindrical nanostructure with a diameter of 100–1000 nm and a length of several micrometers [4]. According to the disposition of graphene layers, there are three types of CNFs: ribbon-like CNFs, platelet CNFs and herringbone CNFs [5]. The discovery of CNFs could be dated back to 1889 when CNFs was observed after using the metallic crucible in a carbon-containing atmosphere [6]. Then, CNFs were considered as nuisance for a long time, and the applications of CNFs start in the 1980s when CNFs were mainly employed as the supporting material for catalysts and additives in polymers [7]. The development of CNFs has been significantly promoted after the utilization of electrospinning technique for preparation with polyacrylonitrile as the principal precursor. Most of investigations on CNFs are focused on catalyst support materials, applied in the fields of supercapacitors [8] and fuel cells [9], etc. Noteworthy, CNFs and related composites with superior properties have efficiently

Nanocarbons for Electroanalysis, First Edition.
Edited by Sabine Szunerits, Rabah Boukherroub, Alison Downard and Jun-Jie Zhu.

enhanced the development of modern electroanalysis, serving as the high-performance electrode materials for the fabrication of electrochemical biosensors. Except for the large surface area, high conductivity and excellent biocompatibility, the most important advantage of CNFs is that their whole surface area could be activated and functionalized which would facilitate the immobilization of nanoparticles and biomolecules [10].

This chapter offers an overview on the electroanalysis application of CNFs and related composites. The first section provides a summary of techniques for the preparation of CNFs and descriptions of several novel CNFs composites, including nitrogen-doped CNFs and metal nanoparticles-loaded CNFs. Consequently, the frequently used electrochemical technologies for electroanalysis are introduced. Then, main part of the chapter is an overview of the applications of CNFs and related composites in the field of electrochemical biosensors which are subdivided to non-enzymatic biosensors, enzyme-based biosensors and immunosensors. Lastly, the future challenges of CNFs-based electrochemical biosensors are discussed.

2.2 Techniques for the Preparation of CNFs

Many state-of-the-art techniques have been developed for the fabrication of CNFs, such as hot filament-assisted sputtering [11], co-catalyst deoxidization method [12], template-assisted method [8], chemical vapor deposition (CVD) [13] and electrospinning methods [14, 15]. For example, Matsushima's group has synthesized CNFs thin films by utilizing a hot filament-assisted sputtering system [11]; Qian's group has proposed a novel co-catalyst deoxidization process to prepare CNFs [12]. Among these methods, the CVD and electrospinning methods are the mostly used techniques for the synthesis of CNFs, and they will be introduced in detail in the following sections.

Chemical Vapor Deposition: Generally, chemical vapor deposition mainly comprises four steps, including adsorption, desorption, evolution and incorporation of vapor species at the substrates [16]. The energy source for CVD could be thermal, photoexcitation and plasma, known as thermal CVD (TCVD), photoexcitation CVD and plasma-enhanced CVD (PECVD), respectively.

Taking the fabrication of CNFs by catalytically controlled TCVD for example, the apparatus is illustrated in Figure 2.1a. The precursor containing methane (CH_4), ammonia

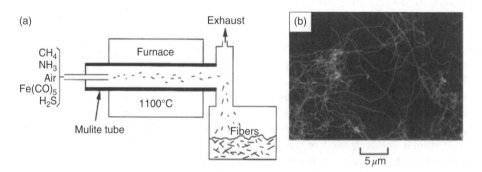

Figure 2.1 (a) Apparatus illustration for the synthesis of CNFs by TCVD; (b) SEM image of CNFs prepared by TCVD. *Source:* Tibbetts 2007 [17]. Reproduced with permission of Elsevier.

(NH3), air, iron pentacarbonyl ($Fe(CO)_5$) and hydrogen sulfide (H_2S) is inlet into the furnace tube at a high temperature of 1100° C. During the reaction, iron particles derived from the $Fe(CO)_5$ decomposition serve as the catalyst, while the presence of sulfur could further increase their catalytic activity; then, CNFs grow on the surface of catalyst. The SEM image of as-obtained CNFs is shown in Figure 2.1b [17]. These methods could be employed for the production of high quality CNFs, however, the residual of metallic catalysts as well as relatively high cost have limited their application in CNFs preparation.

Electrospinning Technique: Electrospinning has been utilized for the preparation of polymer fibers with diameters ranging from several nanometers to micrometers. The electrospinning method could produce continuous nanofibers with a relatively simple and low-cost process. Furthermore, the structure and properties of fibers could be monitored by differing the experimental conditions for electrospinning. As shown in Figure 2.2a, typical electrospinning apparatus consists of high-voltage source, syringe and collector [18]. During the electrospinning, the pendant drop of polymer solution at the tip of syringe is highly electrified by the applied high voltage, and solution jet would be ejected once the electric force surpasses the surface tension of the droplet. After solvent evaporation, the jet transforms to nanofibers and deposits on the collector to form a nonwoven mat [19]. Many parameters could affect the structures and properties of the resulting nanofibers, such as the properties of polymer solution (viscosity, conductivity, etc.) [20], intensity of electric field [21] and flow rate [22].

The development of electrospinning has efficiently facilitated the fabrication of CNFs. Many polymers with carbon backbone have been utilized as precursors for the preparation of CNFs, such as polyvinyl alcohol [23], polyvinylidene fluoride [24], and lignin [25]. Among these polymers, PAN is most frequently used as precursor for electrospun

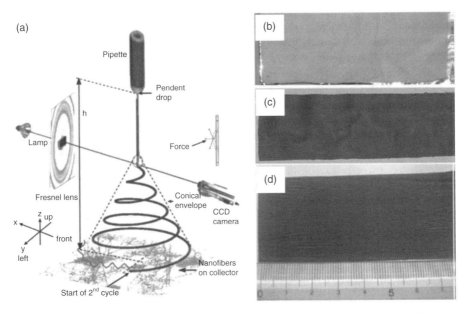

Figure 2.2 (a) Illustration of electrospinning process (Reprinted with permission from [18]). Photographs of (b) PAN nanofiber web, (c) web after stabilization and (d) web after carbonization. *Source:* Kim 2006 [27]. Reproduced with permission of John Wiley and Sons.

CNFs. For PAN-based CNFs, the stabilization and carbonization processes are both required to transform polymer nanofiber to CNFs, and the nanofiber diameter would decrease significantly for the shrinkage and weight loss during the thermal treatments [26]. As shown in Figure 2.2b–d, distinguishable color change could be observed during the thermal treatment of PAN nanofibers that the nanofibers web shows dark-brown and changes into black after the stabilization and carbonization [27]. Most of the electrospun CNFs composites exhibit a unique self-supported structure, and the resulting CNFs films could be directly used as electrodes free of any treatment, which could simplify the preparation of electrode and help to maintain the attracting properties.

2.3 CNFs Composites

The investigation of CNFs has for a long time been concentrated on the supporting materials for catalysts, particularly the metal nanoparticles-loaded CNFs composites, which take advantages of the large surface area and superior electrical conductivity of CNFs. In recent years, the excellent properties of nitrogen-doped carbon materials attracted enormous attention, and nitrogen-doped carbon nanofibers (NCNFs) have been produced for catalysis and biosensing applications. The following section would offer a brief introduction on the NCNFs and metal nanoparticles-loaded CNFs composites.

2.3.1 NCNFs

Doping nitrogen atom into graphite matrix has been verified to be an efficient approach to improve the properties of carbon nanomaterials, such as the catalytic activity toward oxygen reduction. Generally, there are two methods to fabricate NCNFs, including direct carbonization of nitrogen-rich precursor and thermal treatment of CNFs in a nitrogen-containing atmosphere. Huang's group has produced NCNFs with a high nitrogen content of 13.93 wt.% by direct pyrolyzation of polypyrrole nanofibers, and the resultant NCNFs showed large rate capability and superior ability as an anode material in sodium-ion battery [28]. Except for polypyrrole, polyaniline [29], iron(II) phthalocyanine [30] etc. could be directly employed for the preparation of NCNFs by carbonization.

Despite the high content of nitrogen in PAN, NCNFs with high catalytic activity could not be prepared by the carbonization of PAN nanofibers via the common methods. In respect of this challenge, You's group has developed a novel method to prepare NCNFs by carbonizing electrospun PAN nanofibers (Figure 2.3a). Differing from the conventional method, the nitrogen-containing tail gas generated during the decomposition of PAN was reused for surface etching and nitrogen doping by using a capsule shaped reactor (Figure 2.3a). The resultant NCNFs displayed a rough surface and relatively small diameter (Figure 2.3b), and the self-supported NCNFs film with high flexibility could be directly cut to the required shapes for the modification of electrodes (Figure 2.3c) [31].

For the latter method, ammonia (NH3) is frequently used for the thermal treatment of CNFs as nitrogen source. Yu's group employed cellulose as a precursor for CNFs, and doped nitrogen atoms into CNFs by utilizing NH_3 as a carrier gas (Figure 2.4a). The obtained NCNFs retained three-dimensional network of cellulose with high nitrogen content of 5.8 at % and surface area of 916 $m^2 g^{-1}$, resulting in superior oxygen reduction activity (Figure 2.4b–d) [32].

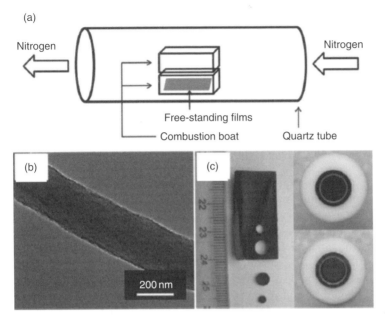

Figure 2.3 (a) Preparation of NCNFs in a capsule-shaped device; (b) TEM image of NCNFs; (c) photograph of NCNFs films for electrode modification. *Source:* Liu 2013 [31]. Reproduced with permission of the Royal Society of Chemistry.

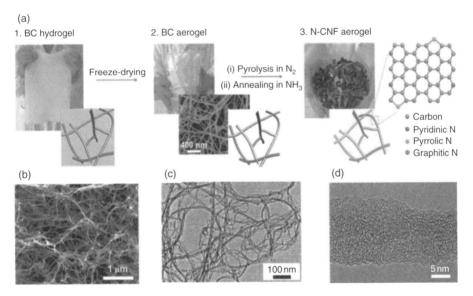

Figure 2.4 (a) The production process of NCNFs from cellulose; (b) SEM, (d) TEM and (d) HRTEM images of NCNFs. *Source:* Liang 2015 [32]. Reproduced with permission of Elsevier. (*See color plate section for the color representation of this figure.*)

2.3.2 Metal nanoparticles-loaded CNFs

Due to large surface area and excellent conductivity, CNFs have been extensively investigated as supporting material for many important catalysts, particularly the noble metal-based catalysts. For example, platinum nanoparticles-loaded CNFs (Pt/CNFs) could be simply synthesized at room temperature by using formic acid as the reducing agent, and the size and loading content of platinum could be simply controlled by varying the molar ratio in the precursor [33]. Loaiza *et al.* prepared CNFs-Pt nanoparticle hybrids by microwave-assisted heating polyol reduction, and the resultant composites could be used for the preparation of screen printed biosensors to detect lactate in wines and ciders [34].

For the preparation of metal nanoparticles-loaded CNFs, most of the studies are focused on the chemical reduction method; however, the relatively low reproducibility may limit the large-scale production. To solve this problem, You's group has achieved the controllable preparation of PdM (M: Ni, Co) nanoparticles-loaded CNFs by a one-step approach. Taking the PdCo/CNFs for example, metal precursor-containing PAN nanofibers ($Co(acac)_2/Pd(acac)_2/PAN$ nanofibers) was firstly prepared by electrospinning, and then PdCo/CNFs could be obtained after thermal treatment. The ratio of Pd and Co in PdCo/CNFs could be controlled by changing the molar ratio of $Co(acac)_2/Pd(acac)_2$ in the precursor. PdCo nanoparticles of all the resultant PdCo/CNFs exhibited uniform morphology with high dispersion on the surface of CNFs, and more importantly, the PdCo nanoparticles were embedded in CNFs which may enhance the stability compared with that obtained by chemical reduction [35].

2.4 Applications of CNFs for electroanalysis

2.4.1 Technologies for electroanalysis

Electrochemical related methods are efficient technologies for the micrometric analysis due to the high sensitivity and selectivity. Here, the introduction is focused on the applications of voltammetric techniques, amperometric techniques and electrochemical impedance spectroscopy in electrochemical analysis.

Voltammetric and amperometric techniques: Voltammetric and amperometric techniques are performed by measuring the current at working electrodes with a setting potential. For voltammetric techniques, the potential applied on electrodes is scanned in the set range during the test, and the measured current is related to the concentration of analyte. The critical advantage of voltammetric techniques is their wide dynamic range, and the frequently used voltammetric methods for electroanalysis include cyclic voltammetry (CV), linear sweep voltammetry (LSV), differential pulse voltammetry (DPV) and square-wave voltammetry (SWV) [36]. Generally, voltammetric methods could be applied for low-level quantitation. For amperometric techniques, a constant potential is applied at the electrode during the test, which differs from that for voltammetry. Typically, the applied potential should be the characteristic potential of analyte which results in a high selectivity [37]. During the testing, the current changes with the process of electrochemical reaction occurring at the electrode, and its value relates with the concentration of electroactive species in solution. Amperometric detection is frequently utilized for electrochemical biosensors for the low detection limit.

Electrochemical impedance spectroscopy: Electrochemical impedance spectroscopy (EIS) is utilized for the evaluation of the resistive and capacitive properties of materials by monitoring the current upon a voltage perturbation with small amplitude [38]. The resistive and capacitive components of impedance could be estimated by the in-phase and out-of-phase currents, respectively. EIS is commonly used to monitor the fabrication of biosensor as well as the recognition process at the biosensors.

2.4.2 Non-enzymatic biosensors

Non-enzymatic biosensors play a critical role in the field of electroanalysis because of their simple fabrication, high sensitivity and good long-term stability. Due to the abundant active sites and high conductivity, CNFs and related composites, mainly NCNFs and metal nanoparticles-loaded CNFs, have been extensively utilized for the fabrication of non-enzymatic electrochemical biosensors. CNFs-based non-enzymatic biosensors toward the detection of dopamine, hydrogen peroxide and glucose are summarized in Table 2.1.

Dopamine: As one of critical catecholamine neurotransmitters, dopamine (DA) is of great importance in human central nervous, metabolism and renal system, and the disorder of DA level may be related to many serious diseases, such as schizophrenia [39]. Despite the simplicity of electrochemical methods, the electrochemical detection of DA is always interfered by ascorbic acid (AA), uric acid (UA) and serotonin (5-HT) which have the similar potentials with that of DA [40, 41].

In recent years, many investigations have demonstrated the superior analysis properties of CNFs-based biosensors for the determination of DA in the presence of high concentration of AA, UA or 5-HT. Mesoporous carbon nanofibers (MCNFs) with an average diameter of 90 nm were prepared using mesoporous silica as the starting template. The MCNFs-based non-enzymatic biosensor showed high sensitivity and selectivity for simultaneous detection of DA, AA and UA with detection limits of 0.02, 50 and 0.2 μM, respectively, which was ascribed to the abundant edge-plane-like defects as well as large surface area of MCNFs [42]. Koehne's group has fabricated a high-performance biosensor for the simultaneous detection of DA and 5-HT using an array of vertically aligned CNFs (VACNFs). The unique nanofiber structure and abundant active sites allowed the VACNFs-based biosensor to efficiently distinguish DA, 5-HT and AA in a mixture. The detection limits toward DA and 5-HT were 50 nM and 250 nM, respectively [43].

To accurately control the electroactive surface area (ESAs) of the fabricated biosensors, the continuous CNFs, which were firmly attached onto a conductive substrate free of binders, were prepared by electrospinning PAN nanofibers on substrates followed by carbonization (Figure 2.5a–c). The resultant CNFs with good mesh integrity and high density of electronic states (DOS) could be directly utilized as the sensors, while the ESAs were monitored by varying the electrospinning deposition time (Figure 2.5d–e). The CNFs biosensor displayed high sensitivity as well as selectivity toward the determination of DA; more importantly, the dynamic range of the biosensor could be controlled by changing the ESA of CNFs [44].

To achieve higher catalytic activity, heteroatom-doped CNFs and metal nanoparticles-loaded CNFs have been constructed for electrochemical biosensing applications. NCNF has been prepared by combining electrospinning and carbonization, using the

Table 2.1 A summary of CNFs-based non-enzymatic biosensors toward dopamine, hydrogen peroxide and glucose.

Analyte	Biosensor	Linear range	LOD	Ref.
	Mesoporous CNFs-modified pyrolytic graphite electrode	0.05 – 30 μM	0.02 μM	42
	Array of vertically aligned CNFs	1 – 10 μM	0.05 μM	43
		0.1 – 0.5 μM		
	Electrospun CNFs	0.2 ..M – 0.70 M	0.07 μM	44
Dopamine	Nitrogen-doped CNFs	1 – 10 μM	0.50 μM	45
		10 – 200 μM		
	Ag–Pt bimetallic nanoparticles loaded nanoporous CNFs	10 – 500 μM	0.11 μM	46
	Electrospun graphene/polyaniline/ polystyrene nanofiber	0.0001 – 100 μM	0.05 nM	47
	Coaxial carbon fiber/ZnO nanorod fibers	5 – 70 μM	0.07 μM	48
	Nitrogen-doped carbon nanoparticles-embedded CNFs	5 μM – 27mM	1.5 μM	50
	CNFs decorated with platinum nanoparticles	10 μM – 9.38 mM	1.9 μM	51
		9.38 – 74.38 mM		
Hydrogen peroxide	CNFs decorated with platinum nanoparticles	10 μM – 15 mM	3.4 μM	52
	Pt nanoparticles/CNFs	5 μM – 15 mM	1.7 μM	53
	PdCo nanoparticle–embedded CNFs	0.4 – 30 μM	0.2 μM	54
		30 – 400 μM		
	Manganese dioxide nanoparticles loaded CNFs	10 μM – 15 mM	1.1 μM	55
	Nickel-doped carbon nanofibers	0.125 – 12.73 μM	0.05 μM	57
	Pd–Ni alloy nanoparticle/carbon nanofiber (Pd–Ni/CNF)	0.1 μM – 5.4 mM	0.06 μM	60
	Cu-Co-Ni nanostructures electrodeposited on CNFs	0.01 – 4.30 mM	3.05 μM	61
	Ccupric oxide nanoparticles/CNFs	0.5 μM – 11mM	0.2 μM	62
Glucose	CuO nanoneedle/graphene/CNFs	1 – 5.3 mM	0.1 μM	63
	Nanorod-aggregated flower-like CuO/ carbon fiber fabric	0.27 μM – 0.96 mM	0.27 μM	64

(Continued)

Table 2.1 (Continued)

Analyte	Biosensor	Linear range	LOD	Ref.
	CuO nanorods dispersed hollow carbon fibers	0.005 – 0.8 mM 0.8 – 8.5 mM	0.1 µM	65
	Nickel(II) hydroxide nanoplates/CNFs	0.001 – 1.2 mM	0.76 µM	66
	Carbon Nanofibers/Co(OH)$_2$	10 µM – 0.12 mM	5 µM	67

mM, mmol/L; µM, µmol/L;

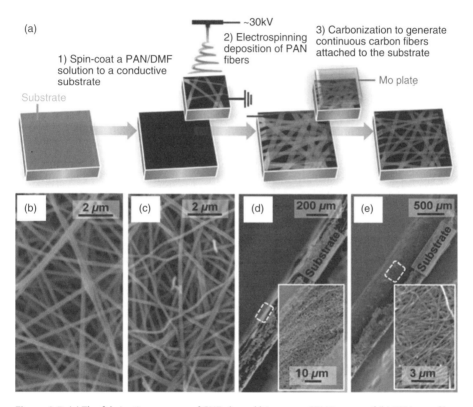

Figure 2.5 (a) The fabrication process of CNFs-based biosensor; SEM images of (b) PAN nanofibers and (c) CNFs; Cross-sectional SEM images of biosensors fabricated for different deposition times: (d) 12 h, (e) 68 h. *Source*: Mao 2014 [44]. Reproduced with permission of the American Chemical Society. (*See color plate section for the color representation of this figure.*)

nitrogen-containing tail gas generated during the carbonization for surface etching to introduce nitrogen atom. The NCNFs displayed superior catalytic activity toward the oxidation of DA, AA and UA; a high selectivity toward the simultaneous detection of DA, AA and UA was observed at NCNFs and large peak to peak potential separations for AA-DA (0.277 V) and DA-UA (0.124 V) were achieved. The limit of detection (LOD) of NCNFs-based biosensor was 0.5, 50 and 1 µM for DA, AA and UA, respectively, implying the superior sensitivity [45]. For metal nanoparticle-loaded CNFs, Liu's group has successfully prepared Ag-Pt bimetallic nanoparticles-loaded electrospun nanoporous CNFs (Ag-Pt/pCNFs) for the fabrication of DA biosensor. Taking advantage of synergistic effect at the interface of Ag-Pt binary structure, the Ag-Pt/pCNFs based biosensor showed excellent sensitivity and selectivity for the detection of DA with a linear concentration range of 10–500 µM and a LOD of 0.11 µM [46]. Furthermore, many other composites with novel structures or ingredients were prepared and employed for the determination of DA, such as electrospun graphene/polyaniline/polystyrene nanofiber [47].

Besides, a nanoelectrode array (NEA) formed by VACNFs has been extensively studied as a sensing element in many important devices. Marsh *et al.* have developed an intriguing device for the simultaneous detection of DA and O_2 by combining the patterned CNFs NEA with wireless instantaneous neurotransmitter concentration sensor. Their results showed that CNFs nanoelectrode array could enhance the spatial resolution for neurochemical detection, which may be beneficial to elucidate the pathophysiology of disorders of the nervous system and the mechanism of deep brain stimulation [49].

Hydrogen peroxide: The detection of hydrogen peroxide (H_2O_2) plays a critical role in various fields, such as pharmacy, food and environmental protection. In recent years, H_2O_2 biosensors based on CNFs composites mainly employed nitrogen-doped CNFs (NCNFs) and metal nanoparticles-loaded CNFs (Pt/CNFs etc.).

You's group has prepared nitrogen-doped carbon nanoparticles-embedded CNFs (NCNPFs) by electrospinning/carbonization for the evaluation of H_2O_2. The polypyrrole nanoparticles (PPy NPs) were embedded in the nanofibers by electrospinning as the nitrogen source, and the characteristic of NCNPFs could be controlled by varying the amount of PPy NPs. The as-obtained NCNPFs with self-supported structure and high flexibility could be directly used as electrode free of any treatment. The optimum NCNPFs based H_2O_2 biosensor exhibited a high sensitivity of 383.9 µA µM^{-1} cm^{-2} with a superior selectivity and reproducibility (Figure 2.6) [50].

Metal nanoparticles-loaded CNFs, mostly Pt/CNFs with different structures and morphologies have been fabricated for the determination of H_2O_2 [51, 53]. Generally, Pt nanoparticles could be directly anchored on the CNFs, or loaded on PAN nanofibers followed by carbonization to obtain Pt/CNFs. For the later method, Pt nanoparticles-loaded PAN nanofibers could be synthesized by electrospining Pt nanoparticles-containing precursor, followed by carbonization to prepare Pt/CNFs [53]. Besides Pt/CNFs, bimetal nanoparticles-loaded CNFs, such as PdCo/CNFs [54], and manganese dioxide nanoparticles-loaded CNFs (MnO$_2$/CNFs) were also prepared for the sensitive detection of H_2O_2 [55].

Glucose: The glucose biosensors have been extensively investigated in many fields, such as food and clinical. Taking clinical for example, the detection of glucose is of great importance in diagnosis of diabetes mellitus, which has been a worldwide health problem [56]. Various CNFs composites, mainly the metallic nanostructure and metallic oxide nanostructure-loaded CNFs, have been employed for the fabrication of nonenzymatic glucose sensors.

For metal nanostructure/CNFs composites, Ni nanoparticles-loaded CNFs (Ni/CNFs) have attracted considerable attention for the construction of non-enzymatic

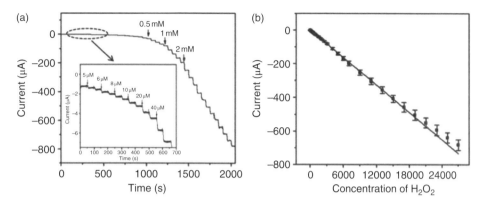

Figure 2.6 (a) *I-t* curve of NCNPFs electrode in 0.1 M PBS (pH 7.4) with the successive addition of H2O2. Inset: the magnified *I-t* curve with low concentration of H_2O_2. (b) Calibration curve for the detection of H_2O_2 in Figure 2.6a. *Source:* Zhanga 2016 [50]. Reproduced with permission of Elsevier.

glucose sensors. Li *et al.* have prepared Ni/CNFs by carbonization of electrospun $Ni(NO_3)_2$-loaded polyvinyl alcohol nanofibers. The resultant Ni/CNFs exhibited an average diameter of 116 nm, and Ni nanoparticles with an average diameter of 64 nm were observed on the surface as well as inside of the CNFs. Interestingly, Ni/CNFs modified electrodes showed an electrochemical behavior as microelectrodes which was ascribed to the continuous and long nanofibers. Taking advantages of large specific surface area ($286.12 \text{ m}^2 \text{ g}^{-1}$) and high electrochemical activity, Ni/CNFs displayed good sensitivity toward the determination of glucose with a detection limit as low as 0.05 µM [57].

Bimetallic nanoparticles could show higher electrocatalytic activity, selectivity, and stability in comparison with monometallic nanoparticles [58]. By the combination of electrospinning and thermal treatment, CuCo, FeCo, NiCo and MnCo bimetallic nanoparticles-loaded CNFs were prepared. In contrast to nanoparticles adsorbed on CNFs, the bimetallic nanoparticles synthesized by electrospinning were anchored and embedded into the CNFs according to the characterization results. The fabricated biosensors showed different analysis performances toward the determination of glucose, and the sensitivity decreases as following: CuCo/CNFs > FeCo/CNFs > NiCo/CNFs > Co/CNFs > MnCo/CNFs. The CuCo/CNFs displayed the best analysis ability (linear range: 0.02–11 mM; detection limit: 1.0 µM), which was attributed to the unique 3D network films as well as the synergistic effect of the Co(III)/Co(IV) and Cu(II)/Cu(III) redox couples [59]. Moreover, the bimetallic nanoparticles could not only decrease the utilization of noble metal (Pt, Pd), but also enhance the electrochemical properties (Figure 2.7a). You's group has prepared PdNi alloy nanoparticle/carbon nanofibers (PdNi/CNFs) with different ratios of Pd and Ni by electrospinning polyacrylonitrile/Pd(acac)$_2$/Ni(acac)$_2$ nanofiber with corresponding ratios of metallic precursors, followed by carbonization. The advantages of this approach are as follows: 1) The characteristic properties of nanoparticles (size, composition, and alloy homogeneity) could be tailored simply by changing the composition of electrospinning solution or carbonization process; 2) The obtained PdNi nanoparticles were firmly embedded in CNF with uniform dispersion. The electrochemical test results

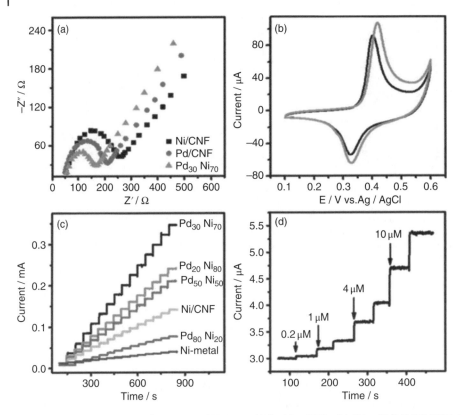

Figure 2.7 (a) Nyquist plots of Ni/CNFs, Pd/CNFs and Pd30Ni70/CNFs; (b) CVs of Pd30Ni70/CNFs in the presence (red) and absence (black) of glucose; (c) *I-t* curves of different electrodes with the addition of glucose; (d) *I-t* curve of Pd30Ni70/CNFs with the addition of varying concentrations of glucose. *Source:* Guo 2014 [60]. Reproduced with permission of the American Chemical Society. (*See color plate section for the color representation of this figure.*)

revealed that the electrocatalytic activity of these composites toward glucose increased in the following order:

$$Ni-metal < Pd_{80}Ni_{20}/CNFs < Ni/CNFs < Pd_{50}Ni_{50}/CNFs < Pd_{20}Ni_{80}/CNFs < Pd_{30}Ni_{70}/CNFs.$$

The $Pd_{30}Ni_{70}$/CNFs-based glucose sensor showed a detection limit of ca. 60 nM with a detection range from 0.1 μM to 5.4 mM (Figure 2.7b–d). When utilized inflow system, liquid chromatography-electrochemical system and flow-injection analysis, $Pd_{30}Ni_{70}$/CNFs electrode displayed high sensitivity to the sugar mixture (glucose, fructose, sucrose and maltose). The detection limits of $Pd_{30}Ni_{70}$/CNFs for glucose, fructose, sucrose, and maltose were 7, 9, 15, and 20 nM, respectively. The superior analysis properties of $Pd_{30}Ni_{70}$/CNFs were ascribed to the unique electronic structure, appropriate surface Ni content in PdNi alloy and uniform dispersion of PdNi nanoparticles in CNFs [60].

In recent years, metal oxide materials, such as CuO and NiO have been investigated for the fabrication of high-performance and low-cost nonenzymatic glucose biosensors.

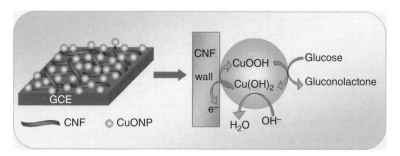

Figure 2.8 The mechanism of CuO/CNFs-based glucose biosensor. *Source:* Zhang 2013 [62]. Reproduced with permission of Elsevier. (*See color plate section for the color representation of this figure.*)

Zhang *et al.* fabricated a novel glucose biosensor based on the cupric oxide nanoparticles-loaded carbon nanofibers (CuO/CNFs). The CuO/CNFs was prepared via a simple precipitation procedure; during glucose biosensing, the glucose was oxidized to glucolactone by the Cu(III) derived from CuO which served as an electron transfer mediator (Figure 2.8). The determination of glucose could be achieved through the analysis of the change of redox current for Cu(III)/Cu(II). The CuO/CNFs-based biosensors exhibited high electrocatalytic activity for the oxidation of glucose with a detection limit of 0.2 µM [62]. For the enhanced analysis properties, CuO nanostructures with abundant electroactive sites were synthesized to fabricate novel glucose biosensors. Xu *et al.* prepared a novel nanorod-aggregated flower-like CuO loaded carbon fiber fabric (CFF) by a hydrothermal approach and investigated its ability for the determination of glucose. Taking the advantages of high surface area and abundant active sites of CFF, CuO nanorod with an average diameter of 10 nm aggregated on the surface of CFF to form a flower-like morphology. The as-obtained CuO/CFF exhibited high flexibility which could facilitate the construction of biosensors. The electrochemical tests revealed the satisfactory catalytic activity of CuO/CFF for glucose oxidation. The CuO/CFF-based biosensor displayed a low detection limit of 0.27 µM with high reproducibility and long-term stability [64].

Besides CuO, nickel(II) hydroxide (NiOOH) was also utilized for the fabrication of high-performance nonenzymatic glucose biosensors. The plate $Ni(OH)_2$-loaded CNFs ($Ni(OH)_2$/CNFs) has been synthesized through a hydrothermal route. For $Ni(OH)_2$/ CNFs, $Ni(OH)_2$ nanoplates with the size of 200 nm aggregated on the surface of CNFs. For the detection of glucose, NiO(OH) species generated under alkaline conditions from $Ni(OH)_2$ could oxidize the glucose to glucolactone. The fabricated glucose biosensor based on $Ni(OH)_2$/CNFs showed high sensitivity and selectivity toward glucose detection with a wide linear range (0.001 to 1.2 mM) and a detection limit of 0.76 µM [66].

Heavy Metal Ions: CNFs composites have been utilized for the fabrication of sensors toward the sensitive detection of heavy metal ions, such as Hg^{2+}, Pb^{2+} and Cd^{2+} [68–70].

Zhao *et al.* has prepared a CNFs/Nafion-based sensor for the evaluation of Pb^{2+} and Cd2+ via anodic stripping voltammetry (ASV). The results revealed that the introduction of CNFs could significantly enhance the sensitivity toward the determination of Pb^{2+} 8-fold, and the LOD of this sensor for the simultaneous detection of Pb^{2+} and Cd^{2+} were 0.9 and 1.5 nM, respectively (Figure 2.9) [69]. As for Hg^{2+}, Liao *et al.* constructed

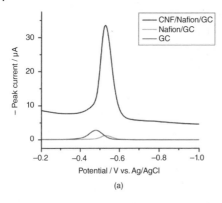 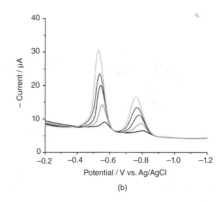

(a)   (b)

Figure 2.9 (a) Stripping voltammograms of CNF/Nafion/GCE, Nafion/GCE, and GCE in 0.1 M acetate buffer containing 0.5 μM Pb^{2+}; (b) stripping voltammograms of CNF/Nafion/GCE in 0.1 M acetate buffer with different concentration of Pb^{2+} and Cd^{2+}. *Source:* Zhao 2015 [69]. Reproduced with permission of the American Chemical Society. (*See color plate section for the color representation of this figure.*)

a Hg^{2+} sensor by using bis(indolyl)methane and mesoporous carbon nanofiber (MCNFs). Noteworthy, the MCNFs capability for adsorbing Hg^{2+} improved the sensitivity of the sensor. Using the ASV technique, the sensor allowed to quantitatively determine Hg^{2+} with a detection range of 5–500 nM and a LOD of 0.3 nM [70].

Other Biosensors: Besides the applications mentioned above, many CNFs composites were also investigated for the fabrication of non-enzymatic biosensors toward the detection of environmental contaminants (catechol [71, 72], hydroquinone [73] and ibuprofen [74] etc.), drug (insulin [75], nimodipine [76] and β-blockers [77] etc.) and so on.

2.4.3 Enzyme-based biosensors

The utilization of enzyme-based biosensors has extensively been promoted in the last decade due to the simple and inexpensive construction as well as rapid analysis and ease of regeneration. The enzyme-based electrochemical biosensors could be prepared by the modification of electrodes with biological enzymes. For enzyme-based biosensors, the enzymes are of great importance due to the fact that the analytical properties of these biosensors, particularly the sensitivity and selectivity, strongly depend on the activity and stability of the immobilized enzymes [78]. However, the activity of immobilized enzyme on the electrode could gradually decrease, leading to the poor lifespan of the biosensor [79]. Thus, the supporting materials which help to retain the activity of enzymes are highly desirable for the construction of enzyme-based biosensors.

CNFs shows many important properties, such as large functionalized surface area and surface-active groups-to-volume ratio, which could benefit the anchor of enzymes on the surface (Figure 2.10). The selective immobilization could be achieved by controlling the type of functional groups on CNFs [80]. On the other hand, the DOS of supporting materials near the Fermi level could strongly influence the direct electron transfer (DET) of enzymes. Hatton's group has developed a novel approach to prepare CNFs with various DOS which could modulate electron transfer kinetics and efficiencies for Cytochrome C and horseradish peroxidase [81]. Except for immobilization

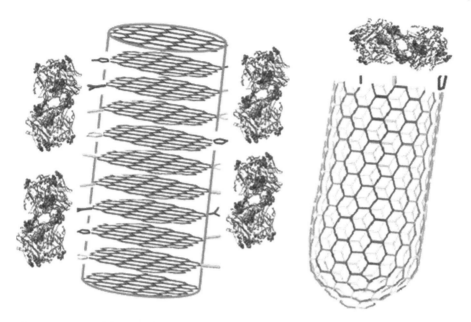

Figure 2.10 Immobilization of glucose oxidase on CNFs and single-walled CNTs. *Source:* Vamvakaki 2006 [10]. Reproduced with permission of the American Chemical Society. (*See color plate section for the color representation of this figure.*)

matrixes, CNFs also serves as good transducers for electrochemical signal due to the superior conductivity. Up to date, CNFs have been utilized to fabricate many biosensors serving as the high-performance supporting matrix for enzymes (Table 2.2).

Among these investigations, glucose oxidase GOx-based biosensors have been most frequently studied for the determination of glucose, which is of great importance in glucose detection for diabetes control and biofuel cells [82]. Chaniotakis's group has employed three different types of CNFs and single-walled CNTs as the supporting matrix for GOx to construct the glucose biosensors. The results demonstrated that the CNFs-based biosensors showed higher sensitivity and long-term stability in comparison with those fabricated by single-walled CNTs. They ascribed the superior analytical performance to the large functionalized surface area and abundant active sites of CNFs, which could benefit the adsorption and retain the bioactivity of GOx [10].

The oxidation of glucose catalyzed by enzyme GOx could produce H_2O_2, and monitoring the amount of H_2O_2 in this process could be utilized for the indirect detection of glucose. Hou's group has developed glucose biosensor by using Prussian blue (PB) nanostructures/carboxylic group-functionalized carbon nanofiber nanocomposites (PB/FCNFs) as supporting matrix for GOx. They employed the PB as the electron-transfer mediator to decrease the reduction potential of H_2O_2. The PB/FCNFs were prepared by controllable growth of PB nanostructures on the surface of FCNFs which exhibited high sensitivity toward the detection of H_2O_2. Then, the glucose biosensor was constructed by the immobilization of GOx on PB/FCNFs (GOx/PB/FCNFs). The electrochemical tests revealed that the as-obtained GOx/PB/FCNFs exhibited high sensitivity and stability for the determination of glucose with a detection limit of 0.5 μM and a wide linear range from 0.02 to 12 mM [82].

Table 2.2 A summary of enzyme-based biosensors fabricated by CNFs composites.

Biosensor	Analyte	Linear range	LOD	Ref.
GOx/prussian blue/ functionalized CNFs	Glucose	0.02–12 mM	0.5 μM	82
GOx//nitrogen-doped CNFs	Glucose	0.2–1.2mM	0.06 mM	84
GOx/nitrogen-doped carbon nanospheres@CNFs	Glucose	12 – 1000 μM	2 μM	85
GOx/palladium/helical CNFs	Glucose	0.06–6.0 mM	0.03 mM	86
Polydopamine-laccase-nickel nanoparticle/CNFs	Catechol	1 μM – 9.1 mM	0.69 μM	87
Laccase/graphene/CNFs	Catechol	0.1 μM – 2.86 mM	0.02 μM	88
Laccase/ZnO/CNFs	Hydroquinone	0.5 – 2.06 μM	9.50 nM	89
Lactate oxidase/Pt/ graphitized CNFs	Lactate	0.1 μM – 5.4 mM	0.06 μM	90
Lactate oxidase/Pt/CNFs/ diallyldimethylammonium) chloride	Hydrogen peroxide	0.125 – 12.73 μM	0.05 μM	91
Horseradish peroxidase/ CNFs/polyaniline/Au	Hydrogen peroxide	0.01–4.30 mM	0.18 μM	92
Hemoglobin/Au/helical CNFs	Hydrogen peroxide	1 – 3157 μM	0.46 μM	93
Hemoglobin/ZnO/CNFs	Hydrazine	19.8 μM – 1.71 mM	0.1 μM	94

mM, mmol/L; μM, μmol/L;

The introduction of nitrogen into the carbon nanomaterials has proven to be an efficient method to enhance the properties of carbons, such as biocompatibility and catalytic activity [83]. You's group has developed a facile approach for the production of free-standing nitrogen-doped carbon nanofibers film (NCNFs). Taking the advantage of free-standing structure with high flexibility, the resultant NCNFs film could be cut to the shapes required, which would efficiently simplify the biosensors fabrication. As the supporting matrix for GOx, the NCNFs could benefit to immobilize and retain the activity of GOx, resulting in the fast electron transfer between enzyme and electrode. The as-fabricated GOx/NCNFs based glucose biosensor showed a linear range of 0.2–1.2 mM at −0.42 V with a detection limit of 0.06 mM. Moreover, these biosensors demonstrated the superior selectivity and stability toward the detection of glucose [84]. Furthermore, the same group developed another glucose biosensor based on the direct electron transfer between the enzyme GOx and nitrogen-doped carbon nanospheres@ carbon nanofibers (NCNSs@CNFs) composite film. The introduction of carbon nanospheres derived from polypyrrole nanospheres served as the nitrogen source during the carbonization and increased the surface area of the obtained composites, leading to high nitrogen content and abundant active sites for GOx immobilization. Electrochemical experiments revealed that the NCNSs@CNFs could realize a fast direct electron transfer between the electrode and GOx in the biosensor. The as-fabricated biosensors displayed high sensitivity, selectivity and long-term stability toward the electrochemical determination of glucose, providing a low detection limit of 2 μM [85].

Besides, CNFs have also been employed as the supporting matrix for some other enzymes to construct high-performance electrochemical biosensors, such as laccase [87–89], lactate oxidase [89–91], horseradish peroxidase [92] and hemoglobin [93, 94].

Laccase (Lac) could catalyze the oxidation of phenolic compounds in the presence of oxygen. Lac-based electrochemical biosensors are always fabricated for the detection of phenolic species which are harmful for human body. Li *et al.* fabricated a novel phenolic electrochemical biosensor based on nickel nanoparticles loaded CNFs (Ni/CNFs). The Lac was anchored onto the surface of Ni/CNFs by polydopamine (PDA) during the process of dopamine oxidation (Figure 2.11). Their experiments demonstrated the superior bio-compatibility of Ni/CNFs which was beneficial for the anchor of Lac as well as the fast electron transfer between Lac and electrode. The PDA/Lac/Ni/CNFs-based biosensors displayed enhanced analysis properties toward the detection of catechol, achieving a detection limit of 0.69 μM. Moreover, the resultant biosensor could be employed for the estimation of catechol in real samples [87]. To achieve higher conductivity and surface area, graphene has been immobilized on the surface of CNFs for the enhancement of graphitization and conductivity. The graphene loaded CNFs (G/CNFs) was

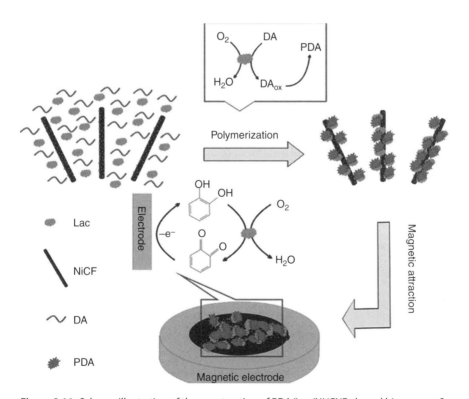

Figure 2.11 Scheme illustration of the construction of PDA/Lac/Ni/CNFs-based biosensors. *Source:* Li 2014 [87]. Reproduced with permission of the Royal Society of Chemistry. (*See color plate section for the color representation of this figure.*)

prepared by adsorption and chemical reduction of graphene oxide on CNFs. As the supporting matrix for Lac, the G/CNFs composites could realize the direct electrochemistry of Lac between enzyme and electrode. For the detection of catechol, the as-fabricated biosensor achieved a detection limit of 20 nM and over a linear range of 0.1 to 2.86 μM with good selectivity and stability [88].

The CNFs have also been investigated for the direct electrochemistry of LOx using platinum nanoparticles decorated CNFs (Pt/CNF) as a supporting matrix. Combining with screen-printed carbon electrodes, the fabricated biosensors could facilitate the estimation of lactate for the practical applications. The linear range of the biosensor was from 25 to 1500 mM with a detection limit of 11 mM. Meanwhile, the selectivity and stability of the biosensor were also satisfactory. Importantly, the as-fabricated biosensor was successfully used for the detection of lactate in sweat and blood samples in a sport tests [91].

2.4.4 CNFs-based immunosensors

Immunosensors are affinity ligand-based biosensors, and their detection mechanism is based on the molecular recognition between antigens and corresponding antibodies [95]. Recent years, many studies have been focused on electrochemical immunosensors because of their low cost, high sensitivity and simple fabrication. Among these investigations, CNFs have been considered as a suitable platform to construct electrochemical immunosensors due to the large surface area for antibody immobilization [96]. CNFs-based electrochemical immunosensors mainly employed differential pulse voltammetry, linear sweep voltammogram as well as electrochemiluminescent techniques for the detection of antigens. The immunosensors based on CNFs composites have been summarized in Table 2.3.

Table 2.3 A summary of immunosensors based on CNFs composites.

Biosensor	Analyte	Linear range	LOD	Ref.
CNFs	Recombinant bovine somatotropin	1pg/mL-10ng/mL	1pg/mL	96
CNFs	Microcystin-LR	$0.0025 - 5\,\mu g/L$	1.68 ng/L	97
CNFs	Protein	$1\,\mu M - 9.1$ mM	$0.69\,\mu M$	98
CNFs forests	Plasmodium falciparum histidine rich protein-2	0.025 to 10 ng/mL	0.025 ng/mL	99
CNFs nanoelectrode arrays	Human cardiac troponin-I (cTnI)	0.25–1.0 ng/mL 5.0–100 ng/mL	0.2 ng/mL	100
CNFs-based SPE	miRNA-34a target RNA	$25 - 100\,\mu g/mL$	$10.98\,\mu g/mL$	103
CNTs functionalized nanofibers	$^-$-Fetoprotein	0.1 pg/mL-160 ng/mL	0.09pg/mL	104
CNFs-polyamidoamine dendrimer	IgG	0.06–6.0 mM	0.5 fg/mL	105

mM, mmol/L; μM, μmol/L;

CNFs have been explored for the construction of label-free immunosensors by DPV, EIS and ECL. CNFs screen-printed electrode (CNFs SPE)-based electrochemical immunosensor was fabricated by Lim and Ahmed for the sensitive detection of recombinant bovinesomatotropin (rBST). In this work, CNFs SPE was employed as a substrate for the immunosensor which could increase the current signal response compared with CNTs; the immobilization of the antibody was achieved by binding boronic acid to an oligosaccharide moiety on the antibody which could help to retain the biological activity of the antibody. The response signal was the decrease of DPV peak current of $Fe(CN)_6^{3-/4-}$ due to the formation of immunocomplex. The fabricated immunosensor showed high analytical performance toward the determination of rBST with a linear range of 1 pg mL^{-1}–10 ng mL^{-1} and a detection limit of 1 pg mL^{-1} [96].

Chen's group fabricated a non-enzymatic immunosensor based on CNFs and gold nanoparticles (Au NPs) to detect microcystin-LR (MC-LR). The MC-LR antigen was immobilized on carboxylic group functionalized CNFs; then, by the addition of target MC-LR and its antibody, MC-LR would compete with MC-LR antigen to generate antibody-antigen immune complex. Next, Au NPs labeled secondary antibodies bound with the immune complex, and the signal for the detection of MC-LR was recorded by the oxidation of Au NPs via DPV (Figure 2.12). The CNF-based immunosensor displayed a linear range of 0.0025–5 µg L^{-1} and a LOD of 1.68 ng L^{-1}. Detailed experiments analysis revealed that the introduction of CNFs increased the conductivity as well as accelerated the electron transfer, resulting in the high-performance immunosensor for MC-LR detection [97].

Koehne's group prepared a CNFs NEA-based label-free immunosensor to detect human cardiac troponin-I (cTnI) via EIS technique. Anti-cTnI Ab was directly

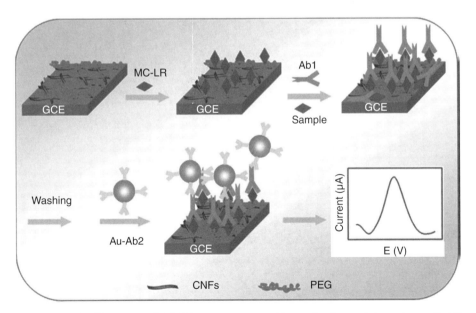

Figure 2.12 The fabrication of MC-LR immunosensor and its application to detect microcystin-LR. *Source:* Zhang 2016 [97]. Reproduced with permission of Elsevier. (*See color plate section for the color representation of this figure.*)

immobilized on CNFs by covalent binding, and the modification and detection were both monitored by EIS. The changes in electrical capacitance or resistance to charge transfer derived from the formation of Ab–Ag were recorded for the evaluation of human-cTnI. The immunosensor was able to sensitively detect human-cTnI with a LOD of 0.2 ng mL^{-1} [100]. Furthermore, CNFs NEA was utilized for the fabrication of a biosensor to investigate the kinetics of phosphorylation and dephosphorylation of surface-attached peptides by real-time electrochemical impedance spectroscopy technique [101]. Besides, CNFs NEA was also used for the activity evaluation of proteases which is related to many important cellular processes [102].

Noteworthy, electrochemiluminescence (ECL) has been extensively studied to fabricate high-performance immunosensors. Dai *et al.* developed a novel immunosensor by using carbon nanotube nanofibers (CNTs@PNFs) for the evaluation of the interaction between antibody and antigen in vitro. For the immunosensor construction, CNTs@PNFs were employed for the enhancement of ECL emission, whilst the α-Fetoprotein (AFP) antibody serving as recognition biomolecule was anchored on CNTs@PNFs by electrostatic interaction (Figure 2.13). The detection of AFP could be achieved through the analysis of ECL change induced by the immunoreaction; a linear range of 0.1 pg mL^{-1}–160 ng mL^{-1} and a detection limit of 0.09 pg mL^{-1} were achieved [104].

Besides the label-free immunosensors, CNFs have also been used for the fabrication of electrochemical immunosensors with labels. Taking Ma's report as example, they have constructed an electrochemical immunosensor with sandwich structure for

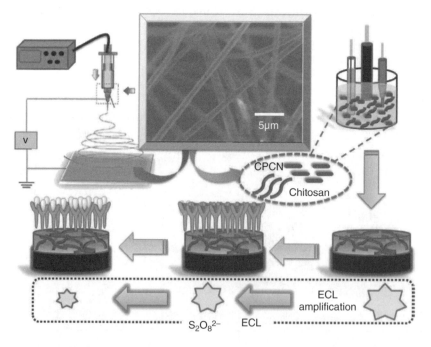

Figure 2.13 The fabrication process of ECL immunosensors based on CNTs@PNFs. *Source:* Dai 2014 [104]. Reproduced with permission of Elsevier. (*See color plate section for the color representation of this figure.*)

immunoglobulin G (IgG) detection using Au@Ag nanorods (Au@Ag NRs) and CNFs. For this sensor, antibodies were immobilized on CNFs-polyamidoamine dendrimer (CNFs-PAMAM), and anti-human IgG was anchored on Au@Ag NRs as trace tag. Due to the superior conductivity of CNFs, the introduction of CNFs-PAMAM in the system could efficiently increase the peak current of Ag in Au@Ag NRs which was proportional to IgG concentration. This immunosensor could be used for the detection of IgG with high sensitivity (LOD: 0.5 fg mL^{-1}) and acceptable stability [105].

2.5 Conclusions

Electrochemical biosensors play an important role in many critical fields, such as clinical detection and food-quality monitoring. Taking advantages of high conductivity, large surface area and superior catalytic activity, carbon nanofibers composites have been extensively utilized to construct high-performance electrochemical biosensors. The investigations of carbon nanofibers on electroanalysis have acquired significant advances in recent years; however, there are still many challenges in this field. Up to date, most of the investigations have been performed in a laboratory setting, and more attentions are supposed to be paid on the practical applications, such as carbon nanofiber-based electroanalysis devices (screen-printed electrode etc.). Meanwhile, the applications of carbon nanofibers for electroanalysis could be expanded in more emerging fields (single molecule analysis etc.), and the analysis properties of carbon nanofibers based biosensor, particularly sensitivity and selectivity, would be enhanced for more accurate estimation of analyte in specific situations. The continuous investigations on the carbon nanofibers may offer feasible strategies to solve these issues.

References

1 S. Iijima, Helical microtubules of graphitic carbon. Nature, 1991, 354, 56–58.
2 A.K. Geim, K.S. Novoselov. The rise of grapheme. Nature Materials, 2007, 6, 183–191.
3 D. Kasuya, M. Yudasaka, K. Takahashi, *et al*. Selective production of single-wall carbon nanohorn aggregates and their formation mechanism. Journal of Physical Chemistry B, 2002, 106, 4947–4951.
4 L.F. Zhang, A. Aboagye, A. Kelkar, *et al*. A review: carbon nanofibers from electrospun polyacrylonitrile and their applications. Journal of Materials Science, 2014, 49, 463–480.
5 P. Serp, M. Corrias, P. Kalck. Applied Catalysis A: General, Carbon nanotubes and nanofibers in catalysis 2003, 253, 337–358.
6 T.V. Hughes, C.R. Chambers. Manufacture of Carbon Filaments, U.S. Patent 1889, 405, 480.
7 K.P. Dejong, J.W. Geus. Carbon nanofibers: Catalytic synthesis and applications. Catalysis Reviews, 2000, 42, 481–510.
8 W. Li, F. Zhang, Y.Q. Dou, Z.X. *et al*. A self-template strategy for the synthesis of mesoporous carbon nanofibers as advanced supercapacitor electrodes. Advanced Energy Materials, 2011, 1, 382–386.

9 J.S. Guo, G.Q. Sun, Q. Wang, *et al.* Carbon nanofibers supported Pt–Ru electrocatalysts for direct methanol fuel cells. Carbon, 2006, 44, 152–157.

10 V. Vamvakaki, K. Tsagaraki, N. Chaniotakis. Carbon nanofiber-based glucose biosensor. Analytical Chemistry, 2006, 78, 5538–5542.

11 Y. Matsumoto, M.T. Oo, M. Nakao, *et al.* Preparation of carbon nanofibers by hot filament-assisted sputtering. Materials Science and Engineering: B, 2000, 74, 218–221.

12 G.F. Zou, D.W. Zhang, C. Dong, *et al.* Carbon nanofibers: Synthesis, characterization, and electrochemical properties. Carbon, 2006, 44, 828–832.

13 G.B. Zheng, K. Kouda, H. Sano, *et al.* A model for the structure and growth of carbon nanofibers synthesized by the CVD method using nickel as a catalyst. Carbon, 2004, 42, 635–640.

14 Z.P. Zhou, C.L. Lai, L.F. Zhang, *et al.* Development of carbon nanofibers from aligned electrospun polyacrylonitrile nanofiber bundles and characterization of their microstructural, electrical, and mechanical properties. Polymer, 2009, 50, 2999–3006.

15 Y. Liu, H. Teng, H.Q. Hou, T.Y. You. Nonenzymatic glucose sensor based on renewable electrospun Ni nanoparticle-loaded carbon nanofiber paste electrode. Biosensors and Bioelectronics, 2009, 24, 3329–3334.

16 A.V. Melechko, V.I. Merkulov, T.E. McKnight, *et al.* Vertically aligned carbon nanofibers and related structures: controlled synthesis and directed assembly. Journal of Applied Physics, 2005, 97, 041301.

17 G.G. Tibbetts, M.L. Lake, K.L. Strong, B.P. Ricec. A review of the fabrication and properties of vapor-grown carbon nanofiber/polymer composites. Composites Science and Technology, 2007, 67, 1709–1718.

18 D.H. Reneker, A.L. Yarin, H. Fong, S. Koombhongse. Bending instability of electrically charged liquid jets of polymer solutions in electrospinning. Journal of Applied Physics, 2000, 87, 4531–4547.

19 J. Zeng, X.Y. Xu, X.S. Chen, *et al.* Biodegradable electrospun fibers for drug delivery. Journal of Controlled Release, 2003, 92, 227–231.

20 S. Megelski, J.S. Stephens, D.B. Chase, J.F. Rabolt. Micro-and nanostructured surface morphology on electrospun polymer fibers. Macromolecules, 2002, 35, 8456–8466.

21 J.M. Deitzel, J. Kleinmeyer, D. Harris, N.C. Tan. Controlled deposition of electrospun poly(ethylene oxide) fibers. Polymer, 2001, 42, 261–272.

22 X.H. Zong, K. Kima, D.F. Fang, *et al.* Structure and process relationship of electrospun bioabsorbable nanofiber membranes. Polymer, 2002, 43, 4403–4412.

23 L. Zou, L. Gan, R. Lv, *et al.* A film of porous carbon nanofibers that contain Sn/SnO_x nanoparticles in the pores and its electrochemical performance as an anode material for lithium ion batteries. Carbon, 2011, 49, 89–95.

24 Y. Yang, A. Centrone, L. Chen, *et al.* Highly porous electrospun polyvinylidene fluoride (PVDF)-based carbon fiber. Carbon, 2011, 49, 3395–3403.

25 R. Ruiz-Rosasa, J. Bediaa, M. Lallaveb, *et al.* The production of submicron diameter carbon fibers by the electrospinning of lignin. Carbon, 2010, 48, 696–705.

26 M. Inagaki, Y. Yang, F.Y. Kang. Carbon nanofibers prepared via electrospinning. Advanced Materials, 2012, 24, 2547–2566.

27 C. Kim, K.S. Yang, M. Kojima, *et al.* Fabrication of electrospinning-derived carbon nanofiber webs for the anode material of lithium-ion secondary batteries. Advanced Functional Materials, 2006, 16, 2393–2397.

28 Z.H. Wang, L. Qie, L.X. Yuan, *et al*. Functionalized N-doped interconnected carbon nanofibers as an anode material for sodium-ion storage with excellent performance. Carbon, 2013, 55, 328–334.

29 N. Lu, C.L. Shao, X.H. Li, *et al*. CuO nanoparticles/nitrogen-doped carbon nanofibers modified glassy carbon electrodes for non-enzymatic glucose sensors with improved sensitivity, Ceramics International, 2016, 42, 11285–11293.

30 S. Maldonado, K.J. Stevenson. Direct preparation of carbon nanofiber electrodes via pyrolysis of iron (II) phthalocyanine: electrocatalytic aspects for oxygen reduction. The Journal of Physical Chemistry B, 2004, 108, 11375–11383.

31 D. Liu, X.P. Zhang, Z.C. Sun, T.Y. You. Free-standing nitrogen-doped carbon nanofiber films as highly efficient electrocatalysts for oxygen reduction. Nanoscale, 2013, 5, 9528–9531.

32 H.W. Liang, Z.Y. Wu, L.F. *et al*. Bacterial cellulose derived nitrogen-doped carbon nanofiber aerogel: An efficient metal-free oxygen reduction electrocatalyst for zinc-air battery. Nano Energy, 2015, 11, 366–376.

33 D.W. Wang, Y. Liu, J.S. Huang, T.Y. You. In situ synthesis of Pt/carbon nanofiber nanocomposites with enhanced electrocatalytic activity toward methanol oxidation. Journal of Colloid and Interface Science, 2012, 367, 199–203.

34 O.A. Loaiza, P.J. Lamas-Ardisana, L. Añorga, *et al*. Graphitized carbon nanofiber–Pt nanoparticle hybrids as sensitive tool for preparation of screen printing biosensors. Detection of lactate in wines and ciders. Bioelectrochemistry, 2015, 101, 58–65.

35 D. Liu, Q. H. Guo, H. Q. Hou, *et al*. Pd_xCo_y nanoparticle/carbon nanofiber composites with enhanced electrocatalytic properties. ACS Catalysis, 2014, 4, 1825–1829.

36 P.T. Kissinger, W.R. Heineman. Laboratory Techniques in Electroanalytical Chemistry (eds). Marcel Dekker, New York, NY, USA, 1996.

37 B. R. Eggins, Chemical Sensors and Biosensors. John Wiley and Sons, West Sussex, England, 2002.

38 I.I. Suni. TrAC, Impedance methods for electrochemical sensors using nanomaterials. Trends in Analytical Chemistry, 2008, 27, 604.

39 S.F. Hou, M.L. Kasner, S.J. Su, *et al*. Highly sensitive and selective dopamine biosensor fabricated with silanized graphene. Journal of Physical Chemistry C, 2010, 114, 14915–14921.

40 P. Ekabutr, P. Sangsanoh, P. Rattanarat, *et al*. Development of a disposable electrode modified with carbonized, graphene-loaded nanofiber for the detection of dopamine in human serum. Journal of Applied Polymer Science, 2014, 131, 40858.

41 H.S. Han, H.K. Lee, J.M. You, *et al*. Electrochemical biosensor for simultaneous determination of dopamine and serotonin based on electrochemically reduced GO-porphyrin. Sensors and Actuators B: Chemical, 2014, 190, 886–895.

42 Y. Yue, G.Z. Hu, M.B. Zheng, *et al*. mesoporous carbon nanofiber-modified pyrolytic graphite electrode used for the simultaneous determination of dopamine, uric acid, and ascorbic acid. Carbon, 2012, 50, 107–114.

43 E. Rand, A. Periyakaruppan, Z. Tanaka, *et al*. A carbon nanofiber based biosensor for simultaneous detection of dopamine and serotonin in the presence of ascorbic acid. Biosensors and Bioelectronics, 2013, 42, 434–438.

44 X.W. Mao, X.Q. Yang, G.C. Rutledge, T. A. Hatton. Ultra-wide-range electrochemical sensing using continuous electrospun carbon nanofibers with high densities of states. ACS Applied Materials and Interfaces, 2014, 6, 3394–3405.

45 J.Y. Sun, L.B. Li, X.P. Zhang, *et al*. Simultaneous determination of ascorbic acid, dopamine and uric acid at a nitrogen-doped carbon nanofiber modified electrode. RSC Advances, 2015, 5, 11925–11932.

46 Y.P. Huang, Y.E, Miao, S.S. Ji, *et al*. Electrospun carbon nanofibers decorated with Ag–Pt bimetallic nanoparticles for selective detection of dopamine. ACS Applied Materials and Interfaces, 2014, 6, 12449–12456.

47 N. Rodthongkum, N. Ruecha, R. Rangkupan, *et al*. Graphene-loaded nanofiber-modified electrodes for the ultrasensitive determination of dopamine. Analytica Chimica Acta, 2013, 804, 84–91.

48 C. Yang, B.X. Gu, D. Zhang, *et al*. Coaxial carbon fiber/ZnO nanorods as electrodes for the electrochemical determination of dopamine. Analytical Methods, 2016, 8, 650–655.

49 M.P. Marsh, J.E. Koehne, R. Carbon nanofiber multiplexed array and wireless instantaneous neurotransmitter concentration sensor for simultaneous detection of dissolved oxygen and dopamine. Biomedical Engineering Letters, 2012, 2, 271–277.

50 X.P. Zhanga, D. Liu, B. Yu, T.Y. You. A novel nonenzymatic hydrogen peroxide sensor based on electrospun nitrogen-doped carbon nanoparticles-embedded carbon nanofibers film. Sensors and Actuators B: Chemical, 2016, 224, 103–109.

51 Y. Li, M.F. Zhang, X.P. Zhang, *et al*. Nanoporous carbon nanofibers decorated with platinum nanoparticles for non-enzymatic electrochemical sensing of H_2O_2. Nanomaterials, 2015, 5, 1891–1905.

52 Y. Yang, R.Z. Fu, H.Y. Wang, C. Wang. Carbon nanofibers decorated with platinum nanoparticles: a novel three-dimensional platform for non-enzymatic sensing of hydrogen peroxide. Microchimica Acta, 2013, 180, 1249–1255.

53 T.T. Yang, M.L. Du, H. Zhu, *et al*. Immobilization of Pt nanoparticles in carbon nanofibers: bifunctional catalyst for hydrogen evolution and electrochemical sensor. Electrochimica Acta, 2015, 167, 48–54.

54 D. Liu, Q.H. Guo, X.P. Zhang, *et al*. PdCo alloy nanoparticle–embedded carbon nanofiber for ultrasensitive nonenzymatic detection of hydrogen peroxide and nitrite. Journal of Colloid and Interface Science, 2015, 450, 168–173.

55 X.P. Xiao, Y.H. Song, H.Y. Liu, *et al*. Electrospun carbon nanofibers with manganese dioxide nanoparticles for nonenzymatic hydrogen peroxide sensing. Journal of Materials Science, 2013, 48, 4843–4850.

56 J. Wang. Electrochemical glucose biosensors. Chemical Reviews, 2008, 108, 814–825.

57 L.L. Li, T.T. Zhou, G.Y. Sun, *et al*. Ultrasensitive electrospun nickel-doped carbon nanofibers electrode for sensing paracetamol and glucose. Electrochimica Acta, 2015, 152, 31–37.

58 Y. Shin, I.T. Bae, B.W. Arey, G.J. Exarhos. Facile stabilization of gold-silver alloy nanoparticles on cellulose nanocrystal. Journal of Physical Chemistry C, 2008, 112, 4844–4848.

59 M. Li, L.B. Liu, Y.P. Xiong, *et al*. Bimetallic MCo (M= Cu, Fe, Ni, and Mn) nanoparticles doped-carbon nanofibers synthetized by electrospinning for nonenzymatic glucose detection. Sensors and Actuators B: Chemical, 2015, 207, 614–622.

60 Q.H. Guo, D. Liu, X.P. Zhang, *et al*. Pd–Ni alloy nanoparticle/carbon nanofiber composites: preparation, structure, and superior electrocatalytic properties for sugar analysis. Analytical Chemistry, 2014, 86, 5898–5905.

61 H.Y. Liu, X.P. Lu, D.J. Xiao, *et al*. Hierarchical Cu–Co–Ni nanostructures electrodeposited on carbon nanofiber modified glassy carbon electrode: application to glucose detection. Analytical Methods, 2013, 5, 6360–6367.

62 J. Zhang, X.L. Zhu, H.F. Dong, *et al.* In situ growth cupric oxide nanoparticles on carbon nanofibers for sensitive nonenzymatic sensing of glucose. Electrochimica Acta, 2013, 105, 433– 438.

63 D.Y. Ye, G.H. Liang, H.X. Li, *et al.* A novel nonenzymatic sensor based on CuO nanoneedle/graphene/carbon nanofiber modified electrode for probing glucose in saliva. Talanta, 2013, 116, 223–230.

64 W.N. Xu, S.G. Dai, X. Wang, *et al.* Nanorod-aggregated flower-like CuO grown on a carbon fiber fabric for a super high sensitive non-enzymatic glucose sensor. Journal of Materials Chemistry B, 2015, 3, 5777–5785.

65 M. Li, Z. Zhao, X.T. Liu, *et al.* Novel bamboo leaf shaped CuO nanorod@ hollow carbon fibers derived from plant biomass for efficient and nonenzymatic glucose detection. Analyst, 2015, 140, 6412–6420.

66 L. Zhang, S.M. Yuan, X.J. Lu. Amperometric nonenzymatic glucose sensor based on a glassy carbon electrode modified with a nanocomposite made from nickel(II) hydroxide nanoplates and carbon nanofibers. Microchimica Acta, 2014, 181, 365–372.

67 Q. Wang, Y. Ma, X. Jiang, *et al.* Electrophoretic deposition of carbon nanofibers/ Co(OH)$_2$ nanocomposites: Application for non-enzymatic glucose sensing. Electroanalysis 2016, 28, 119–125.

68 A. Cantalapiedra, M.J. Gismera, J.R. Procopio, M.T. Sevilla. Nanocomposites based on polystyrene sulfonate as sensing platforms. comprehensive electrochemical study of carbon nanopowders and carbon-nanofibers composite membranes. Electroanalysis, 2015, 27, 378 –387.

69 D.L. Zhao, T.T. Wang, D. Han, *et al.* Electrospun carbon nanofiber modified electrodes for stripping voltammetry. Analytical Chemistry, 2015, 87, 9315–9321.

70 Y. Liao, Q. Li, N. Wang, S.J. Shao. Development of a new electrochemical sensor for determination of Hg(II) based on Bis(indolyl)methane/Mesoporous carbon nanofiber/ Nafion/glassy carbon electrode. Sensors and Actuators B: Chemical, 2015, 215, 592–597.

71 J.P. Fu, H. Qiao, D.W. Li, *et al.* Laccase biosensor based on electrospun copper/carbon composite nanofibers for catechol detection. Sensors 2014, 14, 3543–3556.

72 Q.H. Guo, J S. Huang, P.Q. Chen, *et al.* Simultaneous determination of catechol and hydroquinone using electrospun carbon nanofibers modified electrode. Sensors and Actuators B: Chemical, 2012, 163, 179–185.

73 D.W. Li, P.F. Lv, J.D. Zhu, *et al.* NiCu alloy nanoparticle-loaded carbon nanofibers for phenolic biosensor applications. Sensors 2015, 15, 29419–29433.

74 F. Manea, S. Motoc, A. Pop, *et al.* Silver-functionalized carbon nanofiber composite electrodes for ibuprofen detection. Nanoscale Research Letters, 2012, 7, 331.

75 L. Zhang, X.K. Chu, S.M. Yuan, G.C. Zhao. Ethylenediamine-assisted preparation of carbon nanofiber supported nickel oxide electrocatalysts for sensitive and durable detection of insulin. RSC Advances, 2015, 5, 41317–41323.

76 P. Salgado-Figueroa, C. Gutiérrez, J.A. Squella. Carbon nanofiber screen printed electrode joined to a flow injection system for nimodipine sensing. Sensors and Actuators B: Chemical, 2015, 220, 456–462.

77 L. Xu, Q.H. Guo, H. Yu, *et al.* Simultaneous determination of three β-blockers at a carbon nanofiber paste electrode by capillary electrophoresis coupled with amperometric detection. Talanta, 2012, 97, 462–467.

78 E.A. Songa, J.O. Okonkwo. Recent approaches to improving selectivity and sensitivity of enzyme-based biosensors for organophosphorus pesticides: A review. Talanta, 2016, 155, 289–304.

79 N.J. Ronkainen, H.B. Halsall, W.R. Heineman. Electrochemical biosensors, Chemical Society Reviews, 2010, 39, 1747–1763.

80 J.S. Huang, Y. Liu, T.Y. You. Carbon nanofiber based electrochemical biosensors: A review. Analytical Methods, 2010, 2, 202–211.

81 X. . Mao, F. Simeon, G. . Rutledge, T. Hatton. Electrospun carbon nanofiber webs with controlled density of states for sensor applications. Advanced Materials, 2013, 23, 1309–1314.

82 L. Wang, Y.J. Ye, H.Z. Zhu, *et al.* Controllable growth of Prussian blue nanostructures on carboxylic group-functionalized carbon nanofibers and its application for glucose biosensing. Nanotechnology, 2012, 23, 455502.

83 K.P. Gong, F. Du, Z.H. Xia, *et al.* ScienceNitrogen-doped carbon nanotube arrays with high electrocatalytic activity for oxygen reduction. Science, 2009, 323, 760–764.

84 D. Liu, X.P. Zhang, T.Y. You. Electrochemical performance of electrospun free-standing nitrogen-doped carbon nanofibers and their application for glucose biosensing. ACS Applied Materials and Interfaces, 2014, 6, 6275–6280.

85 X.P. Zhang, D. Liu, L.B. Li, T.Y. You. Direct electrochemistry of glucose oxidase on novel free-standing nitrogen-doped carbon nanospheres@carbon nanofibers composite film Science Reports, 2015, 5, 09885.

86 X.E. Jia, G.Z. Hu, F. Nitze, *et al.* Synthesis of palladium/helical carbon nanofiber hybrid nanostructures and their application for hydrogen peroxide and glucose detection. ACS Applied Materials and Interfaces, 2013, 5, 12017–12022.

87 D.W. Li, L. Luo, Z.Y. Pang, *et al.* Novel phenolic biosensor based on a magnetic polydopamine-laccase-nickel nanoparticle loaded carbon nanofiber composite. ACS Applied Materials and Interfaces, 2014, 6, 5144–5151.

88 D.W. Li, G.H. Li, P.F. Lv, *et al.* Preparation of a graphene-loaded carbon nanofiber composite with enhanced graphitization and conductivity for biosensing applications. RSC Advances, 2015, 5, 30602–30609.

89 D.W. Li, J. Yang, J.B. Zhou, *et al.* Direct electrochemistry of laccase and a hydroquinone biosensing application employing ZnO loaded carbon nanofibers. RSC Advances, 2014, 4, 61831–61840.

90 O.A. Loaiza, P.J. Lamas-Ardisana, L. Añorga, *et al.* Graphitized carbon nanofiber–Pt nanoparticle hybrids as sensitive tool for preparation of screen printing biosensors. Detection of lactate in wines and ciders. Bioelectrochemistry, 2015, 101, 58–65.

91 P.J. Lamas-Ardisana, O.A. Loaiza, L. Añorga, *et al.* Disposable amperometric biosensor based on lactate oxidase immobilised on platinum nanoparticle-decorated carbon nanofiber and poly(diallyldimethylammonium chloride) films. Biosensors and Bioelectronics, 2014, 56, 345–351.

92 S.Y. Bao, M.L. Du, M. Zhang, *et al.* Fabrication of gold nanoparticles modified carbon nanofibers/polyaniline electrode for H_2O_2 determination. Journal of the Electrochemical Society, 2014, 161, H816–H821.

93 M.M. Zhai, R.J. Cui, N. Gu, *et al.* Nanocomposite of Au nanoparticles/helical carbon nanofibers and application in hydrogen peroxide biosensor. Journal of Nanoscience and Nanotechnology, 2015, 15, 4682–4687.

94 M. Wu, W. Ding, J.L. Meng, *et al*. Electrocatalytic behavior of hemoglobin oxidation of hydrazine based on ZnO nano-rods with carbon nanofiber modified electrode. Analytical Sciences, 2015, 31, 1027–1033.

95 P.B. Luppa, L.J. Sokoll, D.W. Chan. Immunosensors-principles and applications to clinical chemistry. Clinica Chimica Acta, 2001, 314, 1–26.

96 S.A. Lim, M.U. Ahmed. A carbon nanofiber-based label free immunosensor for high sensitive detection of recombinant bovine somatotropin. Biosensors and Bioelectronics, 2015, 70, 48–53.

97 J. Zhang, Y.F. Sun, H.F. Dong, *et al*. An electrochemical non-enzymatic immunosensor for ultrasensitive detection of microcystin-LR using carbon nanofibers as the matrix. Sensors and Actuators B: Chemical, 2016, 233, 624–632.

98 S. Baj, T. Krawczyk, N. Pradel, *et al*. Carbon nanofiber-based luminol-biotin probe for sensitive chemiluminescence detection of protein. Analytical Sciences, 2014, 30, 1051–1056.

99 E. Gikunoo, A. Abera, E. Woldesenbet. A novel carbon nanofibers grown on glass microballoons immunosensor: a tool for early diagnosis of malaria. Sensors, 2014, 14, 14686–14699.

100 A. Periyakaruppan, R.P. Gandhiraman, M. Meyyappan, J.E. Koehn. Label-free detection of cardiac troponin-I using carbon nanofiber based nanoelectrode arrays. Analytical Chemistry, 2013, 85, 3858–3863.

101 Y.F. Li, L. Syed, J.W. Liu, *et al*Label-free electrochemical impedance detection of kinase and phosphatase activities using carbon nanofiber nanoelectrode arrays. Analytica Acta, 2012, 744, 45–53.

102 L.Z. Swisher, L.U. Syed, A.M. Prior, *et al*. Electrochemical protease biosensor based on enhanced AC voltammetry using carbon nanofiber nanoelectrode arrays. The Journal of Physical Chemistry C, 2013, 117, 4268–4277.

103 A. Erdem, E. Eksin, G. Congur. Indicator-free electrochemical biosensor for microRNA detection based on carbon nanofibers modified screen printed electrodes. Journal of Electroanalytical Chemistry, 2015, 755, 167–173.

104 H. Dai, G.F. Xu, S.P. Zhang, *et al*. Carbon nanotubes functionalized electrospun nanofibers formed 3D electrode enables highly strong ECL of peroxydisulfate and its application in immunoassay. Biosensors and Bioelectronics, 2014, 61, 575–578.

105 L.N. Ma, D.L. Ning, H.F. Zhang, J.B. Zheng. Au@Ag nanorods based electrochemical immunoassay for immunoglobulin G with signal enhancement using carbon nanofibers-polyamidoamine dendrimer nanocomposite. Biosensors and Bioelectronics, 2015, 68, 175–180.

3

Carbon Nanomaterials for Neuroanalytical Chemistry

Cheng Yang and B. Jill Venton

Department of Chemistry, University of Virginia, USA

3.1 Introduction

Brain neuronal communication occurs primarily through the exocytotic release of neurotransmitters into synaptic junctions and the surrounding extracellular fluid. Many debilitating disorders such as Parkinson's disease, Alzheimer's disease, depression, and drug addiction occur due to problems with neurotransmission [1, 2]. Real-time methods for neurotransmitter detection are necessary to understand how the brain works and develop better treatments for these diseases. Extracellular neurotransmitter concentrations change rapidly and understanding the dynamics of neurotransmission is important for clinical research [3]. Electrochemical sensors are currently the gold standard for the rapid detection of neurotransmitters through their redox reactions. The advantages of electrochemical neurotransmitter sensing include a fast response time, high sensitivity and selectivity, as well as high temporal and spatial resolution. For direct detection, target molecules need to be electroactive within the potential window of the interstitial fluid and common targets include biogenic amines and their metabolites, as well as uric acid and ascorbic acid. Using enzyme modified electrodes, many other non-electroactive neurotransmitters can be detected *in vivo* as well.

Carbon-based electrodes are commonly used for neurotransmitter detection because of their low cost, good electron transfer kinetics, and biocompatibility. The high electrocatalytic activity of neurotransmitters on the carbon surface improves the detection sensitivity [4]. Moreover, the oxidation/reduction range at carbon electrodes in water environments is wider than other electrode materials, such as platinum [5], due to a high overvoltage for water oxidation. Ralph Adams and his colleagues realized the utility of electrochemical methods for the study of easily oxidizable neurotransmitters [6]. They first used a carbon paste electrode implanted in the brain of an anesthetized rat, demonstrating carbon-based electrodes could be applied successfully to biological tissue for neurotransmitters detection [6]. In the late 1970s, Ponchon *et al.* introduced the carbon-fiber microelectrode (CFME) [7], which is still routinely used for *in vivo* neurotransmitter detection today.

Nanocarbons for Electroanalysis, First Edition.
Edited by Sabine Szunerits, Rabah Boukherroub, Alison Downard and Jun-Jie Zhu.
© 2017 John Wiley & Sons Ltd. Published 2017 by John Wiley & Sons Ltd.

Carbon nanomaterials (CNs) have been a hot research topic for the past two decades, since the discovery of carbon nanotubes (CNTs), and CNs have been incorporated into electrochemical sensors for neurotransmitter detection. The feature size of the nanomaterials is 1–100 nm and CN-modified electrodes are advantageous because of their large surface-to-volume ratio and large specific surface area. In addition, CNs have enhanced interfacial adsorption properties, better electrocatalytic activity, and fast electron transfer kinetics compared to many traditional electrode materials. In 1996, the first electrochemical application of CNTs was carried out by Britto *et al.* to detect dopamine at multi-walled carbon nanotube (MWCNT) paste electrodes [8]. There are many strategies for incorporating CNs into electrochemical sensors for *in vivo* neurotransmitter measurements. Dip coating and drop casting methods are widely used because of their minimal cost and suitability for preparing electrodes with mixtures of CNs, polymers, and/or metal nanoparticles (NPs). Direct growth of CNs and CNTs fibers draws more attention for biosensing applications because of the homogeneous CN surface and the preferential exposure of the ends of the CNs which have defect sites that can be functionalized with oxygen to provide abundant electroactive sites for neurotransmitters [4]. The easy batch fabrication and high reproducibility of the direct growth of CNs and CNT fiber (macrostructure of CNTs fabricated by wet spinning or solid state process) microelectrodes makes those methods ideal fabrication approaches for single microelectrodes. Carbon-based nanoelectrodes arrays that contain multiple sensing elements allow detection of neurotransmitters at network levels. This chapter covers the pros and cons of different fabrication methods and electrode designs.

Several challenges still exist for neurotransmitter monitoring with CNs. First, while many studies characterize the materials, few studies investigate the correlation between the electrochemical performance and the surface properties. Second, fouling on an electrode surface in tissue is often a serious problem encountered in the electrochemical analysis. While most of the CN-based sensors reported have not been tested in tissue, new studies focusing on bio-fouling and strategies to combat it show that CNs may be advantageous for *in vivo* testing. Third, novel designs of CN-based electrodes with renewable surface are ongoing to enable electrodes to be reused and improve the reproducibility of consecutive measurements.

In this chapter, we cover the recent advances in CN-based sensors for direct neurotransmitter detection, concentrating on sensors for direct electrochemical detection of dopamine, which is a standard analytes to evaluate neurotransmitter detection at electrodes. First, we provide a comparison of strategies to incorporate CNs into electrochemical sensors for neurotransmitter detection. Second, we highlight several new applications and studies to address the remaining challenges for implementation. Overall, CN-based electrodes are advantageous for neurotransmitter detection and advances in fundamental understanding of structure/ function relationships as well as progress towards in tissue use by minimizing biofouling and making reusable sensors will result in better implementation *in vivo*.

3.2 Carbon Nanomaterial-based Microelectrodes and Nanoelectrodes for Neurotransmitter Detection

3.2.1 Carbon Nanomaterial-based Electrodes Using Dip Coating/Drop Casting Methods

Dip coating/drop casting fabrication methods are popular for early stage new CN applications due to their low cost and the synergistic effects when combing CNs with other composites such as conductive polymers and/or metal NPs. Polymers can modify the physical and chemical properties of composites and facilitate CN deposition on the support surface [9]. The incorporation of CNs with polymers helps to disperse CNTs and graphene in aqueous media via non-covalent interactions, such as van der Waals forces, π- π interactions, or adsorption/wrapping of polymer [10]. Better suspensions of CNs have been reported using hemin-graphene oxide-pristine CNTs complexes on glassy carbon electrode (GCE) [11], polyhistidine dispersed MWCNT-modified GCE [12], and acid yellow 9 stabilized MWCNTs on both ITO and GCE [13]. The lowest dopamine limit of detection (LOD) was found at the MWCNT/Polyhistidine/GCE (15 nM) [12]. Several polymers have been reported to modify CNTs and graphene based electrodes such as polypyrrole (PPy) [14], poly (3,4-ethylenedioxythiophene) (PEDOT) [15, 16], poly(vinylferrocene) (PVF)[17], tryptophan [18], β-cyclodextrin (β-CD) [19], poly (3-methylthiophene) (PMT) [20], polyaniline (PANI) [21–23], polyamide 6/poly(allylamine hydrochloride) (PA6/PAH)[24], and chitosan [25–27]. The best DA sensitivity was found on the chitosan grafted graphite on GCE, with an LOD of 4.5 nM because of the high conductivity and bio-compatibility of chitosan [25]. The introduction of polymers with CNs facilitates the dispersion of CNTs and graphene in aqueous media, but polymers have disadvantages as well, such as restricted diffusion, slow temporal resolution, and a cumbersome fabrication that is not always reproducible. Moreover, no work directly compares the performance of different polymers.

Metal NPs associated with CNs have been applied to fields such as energy storage, catalysis, and chemical sensors [28]. Metal NPs perform several functions including increasing the surface area, facilitating heterogeneous electron transfer, enhancing electrical contact between analyte and the CN surface, inhibiting aggregation of graphene nanosheets, and acting as a spacer [29, 30]. CN-based electrodes incorporated with PtNPs [31, 32], AuNPs [33, 34], CuNPs [35], PbNPs [36], as well as RhNPs, PdNPs, and IrNPs [37] have been reported. Tsierkezos *et al.* recently compared different NPs for the first time, examining the detection of dopamine, ascorbic acid, and uric acid on films of nitrogen doped MWCNTs (N-MWCNTs) decorated with RhNPs, PdNPs, IrNPs, PtNPs, and AuNPs [37]. The LODs for dopamine detection improved with the metal NPs and the best LOD was on N-MWCNT/AuNP film (0.3 μM). This comparison work would be a good reference for future bio-applications based on metal NPs incorporated with CNs. The main limitations of metal NPs application are biocompatibility issues and the difficulty of construction.

Several works use dip coating/drop casting methods to test novel CNs such as nitrogen doped carbon nanofibers [38], ordered mesoporous carbon [39] and nanodiamond-derived carbon nano-onions [40]. Ko *et al.* applied thermally oxygenated cup-stacked carbon nanofibers (ox-CSCNFs) modified GCE for dopamine detection [41]. Ox-CSCNFs provide highly ordered graphene edges and oxygen-containing functional groups, which promote fast electron transfer kinetics. Acid oxidized CSCNF-based screen printed electrodes (SPE) increased the sensitivity for dopamine detection because of the mesoporous structure that increased the electroactive surface area [42]. Another important feature of oxygenated CSCNFs is their easy dispersion in water. However, the fabrication and modification process are difficult.

Carbon nanohorns (CNHs) are horn-shaped aggregates of graphene sheets which were first synthesized using laser ablation of pure graphite by Ijima [43]. CNHs assemble dahlia-like aggregates with an average diameter around 80 nm and the high purity of single walled CNHs (SWCNHs) (metal-free) offers a unique opportunity for sensing and in tissue application [44, 45]. Valentini *et al.* used oxidized SWCNH-based SPE for the selective detection of epinephrine in the presence of serotonin and dopamine with LOD of 0.1 μM [46]. The same group reported better electrochemical performance for pristine SWCNHs compared to oxidized SWCNHs for dopamine detection, due to the improved electrical conductivity, electron transfer features, and surface charge of SWCNHs at different degree of oxygen-containing functional groups [47]. However, the sensitivity (LODs of 0.1 μM and 0.4 μM at pristine and oxidized SWCNHs, respectively,) needs to be improved [9].

While most CN coating studies coat a GCE for convenience, this is not practical for *in vivo* studies. Du *et al.* reported graphene flower-modified CFMEs to simultaneously detect dopamine, ascorbic acid, and uric acid [48]. Graphene flowers were prepared via electrochemical reduction of graphene oxide (GO) sprayed on electrode; these graphene flowers homogeneously decorated the surface of the electrode. In comparison to unmodified CFMEs, graphene flower-modified CFMEs exhibited high electrocatalytic activity towards the oxidation of dopamine, ascorbic acid, and uric acid. In terms of the possible future *in vivo* applications, the graphene-flower modified CFME is advantageous due to its micrometer-scaled structure (7 μm in diameter), which would minimize tissue damage caused during insertion.

Zhang's group used a different strategy to coat acupuncture needles with CNs for the detection of neurotransmitters [49, 50] Acupuncture needles are a unique microelectrode platform and are widely used in traditional Chinese medicine. They reported a CNT/PEDOT modified acupuncture needle for *in vivo* detection of serotonin, with LODs of 50 nM and 78 nM in solution and cell medium, respectively [50]. Although the acupucture needles is a promising platform, a systematic study is required to corelate the application of different CNs with their electrochemical performance.

The largest problem with dip coating and drop casting is a non-homogenous coating that can result in agglomerations, low reproducibility, and slow electron transfer [4, 51]. Screen printing is easy for mass production, but the size and geometry limit their application for deep brain measurement. Most of the dip coated/drop casted CN-based electrodes are characterized by differential pulse voltammetry (DPV), which has limited application for real-time *in vivo* measurement due to its slow temporal resolution. However, the low cost of drop casting and dip coating makes it ideal method for early stage electrochemical study of CNs for neurotransmitter detection.

3.2.2 Direct Growth of Carbon Nanomaterials on Electrode Substrates

Direct growth of CNs on electrode substrates is gaining popularity due to the reproducible mass fabrication and the ability to control the geometry and orientation of the CNs. The ends of CNTs have more defect sites that can be functionalized with oxygen containing groups, which are more likely to be the most electrochemically active sites [3, 4, 9, 52]. The functional groups, such as carbonyls, phenols, lactones, and carboxylic acids, selectively adsorb cationic dopamine and repel the anionic ascorbate and uric acid at physiological pH [4]. Direct growth of CNs is beneficial due to its aligned CNs and high density of edge plane sites, instead of randomly distributed CN films by dip coating/drop casting methods [4, 51].

Vertically-aligned CNTs grown on a microelectrode substrate have high sensitivity for detecting neurotransmitters due to the high density of edge plane sites [53]. One strategy is to chemically self-assemble vertically-aligned CNTs (VACNTs) on substrates with a solution deposition method [54]. An alternative strategy is to directly grow CNTs in an aligned manner through chemical vapor deposition (CVD). The Mao group grew VACNTs on carbon fiber supports formed via pyrolysis of iron phthalocyanine as microelectrodes for the detection of ascorbate, DOPAC (3, 4-dihydroxyphenylacetic acid), dopamine, and oxygen in rat brain [55, 56]. Use of the VACNT-CFMEs avoids manual electrode modification and allows easy fabrication of highly selective, reproducible, and stable microelectrodes. Our group reported CNTs directly grown on metal wires fabricated by CVD for sensitive and selective dopamine detection *in vivo* [57]. Instead of carbon fibers, metal wires were used as substrates because the wires lacked electrochemical reactivity to dopamine (except gold and platinum) and had higher conductivity than carbon fibers [4]. This was the first study to directly grow CNTs on small diameter metal wires, instead of planar metal surfaces such as foils, which allows them to be implanted in tissue with minimal damage and high spatial resolution [58]. CNTs grown on niobium substrates were short and dense, and exhibited the best electrochemical response to dopamine using fast-scan cyclic voltammetry (FSCV), including the detection of stimulated dopamine release in anesthetized rats. Since electrophysiology studies often use arrays of metal wires, future experiments could investigate making arrays of the CNTs on metal wires for multiplexed electrochemical experiments.

Carbon nanospikes (CNS) are a novel CN with tapered and spike-like features about 50–80 nm in length [59]. Our group used plasma-enhanced CVD (PECVD) to directly grow a thin layer of CNS on cylindrical metal substrates under a low-temperature growth condition [60]. This is the first time that CNS have been grown on metals or on cylindrical substrates. In comparison with the CNT growth, direct growth of CNS does not need a catalyst or insulation layer for catalyst deposition, resulting in direct electrical contact between the CNS and substrate. CNS-Ta microelectrodes (Figure 3.1a, b) exhibited the fastest electron transfer kinetics and lowest LOD (8 ± 2 nM) to dopamine. CNS-modified electrodes are a promising biosensing material with a similar sensitivity to CNTs but easier growth at low-temperatures and without catalyst or buffer layers.

Diamond-like carbon (DLC) is a metastable form of amorphous carbon that contains a mixture of tetrahedral and trigonal carbon hybridizations fabricated using PECVD [61].

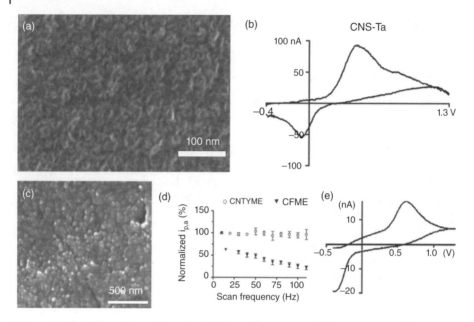

Figure 3.1 Detection of dopamine using FSCV at carbon nanospike (CNS) grown metal microelectrodes (a–b) and CNT yarn microelectrodes (C–E). (a) High magnification of CNS grown on Ta wire. There is uniform coverage of aligned carbon spikes over the entire metal wire. (b) Detection of 1 μM dopamine with CNS-Ta microelectrodes. (c) SEM image of CNT yarn microelectrode surface, shows the ends of individual 30–50 nm diameter CNTs bundled tightly together to form a nanostructured surface. (d) Peak oxidation current at CNTYMEs (circles, n = 10) and CFMEs (triangles, n = 4). Peak currents were normalized to the current at 10 Hz, and error bars represent the standard error of the mean. (e) High-frequency measurements of 1 μM dopamine at CNTYMEs, measured at 2000 V/s and 500 Hz. *Source*: (a, b) Zestos 2015 [60]. Reproduced with permission of Royal Society of Chemistry. (c–e) Jacobs 2014 [72]. Courtesy of ACS Publications.

Silva *et al.* reported the successful growth of DLC thin film on VACNTs (Ti substrate) and its application for dopamine and epinephrine detection, with LODs of 3.9 μM and 4.5 μM, respectively [62]. The VACNT scaffolds create conductive pathways that transport current rapidly to all parts of DLC film. However, the dimensions are on the millimeter scale and the LODs are not yet sensitive enough for monitoring neurotransmitters *in vivo*. Similarly, Sainio *et al.* successfully directly grew MWCNTs on the top of DLC thin film electrodes (Si substrate) [63]. The porous structure of MWCNTs/DLC provides fast electron transfer, reduced double layer capacitance, a wide stable water window (−1.5–2V), and high dopamine sensitivity (LOD of 1.26 ± 0.23 nM). The same group then directly grew carbon nanofibers (CNFs) on DLC thin film (Si substrate), which also exhibited wide water window [64]. Notably, the CNFs/DLC electrodes could detect glutamate without using any enzymes. However, the detection is not direct but depends on the pH shift caused by different amounts of glutamate added into solution. Since the local pH fluctuates [65], this strategy would not be feasible for glutamate detection *in vivo*.

Compared to dip coating/drop casting methods, direct growth of CNs has the advantage of homogeneous coating, large amounts of electroactive sites, and high reproducibility.

The best dopamine detection sensitivity was obtained at the MWCNTs/DLC thin film electrodes (1.26 ± 0.23 nM) [63]. However, this electrode is not feasible for *in vivo* measurement since the dimension of the Si wafer substrate would cause serious tissue damage. In comparison, the application of metal wires (diameter 25 μm) as the substrates in several studies for CNs growth improves the overall conductivity and minimizes tissue damage [55–57]. The direct growth of carbon nanospikes is advantageous due to its low-temperature, catalyst-free, and homogenous growth [60]. Thus, directly growing CNs on electrodes is a promising strategy for reproducibly producing electrodes that could be used *in vivo*.

3.2.3 Carbon Nanotube Fiber Microelectrodes

CNT fibers are a macrostructured carbon material spun from a forest of CNTs [66, 67]. In comparison to randomly structured or tangled CNTs often produced by dip coating or growing, well-aligned CNT fibers have the advantages of a high electroactivity, fast response, high chemical stability, controllable size, and high resistance to dopamine fouling [68]. In particular, fabricating CNT fibers directly into electrodes in a manner similar to CFMEs dramatically simplifies the electrode fabrication process and improves the reproducibility [69]. The first electrochemical application of a CNT fiber sensor was reported in 2003 [70] and CNT fiber microelectrodes have been developed for dopamine detection using DPV [71], chronoamperometry [68], and FSCV [72, 73]. Two distinct routes have been developed for manufacturing CNT fibers: wet-spinning a suspension of CNTs into a bath or direct spinning a fiber from a CNT forest. In solution based wet spinning, CNTs are either dissolved or dispersed in a fluid, extruded out of a spinneret, and coagulated into a solid fiber by extracting the dispersant [74, 75]. Neat CNT fibers can be spun into a polymer bath, such as poly(vinyl alcohol) (PVA) [76] or poly(ethylene)imine (PEI) [77], in which the polymer molecules act as coagulators for CNTs to enhance the intertube interactions. Our group has successfully applied PEI/CNT fiber microelectrodes for *in vivo* monitoring of dopamine and serotonin in brain slice [69]. Thin, uniform fibers were fabricated by wet spinning CNTs into fibers using PEI as a coagulating polymer. In comparison to PVA/CNT fibers, PEI/CNT fibers have better LODs of dopamine (4.7 ± 0.2 nM) and lower overpotential.

An alternative fiber production route employs a solid state process where CNTs are either directly spun as a fiber from the synthesis reaction zone [78] or from a CNT forest grown on a solid substrate [79]. Continuous CNT yarns, generated by drawing one end of the VACNT arrays, are favorable because of the improved conductivity and CNT alignment [80, 81]. Moreover, compared to CNT fibers which are produced with the incorporation of polymers, CNT yarns are made of CNTs only, without impurities. Schmidt *et al.* first used CNT yarn microelectrodes (CNTYMEs) for neurotransmitter monitoring [73]. In this study, the yarns were customizable and tailored for sensitive detection of neurotransmitters. Dopamine concentration fluctuations were successfully detected in acute brain slices using FSCV. Our group utilized a commercially-available CNT yarn fabricated microelectrodes for dopamine detection with LOD of 13 ± 2 nM [72]. We also found the dopamine signal is independent to the repetition rate with FSCV at CNT yarn microelectrodes, an advantage over conventionally used CFMEs which lose sensitivity with increasing repetition rate (Figure 3.1d, e). The high sensitivity with rapid measurement at CNT yarn provides high temporal resolution for neurotransmitters detection and the ability to monitor reaction intermediates and perform neurochemical redox reaction kinetic studies.

To further improve the sensitivity of CNT yarn microelectrodes, our group introduced laser treatment as a simple, reproducible, and efficient surface modification method. At rapid scan repetition frequencies using FSCV, the dopamine LOD was $13 \pm 2\,nM$ [82]. In this work, 15 pulses of a KrF 248 nm pulsed excimer laser was applied, which increased the surface area and oxygen containing functional groups on the CNT yarn surface. CNT yarn microelectrodes were applied for *in vivo* measurement for the first time and maintained high dopamine sensitivity with rapid measurements.

Because of the well aligned CNTs, abundant surface oxide groups, and promising reproducibility, CNT fiber and yarn microelectrodes have already been used for both in tissue and *in vivo* measurements of neurotransmitters [73, 82, 83]. Although several kinds of CNT fibers/yarns and surface modifications methods have been applied, the fundamental understanding of the neurotransmitter reaction mechanism on the surface and the effects of bio-fouling are still elusive and need further study.

3.2.4 Carbon Nanoelectrodes and Carbon Nanomaterial-based Electrode Array

Nanoscale carbon electrodes are useful for localized detection of neurotransmitters at the level of single cells, single vesicles, as well as single synapses. Small animal models such as *Drosophila* and zebrafish have attracted more attention recently due to their homologous neurotransmitters with mammals and easy, fast genetic manipulations [84–86]. The small dimensions of the central nervous systems of *Drosophila* and zebrafish requires better spatial resolution and a small, dagger-like electrode than can penetrate through the tough glial sheath barrier with minimal tissue damage. Nanoelectrodes can be used as single sensors or as an array of densely packed nanometer scaled sensors.

Since the geometric surface area of nanoelectrodes is much smaller than microelectrodes, the application of CNs helps improve the sensitivity with a limited surface area. Li *et al.* used carbon fiber nanometric electrodes for direct monitoring of exocytotic flux from inside an individual synapse in real time [87]. The facile and robust nanoelectrodes were fabricated by flame-etching a glass-pulled microelectrode, which was subsequently etched with a microforge to yield a fine conical tip with diameter less than 100 nm and controlled shaft length (500–2000 nm, Figure 3.2a, b). The real-time monitoring inside neuronal junctions and neuromuscular junctions using amperometry (Figure 3.2c) demonstrated the versatility and importance of this new tool.

Our group reported a small, robust, carbon nanopipette electrode (CNPE) for dopamine detection in *Drosophila* brain [85]. The CNPEs are fabricated by selectively depositing a carbon layer on the inside of a pulled-quartz capillary that is chemically etched to expose the carbon tip. The electrodes had an average diameter of 250 nm and controllable exposed length of 5 to 175 µm. The LOD of $25 \pm 5\,nM$ with FSCV is similar to CFMEs but the size is one order of magnitude smaller in diameter than typical CFMEs. CNPEs were used to detect endogenous dopamine release in *Drosophila* larvae using optogenetic stimulations, which verified the utility of CNPEs for *in vivo* neuroscience studies. This is the first work to combine carbon nanoelectrodes and FSCV together, providing a promising solution for the measurements with both high spatial, temporal resolution and high selectivity.

Neurotransmitter release varies not only within discrete substructures of the brain but also across individual cells [88]. Multi-electrode arrays (MEAs) consolidate multiple sensing elements onto a single device for spatially resolved profiling of neurochemical dynamics as well as simultaneous detection of different analytes [84]. Suzuki *et al.* used planar CNTs-MEA chips for the measurements of dopamine release and electrophysiological responses [89]. These CNT-based MEA chips were fabricated by electroplating bamboo structured MWCNTs on a photolithography-produced indium tin oxide (ITO) electrode (Figure 3.2d–f). The LOD for dopamine was 1 nM using amperometry, and the CNT-MEA chips were used to monitor synaptic dopamine release from mouse striatal brain slices. Similarly, Clark *et al.* reported a functionalized CNT electrode array for dopamine and ascorbic acid detection [90]. These CNT-MEA chips have 64 individually addressable electrodes of oxygen-functionalized CNTs grown on Pt electrodes. The reduced impedance led to improved signal-to-noise ratio. Diner *et al.* applied a nanocrystalline boron doped diamond (BDD) nanoelectrodes array for dopamine detection, combining the advantages of high spatial resolution nanoelectrode arrays with the promising electrochemical properties of BDD [91]. The dopamine detection

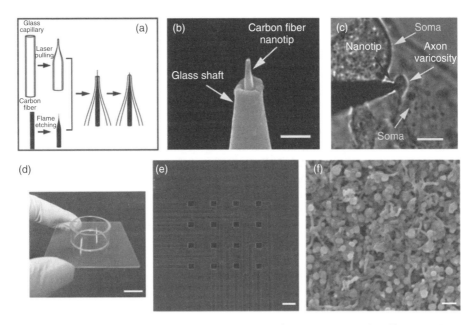

Figure 3.2 Nanoelectrode designs for neurotransmitter detection. (a–c) Carbon fiber nanotip electrodes. (a) Schematic diagram showing main process for fabrication of conical CFNEs. (b) Amplified picture of its tip; the scale bar is 1 μm. (c) Bright-field photomicrographs showing the tip of a sensor inside a synapse between a varicosity of a SCG neuron and the soma of another SCG neuron, scale bars: 5 μm. (d–f) Carbon nanotube-microelectrode arrays. (d) A sample CNT-MEA chip containing 64 electrical recording sites. Scale bar: 1 cm. (e) Phase-contrast image of a CNT-MEA chip showing the spatial configuration of electrodes. Each electrode is 50 μm × 50 μm. Scale bar: 100 μm. (f) SEM image of the CNT-MEA chip surface. Scale bar: 200 nm. *Source*: (a–c) Zhang 2014 [87]. Reproduced with permission of John Wiley and Sons. (d–f) Suzuki 2013 [89]. Reproduced with permission of Elsevier.

sensitivity was $57.9\,nA\,\mu M^{-1}\,cm^{-2}$ and the selectivity for dopamine over ascorbic acid was high due to the negative charge after oxygen-termination.

Carbon-based nanoelectrodes provide robust and sharp tips for neurotransmitter detection at the level of single synapses, vesicles, and single cells *in vivo*. Nanoelectrodes provide high spatial resolution, low background current from double layer charging, and minimal tissue damage. The majority of carbon-based nanoelectrodes consist of protruding carbon materials (e.g. carbon fiber) with the exposed length of several to tens of micrometer; therefore the resolution is not nanometer sized in all dimensions.[92]. Compared to single nanoelectrode systems, carbon-based nanoelectrodes arrays allow detection of neurotransmitters at network levels. However, the stiff and large silicon wafer and bulk metal substrates limit their *in vivo* applications. Van Dersarl *et al.* fabricated a flexible implantable neural probe for electrochemical dopamine sensing [93]. The electrode arrays were fabricated on silicon carrier wafers and then released from the substrate by chemical/electrochemical dissolution, which makes the electrodes array flexible and used for long-term neurotransmitter monitoring *in vivo*. As the development of CNs and new electrodes continues, we expect more attention will be focused on increasing the sensitivity and nanometer-scaled resolution *in vivo*, especially for measurements at single synapses and smaller model organisms.

3.2.5 Conclusions

The dip coating/drop casting method is an ideal method for testing early stage electrochemical performance of novel CNs or combining CNs with polymers and/or metal NPs for neurotransmitter detection. However, disadvantages include the non-homogeneous coating, low reproducibility, and complicated fabrication as well as restricted diffusion and decreased conductivity which jeopardizes electrochemical performance [3]. In comparison, direct growth of CNs on microelectrodes leads to high conductivity, larger amounts of electroactive sites, higher reproducibility, and easier batch fabrication. CNTs fiber/yarn microelectrodes are expected to be a potential alternative to CFMEs for *in vivo* neurotransmitter detection because the highly aligned CNTs lead to controllable electrochemical properties. Moreover, the electrochemical properties of CNT yarn microelectrodes can be controlled by surface treatments, which allows researchers to tune the material depending on the requirements of sensitivity and selectivity to different analytes [82]. The development of flexible electrodes is another future path since most electrodes/arrays are made from stiff silicon or metal substrates, which limit their *in vivo* application [94]. Overall, the best LOD is achieved by the planar CNT-MEA chip with chronoamperometric measurements of dopamine release (1 nM) [89]. The best LOD for microelectrodes is $4.7 \pm 0.2\,nM$, achieved by PEI/CNT fiber microelectrodes [69]. While most of the current experiments have concentrated on dopamine detection, future studies could expand the analytes detected and strive for better sensitivity and selectivity at smaller electrodes for a range of neurochemicals.

In this section, we covered the recent methods to fabricate CNs-based electrochemical sensors for direct neurotransmitter detection and compared strategies for incorporating CNs into sensors. In the next section, we focus on the applications of these CNs based sensors to address remaining challenges for the fundamental understanding of neurotransmitter reaction mechanisms, reducing biofouling, as well as the application of reusable electrode designs.

3.3. Challenges and Future Directions

3.3.1 Correlation Between Electrochemical Performance and Carbon Nanomaterial Surface Properties

Dopamine is a catecholamine with heterogeneous electron transfer that is strongly dependent on surface properties and electrocatalysis [4]. The electrostatic interactions between negatively charged oxide groups on the carbon surface and positively charged dopamine at physiological pH as well as the $\pi-\pi$ stacking between CNs and the catechol significantly influence the redox reactions [95]. The redox processes of dopamine depend on the existence of surface oxygen-containing functional groups, such as carbonyl or carboxylate, which can readily adsorb neurotransmitters and facilitate electron transfer [4, 9]. While CNs can enhance the surface for dopamine redox reactions, correlating the exact properties of the nanomaterials with electrochemistry is challenging.

CNTs and graphene have abundant sp^3-hybridized, edge plane carbons that can be oxidized to provide functional groups compared to conventionally used glassy carbon and carbon fiber [96]. The morphology of CNs and the density of edge plane sites significantly influence the redox reactions. In order to study only the electrochemical properties of CNTs, it is useful to have a substrate that is not sensitive to dopamine [97]. For example, the direct growth of aligned CNTs on metal wires (described in section 3.2.2) was useful for characterizing the electrochemistry of the ends of CNTs [57]. The morphology of CNTs grown on Nb (forest-like) is denser and shorter than CNTs grown on other substrates (spaghetti-like), and the ends of the tubes are likely to have exposed defects sites for electron transfer. In comparison, the diffusion of dopamine may be restricted between the longer CNTs in spaghetti-like CNT grown on other substrates. The dopamine redox reaction is more reversible at CNT electrodes than at carbon fibers, due to similar adsorption energies at the CNT surface, which gives bigger reduction peaks compared to CFMEs. Similarly, Muguruma *et al.* reported an electrochemical study of the synthetic pathway of dopamine (DA), dopamine-o-quinone (DAQ), leucodopaminechrome (LDAC), and dopaminechrome (DAC) at an electrode made of long-length (hundred microns) CNTs dispersed in solution with surfactant cellulose [98]. The electrically conducting networks of long-length CNTs had long π-electron networks and were denser than those of normal-length CNT electrode resulting in reduced activation potential for oxidation and efficient propagation of the electron produced by LDAC $\rightleftharpoons$ DAC reaction. The observation of predominant second redox couple (LDAC $\rightleftharpoons$ DAC) at long-length CNTs could be used as a quantitative and selective detection of dopamine just as in the conventional strategy of direct oxidation of dopamine (DA $\rightleftharpoons$ DOQ). These two works show that CNT alignment, length, and density affect both sensitivity and the dopamine redox reaction products.

Surface properties of CN based electrodes also influence their temporal resolution with FSCV measurements. FSCV is the preferred *in vivo* electrochemical method because the CV fingerprint aids in analyte identification and the typical 100 ms temporal resolution allows neurotransmitter measurements on a subsecond time scale [5, 99–101]. At traditional CFMEs, dopamine signal decreases with increasing scan repetition frequency and the 10 Hz repetition frequency is a compromise between temporal resolution and sensitivity [102, 103]. While methods to improve temporal resolution by raising the scan rate have been explored, the oxidation current still decreases with increasing scan repetition frequency [104, 105]. Our group has recently discovered that at CNTYMEs (described in section 3.2.3), the dopamine current does not change significantly with increasing scan repetition frequency: at a scan rate of 2000 V/s, dopamine can be detected, without any loss in sensitivity, with scan frequencies up to 500 Hz [72]. Thus, the temporal resolution is four times faster without a decrease in signal. The cause of this increased temporal resolution is differences in dopamine and DOQ adsorption and desorption kinetics; at CNTYMEs, the rates of desorption for dopamine and DOQ are almost identical, while at CFMEs, the rate of desorption for DOQ is over an order of magnitude higher than that for dopamine. Therefore, CNT morphology and roughness affect the temporal resolution of measurements and CNTYMEs enable high speed measurements with high sensitivity, since there is no compromise in current with higher sampling rates.

Surface roughness is another important factor affecting the dopamine redox reaction. In our recent work, laser-etched CNT yarn microelectrodes (described in section 3.2.3) were used for the *in vivo* detection of dopamine [82]. The trend for frequency independence at laser-etched CNTYMEs was better than unmodified CNTYMEs which were much better than CFMEs. This matched the trend of surface roughness; the mean surface roughness of the laser-etched CNT yarn is significantly larger than CFMEs, on the same order of magnitude as the dopamine diffusion distance (Figure 3.3).

Thus, dopamine and DOQ are more likely to get trapped near the surface, leading to better reversibility. Additionally, the main reason that the signal decreases with increasing scan repetition frequency for CFMEs is that DOQ desorbs and diffuses away from the electrode. At the rough, laser-etched CNTYME surface, the DOQ would remain close to the surface if it desorbs and thus could more easily absorb again, and be reduced back to dopamine. The response at the electrodes surface would be similar to that in a

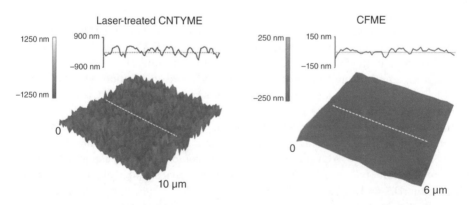

Figure 3.3 Representative three-dimensional laser confocal profile image of a laser treated CNT yarn microelectrode and a CFME. Insets indicate the associated line plot from each sample. Note the scales are different for the different panels. *Source*: Yang 2016 [82]. Reproduced with permission of the American Chemical Society.

thin layer cell, without influencing the temporal response remarkably. Similar results were obtained at PEDOT/graphene oxide coated CFMEs, in which the surface roughness modified by the coating significantly affected the dopamine redox reaction reversibility [106]. Thus, as new surfaces are designed with CNs, crevices that are on the order of the diffusion length can lead to enhanced temporal resolution due to thin layer cell effects.

A better understanding of the correlation between surface properties CNs, including alignment, length, edge plane sites, surface functionalization, and surface roughness, and the electrochemical performance of neurotransmitters will allow researchers to tailor the electrochemistry by applying different CNs or surface modifications. The better aligned CNTs on electrode surface provide higher conductivity and more electroactive sites for dopamine compared to non-oriented, spaghetti-like CNTs. Long-length CNTs form denser conducting networks than the normal-length CNTs, facilitating detection of the second dopamine redox couple, which would be used as a quantitative and selective detection of dopamine. Surface roughness on the order of dopamine diffusion length momentarily traps dopamine and allows higher temporal resolution measurements than at conventional CFMEs. In addition, most sensor designs are based on maximizing exposure of edge plane defects [107], but several recent studies demonstrate there is also electrochemical activity on the side walls of CNTs [107–110]. Therefore, the extent to which CNT sidewalls are electrocatalytic for dopamine is still not well understood and future studies are needed.

To date, almost all studies have focused on dopamine but future research is required to understand the interaction between other electroactive neurotransmitters/neurochemicals (e.g. adenosine, epinephrine, ascorbic acid) and CNs other than CNTs (e.g. graphene, fullerenes, boron-doped diamond, diamond-like carbon). A better fundamental understanding of how CN properties lead to electrochemical response will enable rational development of future electrodes.

3.3.2 Carbon Nanomaterial-based Anti-fouling Strategies for *in vivo* Measurements of Neurotransmitters

Surface fouling can severely affect the sensitivity and reproducibility of electrochemical sensing *in vivo* and CNs have been proposed as a technique to alleviate surface fouling. Electrode fouling involves the surface passivation by a fouling agent that forms an increasingly impermeable layer on the electrode, inhibiting the direct contact of the analyte with the electrode surface for electron transfer [111]. Two main types of fouling arise: (1) passive biological fouling by tissue species such as proteins and lipids and (2) active fouling by byproducts of the analyte redox reaction that form the insulating layer. The fouling agent can adhere to the electrode surface as a result of hydrophobic, hydrophilic, or electrostatic interactions [4, 112]. For example, the products of serotonin oxidation are very reactive and may form an insulating film on the electrode surface, thus jeopardizing the long-term stability of the electrodes [113, 114]. Commonly used antifouling strategies include the addition of a protective layer or barrier on the CNs substrate (polymeric film [115, 116] or metal nanoparticles [117–119]), electrode surface modifications [120, 121], and electrochemical activation [122, 123]. This section examines recent efforts to study fouling resistance study at CN electrodes and develop CN-based antifouling strategies.

The hydrophilic surface of CN electrodes facilitates their antifouling properties [111]. Electrodes with hydrophobic surfaces promote adsorption of hydrophobic molecules, including aromatic compounds and proteins [111]. These hydrophobic interactions are sufficiently favorable, especially in an aqueous condition, that they are typically irreversible [4]. In contrast, fouling caused by hydrophilic interactions tends to be more reversible [124]. Therefore, many strategies aimed at reducing fouling by the application of CNs target reducing the hydrophobicity of the electrode surface. Xiang *et al.* reported VACNTs grown on carbon-fiber electrodes (described in section 3.2.2) exhibited good resistance against electrode fouling in high concentration (0.5 mM) ascorbate solution, with only 2% signal decrease for 30 min continuous amperometric measurements[55]. Our group found the PEI/CNT fiber microelectrodes (described in section 3.2.3) have promising antifouling properties for serotonin. Prevention of electrode fouling from the metabolite of serotonin, 5-hydroxyindoleacetic acid, is particularly important because its physiological concentration is 10 times higher than that of serotonin.[69]. At the PEI/CNT fiber microelectrode, the peak oxidation current for serotonin remained constant upon incubation in 5-hydroxyindolacetic acid for 2 h and serotonin was successfully measured in brain slices. Similarly, Harreither *et al.* reported pristine PVA/CNT fiber microelectrodes exhibited a high resistance to dopamine fouling [68]. While the electrode lost half of its initial signal in 15 min in a high concentration dopamine solution (1 mM), there was no signal change in 100 μM dopamine for 2 h as the insulating patches grew slower on the CNT fiber surfaces than on CFMEs. Since the basal levels of extracellular dopamine are around 10–30 nM and phasic release during fast burst of firing results in release from 0.1 to 1 μM [3], lack of fouling at low concentrations could make them advantageous *in vivo*. The antifouling properties of both the VACNT carbon fiber, PVA/CNT and PEI/CNT fiber electrodes are likely due to the presence of large amount of functionalized edge plane sites on CNTs which makes the surface hydrophilic.

Boron-doped diamond exhibits a high chemical stability and high resistance to surface fouling, usually at the expense of decreased reaction kinetics in comparison to other CNs [125, 126]. Patel *et al.* compared several different carbon electrodes for their antifouling properties for dopamine, including glassy carbon, oxygen-terminated polycrystalline boron-doped diamond (pBDD), edge plane pyrolytic graphite (EPPG), basal plane pyrolytic graphite (BPPG), and the basal surface of highly oriented pyrolytic graphite (HOPG) [127]. Although pBDD was found to be the least susceptible to surface fouling even at relatively high dopamine concentration, the reaction kinetics were relatively slow. In contrast, the reaction of 100 μM dopamine at pristine basal plane HOPG had fast dopamine fouling kinetics and only minor susceptibility toward surface fouling. The basal sp^2 hybridized carbons promote strong coupling of dopamine, leading not only to fast electron transfer kinetics but also to blocking of the electrode by products of the reaction. Meanwhile, the adsorption of dopamine and its oxidation products at pBDD is much less extensive, which is consistent with the slower electrochemical kinetics and the promising anti-passivation property.

Fouling due to adsorption of biological macromolecules, such as proteins, is also important to minimize for optimal sensor performance because adsorption affects mass transport and electron transfer kinetics [128]. Soluble proteins are often

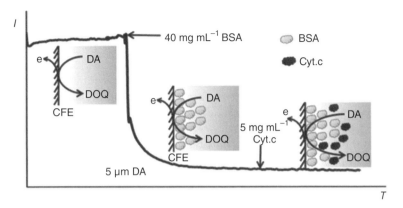

Figure 3.4 Schematic illustration of amperometric response at the electrode/electrolyte interface in different concentrations of protein. *Source*: Liu 2016 [130]. Reproduced with permission of American Chemical Society.

hydrophilic on the surface to interact with the aqueous environment and hydrophobic on the inside to maintain protein folding or the binding of hydrophobic materials [111]. Thus, proteins can foul electrode through both hydrophilic and hydrophobic interactions. Harreither *et al.* reported PVA/CNT fiber electrodes (described in section 3.2.3) were more resistant to fouling than the traditional carbon-fiber electrodes for dopamine detection in the presence of bovine serum albumin (BSA), which accounts for about 50% of total plasma protein content [129]. The presence of albumin reduces the impact of dopamine fouling because the sulfur on the BSA competed with the amine moiety for the nucleophilic binding to the oxidized catechol (a critical step initiating dopamine fouling) and therefore reduced the rate of dopamine polymerization on the electrode surface. Similarly, Liu *et al.* recently reported a new, effective solution for electrode calibration for *in vivo* measurements by pretreating MWCNTS/CFMEs with BSA [130]. The ratio of the sensitivity obtained with the post-calibration to that obtained with pre-calibration containing BSA was about 94%. Essentially, the strategy is to pre-foul the electrodes before implantation, and thus they will not be sensitive to BSA overadsorption or changes in their surface hydrophobicity upon implantation (Figure 3.4).

The electrodes also exhibited a high selectivity for ascorbate against dopamine, DOPAC, uric acid, and serotonin. Chen *et al.* evaluated the impact of the surface charges and morphology of PEDOT modified gold electrodes on the detection of dopamine, ascorbic acid, and uric acid in the presence of proteins, including BSA, lysozyme, and fibrinogen [131]. The adsorption of positively charged lysozyme promoted sensitive detection of ascorbic acid and uric acid, and all protein adsorption lowered the dopamine sensitivity. Although a gold electrode was used in this study, it provides a good reference for the future study focusing on the neurotransmitters detection at CN-based electrodes in the presence of different proteins.

In summary, CNs are being used to change the surface chemistry and improve the surface fouling resistance [132]. While polymer incorporated CN electrodes, such as PEDOT:PSS [115], PEDOT:GO [106] and Nafion-CNT [133], have been proposed to provide fouling resistance for *in vivo*, polymers can affect temporal response and not all are effective [116]. Thus, the future research will likely focus on the promising fouling resistance at CNs such as BDD, HOPG, VACNTs, and CNT fibers. In particular, hydrophilic CN surfaces tend to have better antifouling properties because of the higher reversibility of hydrophilic interactions with the surface than the hydrophobic interactions. Oxidation of the surface of CNs to introduce oxygen-containing functional groups is the easiest method to make the surface more hydrophilic, and oxidized CNs are preferred for *in vivo* neurotransmitters detection because they lead to both fouling resistance and high sensitivity. The surface roughness is another important factor contributing to the fouling resistance, contributing to the observed antifouling properties of superhydrophobic electrodes [134, 135]. Promising methods for future cleaning of adsorbed materials include including applying single anodic/cathodic potentials [123, 136, 137] or a train of pulses to periodically clean the electrode surface [138, 139]. The antifouling mechanism for anodic/cathodic potential relies on altering the amount of oxygen functionalities, which affects surface hydrophilicity [111, 112]. Future studies should focus on how surface properties affect fouling and more studies should work to make measurements in actual tissue, where fouling can truly be evaluated.

3.3.3 Reusable Carbon Nanomaterial-based Electrodes

CN based electrode manufacturing methods are becoming less expensive, more reproducibile, and faster for batch fabrication. However, most of the sensors with CNs are non-reusable and the fouling of analytes/interferences is irreversible, which limits the sensors sustainability and potentially increases the cost for each measurement. To fabricate reusable CN electrodes, several strategies have been applied, including magnetic entrapment, molecularly imprinted electrodes, and TiO_2 nanoparticle based surface self-cleaning. The strategies are important for implantable sensors, but also for future device such as microchip devices, that make continual measurements.

Magnetic entrapment is based on the strong, non-specific adsorption of CNTs to protein–coated magnetic micro-particles (MPs) [132]. The MP/CNT assemblies can then be magnetically captured on cheap and disposable SPEs. Muñoz's group reported carboxylated SWCNT-coated MP modified SPEs for the detection of dopamine (Figure 3.5a), with the LOD of $0.71\,\mu M$ [140].

The magnetic coentrapment approach allows the SPE modified electrodes to be reused at least 15 consecutive times, but the LOD was not improved over other methods and the sensitivity lost with increasing times of consecutive measurements was not reported. Recently, Herrasti *et al.* reported an improved MP-based biosensor for the detection of uric acid [141]. More than ten consecutive cycles of electrode regeneration and nanostructuration were performed without losing electrode performance. However, the sensitivity (LOD of $0.86\,\mu M$) is still not high enough to measure physiological dopamine levels *in vivo*. None of the magnetic entrapment based sensors have been applied for neurotransmitter detection in any real biological samples, which is necessary to investigate possible interferences.

Molecular imprinting has become a powerful tool for the preparation of polymeric materials which can specifically bind a target molecule. Molecularly imprinted polymers (MIPs) are tailor-made for a given target molecules, giving them high chemical selectivity [142], and the complete removal of target neurotransmitters allow

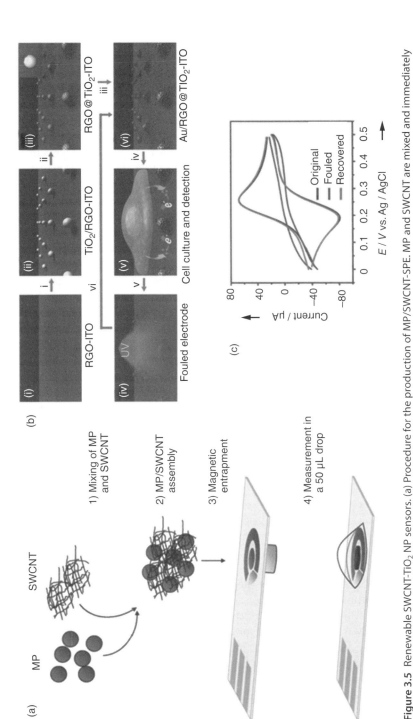

Figure 3.5 Renewable SWCNT-TiO$_2$ NP sensors. (a) Procedure for the production of MP/SWCNT-SPE. MP and SWCNT are mixed and immediately deposited on top of the working electrode. The SPE have been previously modified by placing magnets underneath. (b) Fabrication process of the renewable Au/RGO@TiO$_2$-ITO sensor and cell detection on the sensor. (i) Deposition of TiO$_2$ NPs onto RGO-modified ITO (RGO-ITO) by electrophoresis. (ii) TiO$_2$ NPs wrapped by RGO by electrostatic self-assembly. (iii) Deposition of Au NPs by chemical reduction. (iv) Cell culture and detection on the sensor. (v) Fouled sensor after the cell was detached. (vi) Sensor recovered by photocatalytic cleaning under UV light irradiation. (c) Cyclic voltammograms of 1 mm K$_3$[Fe(CN)$_6$] in 1 m KCl on the Au/RGO@TiO$_2$-ITO electrode before fouling (black), after fouling by 5-HT (blue), and after recycling by UV light irradiation. *Source:* Baldrich 2011 [140]. Reproduced with permission of Elsevier.

MIP-based electrode to be reused. CNs provide enhanced conductivity, increased surface to volume ratio, and maximized porosity in the MIPs. CNs also enhance binding to shorten the incubation and extraction time, improving the temporal resolution of detection. Qian *et al.* reported an ultrasensitive dopamine sensor based on molecularly imprinted oxygen containing polypyrrole (PPy) coated CNTs, in which a novel PPy is produced via polymerization with the incorporation of oxygen-containing groups on PPy backbone [143]. Dopamine was imprinted on PPy via π-π stacking between aromatic rings and hydrogen bonds between amino groups of DA and oxygen-containing groups of PPy. The sensor exhibited a high sensitivity of 16 µA µM and a low detection limit of 0.01 nM for dopamine. After 20 cycles of CV in 0.5 M KOH, the current response was nearly zero, which demonstrates the complete removal of template molecules. Although the concentration of dopamine in human serum and urine samples was determined in this work, the sensitivity for consecutive measurements were not tested. In addition, an incubation time of two minutes is required for stable signal, which is improved compared to previous work [144], but still would limit temporal resolution for real-time measurements.

Photocatalysts such as TiO$_2$ and ZnO can decompose organic molecules by generating oxygen radicals after ultraviolet irradiation without altering the surface morphology and structure [145, 146]. Photocatalytic cleaning methods have therefore been tested for feasibility with CN-based sophisticated sensors, and TiO$_2$ NP-decorated CN-based sensors could be cleaned simply by ultraviolet light irradiation and re-used indefinitely [147]. Xu *et al.* used a sensor fabricated by controllable assembly of reduced graphene oxide (RGO) and TiO$_2$ to form a sandwich structure, followed by deposition of Au NPs onto the RGO shell. The encapsulated TiO$_2$ ensures an excellent photocatalytic cleaning property (Figure 3.5b) [148]. 5-HT was used as a model passivation molecule, which caused a signal drop by accumulation of a non-electrochemically active oxidation product on the electrode surface. Thirty minutes of UV light irradiation after passivation recovers the sensitivity to 90% (Figure 3.5c). The application of this renewable microsensor for sensitive detection of nitric oxide release from cells demonstrated the photocatalytic cleaning of absorbed polymers, proteins, and cell culture medium. Another advantage of TiO$_2$ is the negative charge at physiological pH, which would repel other negatively charged interferences such as uric acid and ascorbic acid, and enhance the attraction of cationic neurochemicals (dopamine, adenosine, etc.) [149]. The combination of photocatalyst and CNs provides a versatile and efficient way for sensitive and renewable sensors for neurotransmitters detection.

Research into reusable electrodes is just beginning and will lead to electrodes that can be used either multiple times *in vivo* or as detectors for separations or in microfluidic devices, where constant changing of the sensor is not possible. The magnetically entrapped SWCNTs electrodes did not lose sensitivity with 15 consecutive measurements in ferrocyanide solution but more testing is needed for other interferences and surface fouling agents [140, 141]. In comparison, the RGO and TiO$_2$ NP electrodes lose some sensitivity over time, but the surface can be cleaned with UV light treatment [148]. However, UV light treatment cannot be performed in the brain so this would only clean electrodes between samples. For the neurotransmitter monitoring at molecularly imprinted electrodes, the measurement is not real time (30 s–10 min dopamine adsorption time was required), which limit the future applications. Overall, because of the large surface area, high conductivity of CNs, the self-cleaning property, TiO$_2$/CNs electrodes is the most favorable strategy for making reusable electrodes.

3.4   Conclusions

While CN electrode studies traditionally focused on CNTs, research has now been extended into many types of materials such as graphene and its derivatives. Newer carbon materials are being used such as carbon nanospikes [60], carbon nanohorns [46, 47] and CNT yarns [72, 73, 82]. Instead of a one-pot approach simply combining CNs with polymers and metal NPs, newer methods that allow growth of CNs directly on the electrode substrate [55–57, 62, 64] or fabrication of electrodes solely from CNs are useful for understanding the electrochemical properties of sensors of the CNs [68, 71–73]. As the number of materials grow, it is important for studies to compare different types of CNs under the same conditions. While a few studies examined the differences between CNs, most studies concentrate on a single electrode design, and it is difficult to evaluate which nanomaterial is most effective for an application. In particular, future studies that focus on identifying structural properties that underlie electrochemical performance will be critical for more rational design of CN-based electrodes. Dopamine redox reactions depend on the surface oxygen-containing functional groups, surface roughness, and π–π stacking reversibility [9, 95, 122, 150–152]. The fundamental electrocatalytic activity of CNs and the relationship between functional groups and defects on CNs are not well understood as some scanning electrochemical microscopy experiments suggest that the side walls of CNTs are also electrochemically active [107–110] while most sensor designs are based on maximizing exposure of edge plane defects [107]. Thus, there is a dichotomy between the practical and the fundamental science still needs to be addressed. New methods to assemble CN-based electrochemical sensors will help elucidate the extent to which electron transfer occurs at either edge plane sites or basal plane sites.

Implementing CN-based electrodes routinely *in vivo* is another challenge. The goal is to continuously monitor neurotransmitters *in vivo* and a major issue is surface fouling. Many electrode development studies still do not test CN based electrodes in real samples, so the extent to which different materials affect surface fouling has not been extensively studied. Future antifouling strategies should involve the advantages of CNs, surface modifications, and electrochemical activation methods. Moreover, reusable CN electrodes, especially the combination of CNs and TiO$_2$ NPs which provide a self-cleaning property, will draw more interest for applications with repeated measurements. In particular, electrochemical sensors as detectors of separation methods need to be robust and easily reusable. Researchers need to continue to test their CN based electrodes in tissue as much as possible and to focus on the use with a wide variety of neurotransmitters, some of which cause surface passivation. Advances in fundamental knowledge of new nanomaterials, along with a focus on practical applications in real-world systems, will lead to breakthroughs in both understanding neurotransmission and development of real-time sensing technology. We expect that CN based electrodes will see more routine usage in direct neurotransmitter sensing in the next five years, as their properties are better understood and tailored to *in vivo* measurements.

References

1 Koob, G.F. Alcohol addiction as a reward deficit and stress surfeit disorder. Front Psychiatry, 2013; 4: 72. Published online 2013 Aug 1, DOI: 10.3389/fpsyt.2013.00072 PMCID: PMC3730086

2 Janezic, S., Threlfell, S., Dodson, P.D., *et al.* Deficits in dopaminergic transmission precede neuron loss and dysfunction in a new Parkinson model. Proceedings of the National Academy of Sciences of the United States of America, 110 (42), E4016-4025; 2013.

3 Jacobs, C.B., Peairs, M.J., Venton, B.J. Review: Carbon nanotube based electrochemical sensors for biomolecules. Analytica Chimica Acta, 662 (2), 105–127; 2010.

4 McCreery, R.L. Advanced carbon electrode materials for molecular electrochemistry. Chemical Reviews, 108 (7), 2646–2687; 2008.

5 Robinson, D.L., Venton, B.J., Heien, M.L., Wightman, R.M. Detecting subsecond dopamine release with fast-scan cyclic voltammetry in vivo. Clinical chemistry, 49 (10), 1763–1773; 2003.

6 Kissinger, P., Hart, J., Adams, R.N. Voltammetry in brain tissue - a new neurophysilogical measurement. Brain Research, 55, 209–213; 1973.

7 Ponchon, J.L., Cespuglio, R., Gonon, F., *et al.* Normal pulse polarography with carbon fiber electrodes for in vitro and in vivo determination of catecholamines. Analytical Chemistry, 51 (9), 1483–1486; 1979.

8 Britto, P.J., Santhanam, K.S. V, Ajayan, P.M. Carbon nanotube electrode for oxidation of dopamine. Bioelectrochemistry and Bioenergetics, 41 (1), 121–125; 1996.

9 Yang, C., Denno, M.E., Pyakurel, P., Venton, B.J. Recent trends in carbon nanomaterial-based electrochemical sensors for biomolecules: A review. Analytica Chimica Acta, 887, 17–37; 2015.

10 Mittal, V. Surface modification of nanotube fillers. Wiley-VCH Verlag, Weinheim, Germany, pp. 2–20; 2011.

11 Zhang, Y., Xia, Z., Liu, H., *et al.* Hemin-graphene oxide-pristine carbon nanotubes complexes with intrinsic peroxidase-like activity for the detection of $H2O2$ and simultaneous determination for Trp, AA, DA, and UA. Sensors and Actuators B: Chemical, 188 (2013), 496–501; 2013.

12 Dalmasso, P.R., Pedano, M.L., Rivas, G. Dispersion of multi-wall carbon nanotubes in polyhistidine: characterization and analytical applications. Analytica Chimica Acta, 710 (2012), 58–64; 2012.

13 Kumar, S.A., Wang, S.-F., Yang, T.C.-K., Yeh, C.-T. Acid yellow 9 as a dispersing agent for carbon nanotubes: preparation of redox polymer-carbon nanotube composite film and its sensing application towards ascorbic acid and dopamine. Biosensors and Bioelectronics, 25 (12), 2592–2597; 2010.

14 Si, P., Chen, H., Kannan, P., Kim, D.-H. Selective and sensitive determination of dopamine by composites of polypyrrole and graphene modified electrodes. Analyst, 136 (24), 5134–5138; 2011.

15 Wang, W., Xu, G., Cui, X.T., *et al.* Enhanced catalytic and dopamine sensing properties of electrochemically reduced conducting polymer nanocomposite doped with pure graphene oxide. Biosensors and Bioelectronics, 58, 153–156; 2014.

16 Xu, G., Li, B., Cui, X.T., *et al.* Electrodeposited conducting polymer PEDOT doped with pure carbon nanotubes for the detection of dopamine in the presence of ascorbic acid. Sensors and Actuators B: Chemical, 188 (2013), 405–410; 2013.

17 Şen, M., Tamer, U., and Pekmez, N.Ö. Carbon nanotubes/alizarin red S–poly(vinylferrocene) modified glassy carbon electrode for selective determination of dopamine in the presence of ascorbic acid. Journal of Solid State Electrochemistry, 16 (2), 457–463; 2011

18 Lian, Q., He, Z., He, Q., *et al.* Simultaneous determination of ascorbic acid, dopamine and uric acid based on tryptophan functionalized graphene. Analytica Chimica Acta, 823, 32–39; 2014.

19 Tan, L., Zhou, K.-G., Zhang, Y.-H., *et al.* Nanomolar detection of dopamine in the presence of ascorbic acid at β-cyclodextrin/graphene nanocomposite platform. Electrochemistry Communications, 12 (4), 557–560; 2010.

20 Quan, D.P., Tuyen, D.P., Lam, T.D., *et al.* Electrochemically selective determination of dopamine in the presence of ascorbic and uric acids on the surface of the modified Nafion/single wall carbon nanotube/poly(3-methylthiophene) glassy carbon electrodes. Colloids and Surfaces B, Biointerfaces, 88 (2), 764–770; 2011.

21 Shieh, Y.-T., Tu, Y.-Y., Wang, T.-L., *et al.* Apparent electrocatalytic activities of composites of self-doped polyaniline, chitosan, and carbon nanotubes. Journal of Electroanalytical Chemistry, 704 (2013), 190–196; 2013.

22 Rodthongkum, N., Ruecha, N., Rangkupan, R., *et al.* Graphene-loaded nanofiber-modified electrodes for the ultrasensitive determination of dopamine. Analytica Chimica Acta, 804, 84–91; 2013.

23 Liu, J., Luo, J., Liu, R., *et al.* Micelle-encapsulated multi-wall carbon nanotubes with photosensitive copolymer and its application in the detection of dopamine. Colloid and Polymer Science, 292 (1), 153–161; 2013.

24 Mercante, L.A., Pavinatto, A., Iwaki, L.E.O., *et al.* Electrospun polyamide 6/poly(allylamine hydrochloride) nanofibers functionalized with carbon nanotubes for electrochemical detection of dopamine. ACS Applied Materials and Interfaces, 7 (8), 4784–4790; 2015.

25 Palanisamy, S., Sakthinathan, S., Chen, S.-M., *et al.* Preparation of β-cyclodextrin entrapped graphite composite for sensitive detection of dopamine. Carbohydrate Polymers, 135, 267–273; 2016.

26 Xu, H., Dai, H., Chen, G. Direct electrochemistry and electrocatalysis of hemoglobin protein entrapped in graphene and chitosan composite film. Talanta, 81 (1–2), 334–338; 2010.

27 Singh, A., Sinsinbar, G., Choudhary, M., *et al.* Graphene oxide-chitosan nanocomposite based electrochemical DNA biosensor for detection of typhoid. Sensors and Actuators B-Chemical, 185, 675–684; 2013.

28 Muszynski, R., Seger, B., Kamat, P. V. Decorating Graphene Sheets with Gold Nanoparticles. The Journal of Physical Chemistry C, 112 (14), 5263–5266; 2008.

29 Jackowska, K., and Krysinski, P. New trends in the electrochemical sensing of dopamine. Analytical and Bioanalytical Chemistry, 405 (11), 3753–3771; 2013.

30 Ponnusamy, V.K., Mani, V., Chen, S.-M., *et al.* Rapid microwave assisted synthesis of graphene nanosheets/polyethyleneimine/gold nanoparticle composite and its application to the selective electrochemical determination of dopamine. Talanta, 120 (2014), 148–157; 2014.

31 Guo, D.-J., Qiu, X.-P., Chen, L.-Q., Zhu, W.-T. Multi-walled carbon nanotubes modified by sulfated TiO_2 – A promising support for Pt catalyst in a direct ethanol fuel cell. Carbon, 47 (7), 1680–1685; 2009.

32 Zan, X., Bai, H., Wang, C., *et al.* Graphene paper decorated with a 2D array of dendritic platinum nanoparticles for ultrasensitive electrochemical detection of dopamine secreted by live cells. Chemistry - A European Journal, 22 (15), 5204–5210; 2016.

33 Yuan, D., Chen, S., Yuan, R., *et al.* An ECL sensor for dopamine using reduced graphene oxide/multiwall carbon nanotubes/gold nanoparticles. Sensors and Actuators B: Chemical, 191, 415–420; 2014.

34 Komathi, S., Gopalan, A.I., Lee, K.-P. Nanomolar detection of dopamine at multi-walled carbon nanotube grafted silica network/gold nanoparticle functionalised nanocomposite electrodes. The Analyst, 135 (2), 397–404; 2010.

35 Zheng, S., Huang, Y., Cai, J., Guo, Y. Nano-copper-MWCNT-modified glassy carbon electrode for selective detection of dopamine. International Journal of Electrochemical Science, 8 (2013), 12296–12307; 2013.

36 Tyszczuk-Rotko, K., Sadok, I. The new application of boron doped diamond electrode modified with nafion and lead films for simultaneous voltammetric determination of dopamine and paracetamol. Electroanalysis, 28 (9), 2178–2187; 2016.

37 Tsierkezos, N.G., Othman, S.H., Ritter, U., *et al.* Electrochemical analysis of ascorbic acid, dopamine, and uric acid on nobel metal modified nitrogen-doped carbon nanotubes. Sensors and Actuators, B: Chemical, 231, 218–229; 2016.

38 Sun, J.Y., Li, L., Zhang, X.P., *et al.* Simultaneous determination of ascorbic acid, dopamine and uric acid at a nitrogen-doped carbon nanofiber modified electrode. RSC Advances, 5 (16), 11925–11932; 2015.

39 Martín-Yerga, D., Rama, E.C., Costa-García, A. Electrochemical characterization of ordered mesoporous carbon screen-printed electrodes. Journal of the Electrochemical Society, 163 (5), B176–B179; 2016.

40 Yang, J., Zhang, Y., Kim, D.Y. Electrochemical sensing performance of nanodiamond-derived carbon nano-onions: Comparison with multiwalled carbon nanotubes, graphite nanoflakes, and glassy carbon. Carbon, 98, 74–82; 2016.

41 Ko, S., Tatsuma, T., Sakoda, A., *et al.* Electrochemical properties of oxygenated cup-stacked carbon nanofiber-modified electrodes. Physical Chemistry Chemical Physics, 16 (24), 12209–12213; 2014.

42 Lamas-Ardisana, P.J., Fanjul-Bolado, P., Costa-García, A. Manufacture and evaluation of cup-stacked carbon nanofiber-modified screen printed electrodes as electrochemical tools. Journal of Electroanalytical Chemistry, 775, 129–134; 2016.

43 Iijima, S., Yudasaka, M., Yamada, R., *et al.* Nano-aggregates of single-walled graphitic carbon nano-horns. Chemical Physics Letters, 309 (3–4), 165–170; 1999.

44 Martín, A., Hernández-Ferrer, J., Vázquez, L., *et al.* Controlled chemistry of tailored graphene nanoribbons for electrochemistry: a rational approach to optimizing molecule detection. RSC Advances, 4 (1), 132–139; 2014.

45 Zhu, S., Li, H., Niu, W., and Xu, G. Simultaneous electrochemical determination of uric acid, dopamine, and ascorbic acid at single-walled carbon nanohorn modified glassy carbon electrode. Biosensors and Bioelectronics, 25 (4), 940–943; 2009.

46 Valentini, F., Ciambella, E., Conte, V., *et al.* Highly selective detection of epinephrine at oxidized single-wall carbon nanohorns modified screen printed electrodes (SPEs). Biosensors and Bioelectronics, 59, 94–98; 2014.

47 Valentini, F., Ciambella, E., Boaretto, A., *et al.* Sensor properties of pristine and functionalized carbon nanohorns. Electroanalysis, 28 (10), 2489–2499; 2016.

48 Du, J., Yue, R., Ren, F., *et al.* Novel graphene flowers modified carbon fibers for simultaneous determination of ascorbic acid, dopamine and uric acid. Biosensors and Bioelectronics, 53, 220–224; 2014.

49 Tang, L., Du, D., Yang, F., *et al.* Preparation of graphene-modified acupuncture needle and its application in detecting neurotransmitters. Scientific Reports, 5 11627; 2015.

50 Li, Y.-T., Tang, L.-N., Ning, Y., *et al.* In vivo monitoring of serotonin by nanomaterial functionalized acupuncture needle. Scientific Reports, 6, 28018; 2016.

51 Jacobs, C.B., Vickrey, T.L., Venton, B.J. Functional groups modulate the sensitivity and electron transfer kinetics of neurochemicals at carbon nanotube modified microelectrodes. The Analyst, 136 (17), 3557–3565; 2011.

52 Liu, J. Fullerene Pipes. Science, 280 (5367), 1253–1256; 1998.

53 Álvarez-Martos, I., Fernández-Gavela, A., Rodríguez-García, J., *et al.* Electrochemical properties of spaghetti and forest like carbon nanotubes grown on glass substrates. Sensors and Actuators B: Chemical, 192 (2014), 253–260; 2014.

54 Xiao, N., and Venton, B.J. Rapid, sensitive detection of neurotransmitters at microelectrodes modified with self-assembled SWCNT forests. Analytical Chemistry, 84 (18), 7816–7822; 2012.

55 Xiang, L., Yu, P., Hao, J., *et al.* Vertically aligned carbon nanotube-sheathed carbon fibers as pristine microelectrodes for selective monitoring of ascorbate in vivo. Analytical Chemistry, 86 (8), 3909–3914; 2014.

56 Xiang, L., Yu, P., Zhang, M., *et al.* Platinized aligned carbon nanotube-sheathed carbon fiber microelectrodes for in vivo amperometric monitoring of oxygen. Analytical Chemistry, 86 (10), 5017–5023; 2014.

57 Yang, C., Jacobs, C.B., Nguyen, M.D., *et al.* Carbon nanotubes grown on metal microelectrodes for the detection of dopamine. Analytical Chemistry, 88 (1), 645–652; 2016.

58 Kozai, T.D.Y., Jaquins-Gerstl, A.S., Vazquez, A.L., *et al.* Brain tissue responses to neural implants impact signal sensitivity and intervention strategies. ACS Chemical Neuroscience, 6 (1), 48–67; 2015.

59 Sheridan, L.B., Hensley, D.K., Lavrik, N. V., *et al.* Growth and electrochemical characterization of carbon nanospike thin film electrodes. Journal of the Electrochemical Society, 161 (9), H558–H563; 2014.

60 Zestos, A.G., Yang, C., Jacobs, C.B., *et al.* Carbon nanospikes grown on metal wires as microelectrode sensors for dopamine. The Analyst, 140 (21), 7283–7292; 2015.

61 Robertson, J. Diamond-like amorphous carbon. Materials Science and Engineering: R: Reports, 37 (4–6), 129–281; 2002.

62 Silva, T.A., Zanin, H., May, P.W., *et al.* Electrochemical performance of porous diamond-like carbon electrodes for sensing hormones, neurotransmitters, and endocrine disruptors. ACS Applied Materials and Interfaces, 6 (23), 21086–21092; 2014.

63 Sainio, S., Palomki, T., Rhode, S., *et al.* Carbon nanotube (CNT) forest grown on diamond-like carbon (DLC) thin films significantly improves electrochemical sensitivity and selectivity towards dopamine. Sensors and Actuators, B: Chemical, 211, 177–186; 2015.

64 Sainio, S., Palomäki, T., Tujunen, N., *et al.* Integrated carbon nanostructures for detection of neurotransmitters. Molecular Neurobiology, 52 (2), 859–866; 2015.

65 Dengler, A.K., Wightman, R.M., McCarty, G.S. Microfabricated collector-generator electrode sensor for measuring absolute pH and oxygen concentrations. Analytical Chemistry, 87 (20), 10556–10564; 2015.

66 Vilatela, J.J., and Marcilla, R. Tough Electrodes: Carbon nanotube fibers as the ultimate current collectors/active material for energy management devices. Chemistry of Materials, 27 (20), 6901–6917; 2015.

67 Lu, W., Zu, M., Byun, J.H., *et al.* State of the art of carbon nanotube fibers: Opportunities and challenges. Advanced Materials, 24 (14), 1805–1833; 2012.

68 Harreither, W., Trouillon, R., Poulin, P., *et al.* Carbon nanotube fiber microelectrodes show a higher resistance to dopamine fouling. Analytical Chemistry, 85 (15), 7447–7453; 2013.

69 Zestos, A.G., Jacobs, C.B., Trikantzopoulos, E., *et al.* Polyethylenimine carbon nanotube fiber electrodes for enhanced detection of neurotransmitters. Analytical Chemistry, 86, 8568–8575; 2014.

70 Wang, J., Deo, R.P., Poulin, P., Mangey, M. Carbon nanotube fiber microelectrodes. Journal of the American Chemical Society, 125 (48), 14706–14707; 2003.

71 Viry, L., Derré, A., Poulin, P., Kuhn, A. Discrimination of dopamine and ascorbic acid using carbon nanotube fiber microelectrodes. Physical Chemistry Chemical Physics, 12 (34), 9993–9995; 2010.

72 Jacobs, C.B., Ivanov, I.N., Nguyen, M.D., *et al.* High temporal resolution measurements of dopamine with carbon nanotube yarn microelectrodes. Analytical Chemistry, 86 (12), 5721–5727; 2014.

73 Schmidt, A.C., Wang, X., Zhu, Y., Sombers, L. Carbon nanotube yarn electrodes for enhanced detection of neurotransmitter dynamics in live brain tissue. ACS Nano, 7 (9), 7864–7873; 2013.

74 Li, Z., Liu, Z., Sun, H., Gao, C. Superstructured assembly of nanocarbons: Fullerenes, nanotubes, and graphene. Chemical Reviews, 115 (15), 7046–7117; 2015.

75 Behabtu, N., Young, C.C., Tsentalovich, D.E., *et al.* Strong, light, multifunctional fibers of carbon nanotubes with ultrahigh conductivity. Science, 339 (6116), 182–186; 2013.

76 Vigolo, B., Pénicaud, A., Coulon, C., *et al.* Macroscopic fibers and ribbons of oriented carbon nanotubes. Science, 290 (5495), 1331–1334; 2000.

77 Muñoz, E., Suh, D.S., Collins, S., *et al.* Highly conducting carbon nanotube/ polyethyleneimine composite fibers. Advanced Materials, 17 (8), 1064–1067; 2005.

78 Koziol, K., Vilatela, J., Moisala, A., *et al.* High-Performance Carbon Nanotube Fiber. Science, 318 (5858), 1892–1895; 2007.

79 Zhang, M., Atkinson, K.R., Baughman, R.H. Multifunctional carbon nanotube yarns by downsizing an ancient technology. Science, 306 (5700), 1358–1361; 2004.

80 Mayhew, E., and Prakash, V. Thermal conductivity of high performance carbon nanotube yarn-like fibers. Journal of Applied Physics, 115 (17), 174306; 2014.

81 Jiang, K., Li, Q., and Fan, S. Nanotechnology: Spinning continuous carbon nanotube yarns. Nature, 419 (6909), 801; 2002.

82 Yang, C., Trikantzopoulos, E., Nguyen, M.D., *et al.* Laser Treated carbon nanotube yarn microelectrodes for rapid and sensitive detection of dopamine in vivo. ACS Sensors, 1 (5), 508–515; 2016.

83 Yang, C., Trikantzopoulos, E., Jacobs, C.B., Venton, B.J. Evaluation of carbon nanotube fiber microelectrodes for neurotransmitter detection: correlation of electrochemical performance and surface properties. Analytica Chimica Acta, 965, 1-8; 2017

84 Bucher, E.S., Wightman, R.M. Electrochemical analysis of neurotransmitters. Annual Review of Analytical Chemistry, 8, 239–261; 2015.

85 Rees, H.R., Anderson, S.E., Privman, E., *et al.* Carbon nanopipette electrodes for dopamine detection in drosophila. Analytical Chemistry, 87 (7), 3849–3855; 2015.

86 Traver, D., Paw, B.H., Poss, K.D., *et al.* Transplantation and in vivo imaging of multilineage engraftment in zebrafish bloodless mutants. Nature Immunology, 4 (12), 1238–1246; 2003.

87 Li, Y., Zhang, S.-H., Wang, L., *et al.* Nanoelectrode for amperometric monitoring of individual vesicular exocytosis inside single synapses. Angewandte Chemie International Edition, 53 (46), 12456–12460; 2014.

88 Wightman, R.M., Heien, M.L.A. V, Wassum, K.M., *et al.* Dopamine release is heterogeneous within microenvironments of the rat nucleus accumbens. European Journal of Neuroscience, 26 (7), 2046–2054; 2007.

89 Suzuki, I., Fukuda, M., Shirakawa, K., *et al.* Carbon nanotube multi-electrode array chips for noninvasive real-time measurement of dopamine, action potentials, and postsynaptic potentials. Biosensors and Bioelectronics, 49, 270–275; 2013.

90 Clark, J., Chen, Y., Silva, S.R. Low impedance functionalised carbon nanotube electrode arrays for electrochemical detection. Electroanalysis, 28 (1), 58–62; 2016.

91 Dincer, C., Ktaich, R., Laubender, E., *et al.* Nanocrystalline boron-doped diamond nanoelectrode arrays for ultrasensitive dopamine detection. Electrochimica Acta, 185, 101–106; 2015.

92 Colombo, M.L., Sweedler, J. V., Shen, M. Nanopipet-based liquid-liquid interface probes for the electrochemical detection of acetylcholine, tryptamine, and serotonin via ionic transfer. Analytical Chemistry, 87 (10), 5095–5100; 2015.

93 Van Dersarl, J.J., Mercanzini, A., Renaud, P. Integration of 2D and 3D thin film glassy carbon electrode arrays for electrochemical dopamine sensing in flexible neuroelectronic implants. Advanced Functional Materials, 25 (1), 78–84; 2015.

94 Trikantzopoulos, E., Yang, C., Ganesana, M., *et al.* Novel carbon-fiber microelectrode batch fabrication using a 3D-printed mold and polyimide resin. The Analyst, 141 (18), 5256–5260; 2016.

95 Cao, M., Fu, A., Wang, Z., *et al.* Electrochemical and Theoretical Study of π–π Stacking Interactions between Graphitic Surfaces and Pyrene Derivatives. The Journal of Physical Chemistry C, 118 (5), 2650–2659; 2014.

96 Yang, X., Feng, B., He, X., *et al.* Carbon nanomaterial based electrochemical sensors for biogenic amines. Microchimica Acta, 180 (11–12), 935–956; 2013.

97 Swamy, B.E.K., Venton, B.J. Carbon nanotube-modified microelectrodes for simultaneous detection of dopamine and serotonin in vivo. The Analyst, 132 (9), 876–884; 2007.

98 Muguruma, H., Inoue, Y., Inoue, H., Ohsawa, T. Electrochemical study of dopamine at electrode fabricated by cellulose-assisted aqueous dispersion of long-length carbon nanotube. The Journal of Physical Chemistry C, 120 (22), 12284–12292; 2016.

99 Robinson, D.L., Hermans, A., Seipel, A.T., Wightman, R.M. Monitoring rapid chemical communication in the brain. Chemical Reviews, 108 (7), 2554–2584; 2008.

100 Nguyen, M.D., Lee, S.T., Ross, A.E., *et al.* Characterization of Spontaneous, Transient Adenosine Release in the Caudate-Putamen and Prefrontal Cortex. PLoS ONE, 9 (1), e87165; 2014.

101 Xu, Y., Venton, B.J. Rapid determination of adenosine deaminase kinetics using fast-scan cyclic voltammetry. Physical Chemistry Chemical Physics, 12 (34), 10027–10032; 2010.

102 Bath, B.D., Michael, D.J., Trafton, B.J., *et al.* Subsecond adsorption and desorption of dopamine at carbon-fiber microelectrodes. Analytical chemistry, 72 (24), 5994–6002; 2000.

103 Venton, B.J., Troyer, K.P., Wightman, R.M. Response times of carbon fiber microelectrodes to dynamic changes in catecholamine concentration. Analytical Chemistry, 74 (3), 539–546; 2002.

104 Keithley, R.B., Takmakov, P., Bucher, E.S., *et al.* Higher sensitivity dopamine measurements with faster-scan cyclic voltammetry. Analytical Chemistry, 83 (9), 3563–3571; 2011.

105 Ross, A.E., Venton, B.J. Sawhorse waveform voltammetry for selective detection of adenosine, ATP, and hydrogen peroxide. Analytical Chemistry, 86 (15), 7486–7493; 2014.

106 Taylor, I.M., Robbins, E.M., Catt, K.A., *et al.* Enhanced dopamine detection sensitivity by PEDOT/graphene oxide coating on in vivo carbon fiber electrodes. Biosensors and Bioelectronics, 89, 400–410; 2017.

107 Dumitrescu, I., Unwin, P.R., Macpherson, J.V Electrochemistry at carbon nanotubes: perspective and issues. Chemical Communications, 7345 (45), 6886–6901; 2009.

108 Kim, J., Xiong, H., Hofmann, M., *et al.* Scanning electrochemical microscopy of individual single-walled carbon nanotubes. Analytical Chemistry, 82 (5), 1605–1607; 2010.

109 Byers, J.C., Güell, A.G., Unwin, P.R. Nanoscale electrocatalysis: Visualizing oxygen reduction at pristine, kinked, and oxidized sites on individual carbon nanotubes. Journal of the American Chemical Society, 136 (32), 11252–11255; 2014.

110 Güell, A.G., Meadows, K.E., Dudin, P. V., *et al.* Mapping nanoscale electrochemistry of individual single-walled carbon nanotubes. Nano Letters, 14 (1), 220–22; 2014.

111 Hanssen, B.L., Siraj, S., Wong, D.K. Recent strategies to minimise fouling in electrochemical detection systems. Reviews in Analytical Chemistry, 35 (1), 1–28; 2016.

112 Roeser, J., Alting, N.F.A., Permentier, H.P., *et al.* Boron-doped diamond electrodes for the electrochemical oxidation and cleavage of peptides. Analytical Chemistry, 85 (14), 6626–6632; 2013.

113 Wrona, M.Z., Lemordant, D., Lin, L., *et al.* Oxidation of 5-hydroxytryptamine and 5,7-dihydroxytryptamine. A new oxidation pathway and formation of a novel neurotoxin. Journal of Medicinal Chemistry, 29 (4), 499–505; 1986.

114 Wrona, M.Z., Dryhurst, G. Electrochemical oxidation of 5-hydroxytryptamine in aqueous solution at physiological pH. Bioorganic Chemistry, 18 (3), 291–317; 1990.

115 Yang, X., Kirsch, J., Olsen, E. V., *et al.* Anti-fouling PEDOT:PSS modification on glassy carbon electrodes for continuous monitoring of tricresyl phosphate. Sensors and Actuators, B: Chemical, 177, 659–667; 2013.

116 Singh, Y.S., Sawarynski, L.E., Dabiri, P.D., *et al.* Head-to-head comparisons of carbon fiber microelectrode coatings for sensitive and selective neurotransmitter detection by voltammetry. Analytical Chemistry, 83 (17), 6658–6666; 2011.

117 Muna, G.W., Partridge, M., Sirhan, H., *et al.* Electrochemical detection of steroid hormones using a nickel-modified glassy carbon electrode. Electroanalysis, 26 (10), 2145–2151; 2014.

118 Guo, Q., Liu, D., Zhang, X., *et al.* Pd-Ni alloy nanoparticle/carbon nanofiber composites: Preparation, structure, and superior electrocatalytic properties for sugar analysis. Analytical Chemistry, 86 (12), 5898–5905; 2014.

119 Yadav, S.K., Agrawal, B., Oyama, M., and Goyal, R.N. Graphene modified Palladium sensor for electrochemical analysis of norepinephrine in pharmaceuticals and biological fluids. Electrochimica Acta, 125, 622–629; 2014.

120 Mudrinić, T., Mojović, Z., Milutinović-Nikolić, A., *et al.* Beneficial effect of Ni in pillared bentonite based electrodes on the electrochemical oxidation of phenol. Electrochimica Acta, 144, 92–99; 2014.

121 Zheng, M., Zhou, Y., Chen, Y., *et al.* Electrochemical behavior of dopamine in the presence of phosphonate and the determination of dopamine at phosphonate modified zirconia films electrode with highly antifouling capability. Electrochimica Acta, 55 (16), 4789–4798; 2010.

122 Takmakov, P., Zachek, M.K., Keithley, R.B., *et al.* Carbon microelectrodes with a renewable surface. Analytical Chemistry, 82 (5), 2020–2028; 2010.

123 Kiran, R., Scorsone, E., de Sanoit, J., *et al.* Boron doped diamond electrodes for direct measurement in biological fluids: An In situ regeneration approach. Journal of the Electrochemical Society, 160 (1), H67–H73; 2013.

124 Tang, L., Thevenot, P., Hu, W. Surface chemistry influences implant biocompatibility. Current Topics in Medicinal Chemistry, 8 (4), 270–280; 2008.

125 Sarada, B.V. Electrochemical characterization of highly boron-doped diamond microelectrodes in aqueous electrolyte. Journal of The Electrochemical Society, 146 (4), 1469; 1999.

126 Xu, J., Chen, Q., Swain, G.M. Anthraquinonedisulfonate electrochemistry: a comparison of glassy carbon, hydrogenated glassy carbon, highly oriented pyrolytic graphite, and diamond electrodes. Analytical Chemistry, 70 (15), 3146–3154; 1998.

127 Patel, A.N., Tan, S., Miller, T.S., *et al.* Comparison and reappraisal of carbon electrodes for the voltammetric detection of dopamine. Analytical Chemistry, 85 (24), 11755–11764; 2013.

128 Patel, J., Radhakrishnan, L., Zhao, B., *et al.* Electrochemical properties of nanostructured porous gold electrodes in biofouling solutions. Analytical Chemistry, 85 (23), 11610–11618; 2013.

129 Harreither, W., Trouillon, R., Poulin, P., *et al.* Cysteine residues reduce the severity of dopamine electrochemical fouling. Electrochimica Acta, 210, 622–629; 2016.

130 Liu, X., Zhang, M., Xiao, T., *et al.* Protein pretreatment of microelectrodes enables in vivo electrochemical measurements with easy precalibration and interference-free from proteins. Analytical Chemistry, 88 (14), 7238–7244; 2016.

131 Chen, C.-H., Luo, S.-C. Tuning surface charge and morphology for the efficient detection of dopamine under the interferences of uric acid, ascorbic acid, and protein adsorption. ACS Applied Materials and Interfaces, 7 (39), 21931–21938; 2015.

132 Baldrich, E., Muñoz, F.X. Carbon nanotube wiring: a tool for straightforward electrochemical biosensing at magnetic particles. Analytical chemistry, 83 (24), 9244–9250; 2011.

133 Cahill, P.S., Walker, Q.D., Finnegan, J.M., *et al.* Microelectrodes for the measurement of catecholamines in biological systems. Analytical Chemistry, 68 (18), 3180–3186; 1996.

134 Xue, Q., Kato, D., Kamata, T., *et al.* Electron cyclotron resonance-sputtered nanocarbon film electrode compared with diamond-like carbon and glassy carbon electrodes as regards electrochemical properties and biomolecule adsorption. Japanese Journal of Applied Physics, 51 (9), 90124; 2012.

135 Kato, D., Komoriya, M., Nakamoto, K., *et al.* Electrochemical determination of oxidative damaged DNA with high sensitivity and stability using a nanocarbon film. Analytical Sciences : The International Journal of the Japan Society for Analytical Chemistry, 27 (7), 703; 2011.

136 Duran, B., Brocenschi, R.F., France, M., *et al.* Electrochemical activation of diamond microelectrodes: implications for the in vitro measurement of serotonin in the bowel. The Analyst, 139, 3160–3166; 2014.

137 Stoytcheva, M., Zlatev, R., Gochev, V., *et al.* Amperometric biosensors precision improvement. Application to phenolic pollutants determination. Electrochimica Acta, 147, 25–30; 2014.

138 Agnesi, F., Tye, S.J., Bledsoe, J.M., *et al.* Wireless instantaneous neurotransmitter concentration system-based amperometric detection of dopamine, adenosine, and glutamate for intraoperative neurochemical monitoring. Journal of Nurosurgery, 111 (4), 701–711; 2009.

139 Schmidt, A.C., Dunaway, L.E., Roberts, J.G., *et al.* Multiple scan rate voltammetry for selective quantification of real-time enkephalin dynamics. Analytical Chemistry, 86 (15), 7806–7812; 2014.

140 Baldrich, E., Gómez, R., Gabriel, G., Muñoz, F.X. Magnetic entrapment for fast, simple and reversible electrode modification with carbon nanotubes: application to dopamine detection. Biosensors and Bioelectronics, 26 (5), 1876–1882; 2011.

141 Herrasti, Z., Martínez, F., Baldrich, E. Detection of uric acid at reversibly nanostructured thin-film microelectrodes. Sensors and Actuators B: Chemical, 234, 667–673; 2016.

142 Gu, L., Jiang, X., Liang, Y., *et al.* Double recognition of dopamine based on a boronic acid functionalized poly(aniline-co-anthranilic acid)-molecularly imprinted polymer composite. The Analyst, 138 (18), 5461–5469; 2013.

143 Qian, T., Yu, C., Zhou, X., *et al.* Ultrasensitive dopamine sensor based on novel molecularly imprinted polypyrrole coated carbon nanotubes. Biosensors and Bioelectronics, 58C, 237–241; 2014.

144 Liu, B., Lian, H.T., Yin, J.F., Sun, X.Y. Dopamine molecularly imprinted electrochemical sensor based on graphene–chitosan composite. Electrochimica Acta, 75 (2012), 108–114; 2012.

145 Hisatomi, T., Kubota, J., Domen, K. Recent advances in semiconductors for photocatalytic and photoelectrochemical water splitting. Chemical Society Reviews, 43 (22), 7520–7535; 2014.

146 Wang, H., Zhang, L., Chen, Z., *et al.* Semiconductor heterojunction photocatalysts: design, construction, and photocatalytic performances. Chemical Society Reviews, 43 (15), 5234–5244; 2014.

147 Li, X., Chen, G., Yang, L., *et al.* Multifunctional Au-coated TiO2 nanotube arrays as recyclable SERS substrates for multifold organic pollutants detection. Advanced Functional Materials, 20 (17), 2815–2824; 2010.

148 Xu, J.-Q., Liu, Y.-L., Wang, Q., *et al.* Photocatalytically renewable micro-electrochemical sensor for real-time monitoring of cells. Angewandte Chemie International Edition, 54 (48), 14402–14406; 2015.

149 Pifferi, V., Soliveri, G., Panzarasa, G., *et al.* Photo-renewable electroanalytical sensor for neurotransmitters detection in body fluid mimics. Analytical and Bioanalytical Chemistry, 408 (26), 7339–7349; 2016.

150 Hočevar, S.B., Wang, J., Deo, R.P., *et al.* Carbon nanotube modified microelectrode for enhanced voltammetric detection of dopamine in the presence of ascorbate. Electroanalysis, 17 (5–6), 417–422; 2005.

151 Swamy, B.E.K., Venton, B.J. Carbon nanotube-modified microelectrodes for simultaneous detection of dopamine and serotonin in vivo. The Analyst, 132 (9), 876–884; 2007.

152 Xiao, N., Venton, B. Rapid, sensitive detection of neurotransmitters at microelectrodes modified with self-assembled SWCNT forests. Analytical Chemistry, 84 (18), 7816–7822; 2012.

4

Carbon and Graphene Dots for Electrochemical Sensing

Ying Chen, Lingling Li and Jun-Jie Zhu

School of Chemistry and Chemical Engineering, Nanjing University, China

4.1 Introduction

In recent decades, carbon-based nanomaterials have gained enormous attention in various fields for their low cost, simple synthesis techniques, abundant resources, environmental protection, and numerous excellent physicochemical properties. To date, carbon-based nanomaterials have developed into a prosperous family, including carbon nanotubes, graphene, nanodiamond, fullerene, carbon onion, carbon nanohorn, carbon dots (CDs) and graphene dots (GDs), in which carbon-based dots including CDs and GDs become burgeoning focuses for their unique structures, photoluminescence properties, exceptional physicochemical properties, and biocompatibility. Generally speaking, CDs and GDs are usually considered as two different kinds of carbon nanomaterials due to their distinct morphologies and structures. CDs are classical zero-dimensional (0D) materials with less than 10 nm diameter. They may be amorphous or nanocrystalline with sp^2 carbon clusters [1]. GDs resemble graphene nanosheets with less than 100 nm lateral size. Unlike their cousins, GDs clearly possess a graphene structure inside the dots and are generally of higher crystallinity, regardless of the dot size [1, 2].

Nowadays, CDs and GDs have achieved considerable success for their superior potentials in chemical sensing, biosensing, optical sensing, solar cell, biomedicine, catalysis, and imaging applications [2]. Among these applications, optical sensing and imaging have been intensively studied due to their excellent fluorescent properties, but electrochemical sensing is rapidly rising. CDs and GDs are recognized as optimal candidates for electroanalytical sensor materials, thanks to (1) their excellent electrochemical properties such as good electron mobility which facilitate charge transfer and electrochemical activity, (2) abundant defects and chemical groups on their surface providing the convenience and flexibility for various kinds of functionalisations, (3) their highly tunable chemical and electrical properties, and (4) their great stability as electrode materials [2]. In the present chapter, we will first summarize some recent progress on CDs and GDs based electrochemical sensing from two aspects, one focuses the roles of CDs and GDs acted in the whole sensing system, such as substrates, carriers and signal probes in detail, another one focuses the detection targets with the CDs and GDs based electrochemical sensing. Then we will simply introduce the branches of

Nanocarbons for Electroanalysis, First Edition.
Edited by Sabine Szunerits, Rabah Boukherroub, Alison Downard and Jun-Jie Zhu.
© 2017 John Wiley & Sons Ltd. Published 2017 by John Wiley & Sons Ltd.

electrochemical sensing, that is, electrochemiluminescence (ECL) sensing and photo-electrochemical (PEC) sensing, based on the CDs and GDs.

4.2 CDs and GDs for Electrochemical Sensors

4.2.1 Substrate Materials in Electrochemical Sensing

Owing to their large specific surface area, abundant edge sites, intrinsic catalytic activities, good electron mobility and chemical stability, CDs and GDs are recognized as optimal candidates for electroanalytical sensor materials. Due to the electronic and electrochemical properties, superior biocompatibility, and stable physical properties, CDs and GDs are considered to be the ideal electrochemical substrates, which are sensitive to minute perturbations, rendering them of great potential for sensitive sensing applications. As the substrates, the surfaces of CDs and GDs can be decorated with multifarious functional groups, such as –COOH and –OH groups, allowing a tethering of the target molecule for sensing applications. As follow, some works based on CDs and GDs as the substrate including as the electrocatalyst will be introduced in detail.

4.2.1.1 Immobilization and Modification Function

CDs are often employed as the substrate of the electrochemical sensor for its large specific surface area, unique physicochemical properties, and versatile functionalization with various functional groups, as, generally, carboxyl, hydroxyl, amino and so on. With these, varied molecules could be immobilized on the surface of CDs, such as organic molecules, biomolecules (enzyme, antigen, antibody, protein, DNA, *et al.*), and modified inorganic materials. On the other hand, CDs own strong $\pi-\pi$ conjugation and electrostatic adsorption, contributing to immobilize organic conjugated molecules and materials with charged groups.

Chen *et al.* prepared a CDs/nafion nanocomposite as the substrate of the electrochemical sensor [3]. CDs/nafion could generate an architecture layer to immobilize antibodies (Abs) through generation of a nonconducting immunocomplex towards antigen (Ag) by immunoreactions. The good conductivity of CDs could also promote electron propagation on the electrode to improve the sensitivity and enhance electrochemical response. After capturing the target antigens (MT-3), the electrochemical signal of biosensor showed a linear relationship with the logarithm concentrations of MT-3 from 5 pg ml to 20 ng ml with a detection limit of 2.5 pg ml, much lower than most of previously reports of label-free immunosensors. Similar work as Gao *et al.* reported that a novel PAMAM–CDs/Au nanocrystal nanocomposite was synthesized, exhibiting excellent conductivity, stability and biocompatibility, for immobilization of anti-alpha-fetoprotein (anti-AFP) [4]. The immunosensor showed a linear detection range from 100 fg ml to 100 ng ml with a detection limit of 0.025 pg ml. A novel electrochemical sensor was developed for glucose determination [5]. Nitrogen-doped carbon dots (N-doped CDs) as the substrate could enhance the electrocatalytic activity toward the reduction of O_2, and efficiently promote electron transfer between enzymes and electrodes, and allow the detection at low potential. The as-expected biosensor showed a response at the presence of glucose ranging from 1–12 mm with a detection limit of 0.25 mm.

GDs are also a promising nanomaterial as the substrate of electrochemical biosensing. Similar to CDs, GDs exhibit large specific surface area and outstanding physicochemical properties. Furthermore, GDs are more easily functional by diversified molecules or nanomaterials, and show more powerful π–π conjugation, due to the unique thin-layer structure. Ganesh reported a green color emitting GDs through an acid reflux reaction of graphene oxide (GO) [6]. Such GDs exhibited stable and uniform structure, and possessed surface functional moieties with epoxide, hydroxyl and carboxyl groups. With these, biomolecules could be easily immobilized onto the surface of GDs. In this work, horseradish peroxidase (HRP) enzyme was bound to the GDs surface through amide interaction. The biosensor with HRP as the enzyme probe and GDs as the substrate was further used for hydrogen peroxide (H_2O_2) detection, revealing a well-defined redox peak. A linear variation of H_2O_2 concentration with reduction current of cyclic voltammetry (CV) method was observed from 100 μm to 1 mm. Amperometric studies showed the reduction current increased with the concentration of H_2O_2 and the linear relationship was divided into two parts, ranging from 10–0 μm to 1.3 mm, and from 1.7–2.6 mm. The detection limit are respectively considered to be 530.85 nm and 2.16 μm. Dong *et al.* reported a simple electrochemical sensor for the detection of non-electroactive organophosphorus pesticides (OPs) [7]. GDs with abundant hydroxyl and carboxyl groups were still introduced as the substrate for glassy carbon electrode (GCE) modification. Pralidoxime (PAM) was attached on the GDs via electrostatic adsorption and π–π stacking interaction. PAM is a common used antidotes for OPs, and a nucleophilic substitution reaction could take place between the oxime group on PAM and OPs, leading to the reduction of oxidation current of PAM. The current change was proportional to OPs concentration from 10 pm to 0.5 μm, with a detection limit of 6.8 μm.

4.2.1.2 Electrocatalysis Function

In addition to the inherent physical and chemical properties, CDs and GDs have two significant performances, one is the photoluminescence (PL) property and another one is the strong electrocatalytic performance as a mimic enzyme. In the past decade, CDs and GDs have already considered to be promising nanomaterials in catalysis field, for their quantum size, high surface area, supernormal π–π conjugation structure, abundant functional moieties, excellent conductivity, and exposed marginal texture. In some work, CDs and GDs were introduced as peroxidase-like enzyme for the redox reaction of hydrogen peroxide (H_2O_2), oxygen (O_2), and hydroxyl ions (OH^-), also considered as mimic glucose oxidase. In addition, CDs and GDs can be used for the redox reaction of organic molecules, amino acid, metal ions and related compounds.

1) *Electrocatalyst for hydrogen peroxide reduction*

   H_2O_2 is an important analyte which is excessively used in industry, chemistry and biology fields. It also plays a crucial role in living cells, cancer and disease, metabolism, cell proliferation and apoptosis, redox reaction in vivo. Thus H_2O_2 determination and H_2O_2 sensing is quite important. In present methods for H_2O_2 sensing, electrochemistry is one of the mainstream techniques for its convenience, rapidity, good selectivity and sensitivity. Generally, related proteins or enzymes will be applied in the electrochemical sensing for H_2O_2 detection, such as hemoglobin, myoglobin and horseradish peroxidase. However, both proteins and enzymes will produce some problem as high cost, limited lifetime and rigorous detection conditions, to limit the development of H_2O_2 sensing. Recently, some inorganic nanomaterials such as metal materials, metal oxides, carbon nanomaterials, were designed as peroxidase-like enzyme for the free-enzyme sensing of H_2O_2, replacing related proteins and enzymes. In these, CDs and GDs are novel carbon-based nanomaterials with mimic enzyme activity. Compared to traditional carbon-based nanomaterials, CDs and GDs are more suitable as peroxidase-like enzyme for their quantum size, more exposed marginal texture, more excellent biocompatibility, and higher surface area, leading to higher catalysis activity. More recently, CDs and GDs have already be introduced as substrate and electrocatalyst in H_2O_2 sensing.

Sadhukhan *et al.* reported a sensitive and selective non-enzymatic, metal-free electrochemical H_2O_2 sensor based on CDs as electrocatalytic probe in-situ generated on the surface of graphene sheets [8]. This sensor exhibited an excellent response towards the electrochemical reduction and oxidation of H_2O_2, with a linear relationship between current and H_2O_2 concentration ranging from 10 μm to 1 mm, with a detection limit of 300 nm . Similar to CDs, Umrao *et al.* reported an enzyme-free sensor for H_2O_2 detection [9]. Green and blue luminescent GDs (g-GDs and b-GDs) were introduced as the probes with tunable and switchable functionalities. The b-GDs was the final product with pH-independent blue luminescence at 433 nm. Both g-GDs and b-GDs showed exhibited a remarkable response towards the reduction of H_2O_2, in which b-GDs were better than g-GDs due to fewer oxygen groups on the surface with a faster electron transfer rate. The b-GDs-embedded sensor showed a linear increase of current with H_2O_2 concentration from 1–10 mm with a detection limit of 0.02 mm.

To further improve the enzymatic activity of CDs and GDs, various heterostructure nanocomposites were designed combining GDs and CDs with other nanomaterials to improve the mimic enzymatic activity. These materials include metal related nanomaterials, carbon-based nanomaterials. Jahanbakhshi *et al.* reported a novel CDs nanohybrid combined with silver nanoparticles (AgNPs/CDs) [10]. The CDs was synthesized through hydrothermal treatment of salep and Ag NPs were embedded on the CDs surface by ultra-violate (UV) irradiation. Both Ag NPs and CDs could improve the mimic enzyme activity of the composites. The nanohybrid was introduced in H_2O_2 sensor as substrate and electrocatalyst, showing a superior performance toward H_2O_2 reduction with high stability and sensitivity. The linear range of H_2O_2 determination was from 0.2–27.0 μm with a limit detection of 80 nm. The H_2O_2 detection in living cells was further successfully achieved with human breast adenocarcinoma cell line MCF-7. Yang *et al.* prepared a novel GDs/ZnO hybrid nanofibers (NFs) by electrospun polymer templates [11]. This GDs/ZnO hybrid nanofibers were introduced in a biosensor as the catalytic probes for the intracellular H_2O_2 detection from cancer cells and normal cells after anticancer drugs treatment. The rGO QDs/ZnO hybrid nanofibers were fabricated by electrospinning and hydrothermal growth method. ZnO NFs with good electrochemical activity and biocompatibility make the sensors possess high sensitivities, low detection limits, and fast electron transfer. The hybrid ZnO NFs are also endowed with catalytic performance. GDs were decorated on the surface of ZnO NFs, leading to enhanced catalytical activity, ascribed to the large surface area and the synergistic effect. The as-expected hybrid nanofibers based electrochemical sensor exhibited high electrocatalytic activity of H_2O_2 from prostate cancer cells (PC-3) with (320 ± 12) amol/cell and noncancerous cells (BPH-1) with (210 ± 6) amol/cell. Bai *et al.* reported a novel enzyme-free hydrogen peroxide sensor based on the CDs modified multi-walled carbon nanotubes (MWCNTs) composites (CDs/MWCNTs). MWCNTs have attached excellent physical and chemical properties due to the superior conductivity, large surface area and high stability as the great electrode substrate and catalysts. Pure MWCNTs show low catalytic activity, thus it is necessary to introduce CDs to increase the catalytic sites with the increase of edge plane-like defective sites. CDs/MWCNTs based electrode possessed a significant enhanced electrocatalytic activity for H_2O_2 determination compared to pure MWCNTs and pure CDs. The current response was proportional to H_2O_2 concentration ranging from 3.5 µm to 30 mm with a detection limit of 0.25 µm. The sensor was further applied to monitor the real-time tracking of H_2O_2 released from human cervical cancer cells. In addition to combination with other nanomaterials, the catalytic activity of CDs and GDs could also be enhanced via doping with other elements. Ju *et al.* reported a nitrogen-doped GDs (N-GDs) with enhanced catalytic activity for the reduction of H_2O_2 [12]. They believed that since carbon materials could serve as reducing agents for metal nanoparticles reduction. The chemically doped N atoms will introduce more active sites and new electronic characteristics. Besides, N doping will also support more anchoring sites for the nucleation and growth of metal NPs. Thus, in this work, N-GDs modified with the in situ growth of Au NPs (Au NPs–N-GDs) were designed with significant enhanced catalytic activity as the catalytic probes. The electrochemical sensor based on the AuNPs/N-GDs composites exhibited sensitive and selective detection of H_2O_2 in human serum and live cells. The linear ranged from 0.25 µm to 13.327 mm with a detection limit of 0.12 µm.

Besides H_2O_2, the reaction of O_2 and OH^- was also investigated with enzyme free electrochemical sensor based on CDs and GDs as the catalytic probes. A novel GDs modified γ-MnOOH nanotubes on the surface of three-dimensional (3D) graphene aerogels (γ-MnOOH@GA/GDs) were synthesized as a highly efficient electrocatalyst [13]. The composites were used for oxygen reduction reaction (ORR) on a cathode. The common used ORR catalysts include platinum (Pt), ruthenium (Ru) and their alloys with high cost and limited scarce resources. Instead of Pt and Ru, manganese compounds are much cheaper and rich in resources, with advantages of low toxicity and multiple valence states. Among the manganese compounds, the catalytic activity of γ-MnOOH for ORR is generally higher for its unique composition, structure, and morphology. MnOOH possesses excellent catalytic activity due to the disproportionation of the produced $O_2^{\bullet-}$ into O_2 and HO_2^-, and then into O_2 and OH^- in an alkaline media. Introducing carbon materials into γ-MnOOH could increase the electronic conductivity and enhance the ORR catalytic activity. 3D graphene aerogels are suitable as the substrate with unique structure, large surface area, good conductivity, and enhanced electron and ion transfer to improve the catalytic performance. GDs with abundant oxygen functional groups, strong quantum confinement and edge effects, could generate positive charged active sites for the adsorption of O_2. The edge structure of GDs will also promote the ORR reaction. Thus, in this work, γ-MnOOH@GA/GDs was introduced as the cathode substrate for the reduction of O_2 with a significant performance. The intercalating property of OH^- was investigated by Doroodmand *et al.* through reversible conversion of GDs into CDs by electrochemical techniques [14]. The research showed that OH^- could selectively intercalate the nanocarbons, leading to reversible conversion of GDs to CDs, where OH^- acted as intercalating agent to make the GDs become aggregate CDs. Through this phenomenon, OH^- could be separated by the electro-filtration process, OH^- could also be exchanged with other anionic species, and the concentration of OH^- could be detected by the GDs based electrochemical sensor with an enhanced electrochemical response. The detection limit is up to 1.5 m of OH^-.

2) *Electrocatalysis of organic redox reaction*

In many ways, organic molecules are distributed in numerous fields including industrial manufacture, life movement, medicine, material life, and so on. Thus it is necessary for the determination of these organic molecules, especially those with special significance and value. Various techniques are designed for the determination of organic molecules, including chromatography, fluorescence, mass spectrometry, spectrophotometry and electrochemistry. Electrochemical technique has gained much attention to the determination of organic molecules for its simple operation, low cost and high sensitivity. The detection of organic molecules by electrochemical sensing could be possible, due to the electrochemical activity of them with the inherent redox reaction. In addition, to improve the performance of electrochemical sensing, many materials are introduced for the detection with enhanced electrocatalytic activity, including metal oxides, conducting polymers, carbon-based materials, etc. Among these materials, CDs and GDs, as novel carbon-based nanomaterials developed in recent years, are also applied in the electrochemical sensing for their unique properties. For example, dopamine (DA) is one member of catecholamine family released from the brain to regulate the central nervous, hormonal, and cardiovascular systems of human beings, which could also be artificially produced. Nowadays, DA is widely studied in the clinical fields for its potential relationship with various diseases. Jiang *et al.* reported an electrochemical sensor based on N-doped carbon quantum dots (NCDs) as the substrate and a mimic enzyme for the high sensitive detection of DA [15]. NCDs could improve the electrocatalytic activity of the sensor due to good conductivity, abundant functional groups, large surface area, and conjugation effect with organic molecules. The sensor showed an increase performance with the DA concentration in a linear range from 0 to 1.0 mm with a detection limit of 1.0 nm . Li *et al.* report a sensitive DA sensor with GDs self-assembled monolayers (GDs–NHCH$_2$CH$_2$NH) as the substrate and a mimic enzyme as shown in Figure 4.1 [16]. Similar to CDs, GDs exhibit a superior electrocatalytic activity for organic molecules due to the suitable conductivity, abundant functional groups, large surface area, conjugation effect, quantum confinement and edge effects. The electrochemical sensor based on the GDs–NHCH$_2$CH$_2$NH layer showed a linear range to DA from 1–150 μm with a detection limit of 0.115 μm. There are also some other reports for the determination of DA with the electrochemical sensing based on CDs, GDs, and their related composites. Besides DA, some other small organic molecules were investigated by the CDs and GDs electrochemical sensing, such as uric acid (UA), ascorbic acid (AA), 2,4,6-trinitrotoluene (TNT), etc.

Glucose (Glu) is a polyhydroxy aldehyde, which is considered to be the most important monosaccharide widely distributed in nature. Glu plays an important role in the biology field, especially in the bodily functions of human beings. Glu is basic unit of the energy storage in the blood, to supply the energy for the living cells and

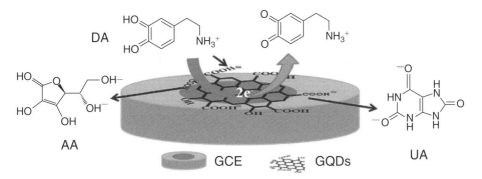

Figure 4.1 The schematic illustration of the reaction mechanism of DA on the GQDs–NHCH$_2$CH$_2$NH/ GCE. *Source*: Li 2016 [16]. Reproduced with permission of Elsevier. (*See color plate section for the color representation of this figure.*)

various organs. Glu also takes part in the metabolism to regulate the balance of body operation. In addition, Glu is also widely applied in the fields of photosynthesis, food and medicine manufacturing. Thus, the sensitive determination of Glu is essential, and electrochemical sensing is one of the detection techniques. Generally, in electrochemical sensing for the detection of Glu, glucose oxidase (GOx) is necessary, which could catalyze the oxidation reaction of Glu. Akhtar *et al.* prepared dendritic NiO@ carbon–nitrogen dots (NiO@NCDs) as the substrate of an effective and sensitive Glu sensor [17]. NiO has attracted great interest for the Glu oxidation with high free-enzyme electrocatalytic performance due to the large surface area, high porosity, and good hydrophilicity. The NCDs could enhance the electrocatalytic activity for free-enzyme Glu determination since they gained large surface area, excessive surface defects, abundant catalytic active sites and good electron transfer and conductivity. The NiO@NCDs based sensor showed a wide linear range of Glu detection ranging from 5 μm to 12 mm with a detection limit of 0.01 μm, with great recyclability, high selectivity, and superior anti-interference.

3) *Electrocatalyst for amino acids detection*

Amino acids, a series of organic compounds containing alkaline amino and acidic carboxyl, are the essential materials for the generation of various proteins. Amino acids are crucial for health, proteins could only be adsorbed by the body after degradation into amino acids, and the generation of proteins also need the presence of relevant amino acids. The balance of nitrogen element in the body should be always kept through the ingestion of proteins and the metabolism of amino acids. Some amino acids will take part in the generation of the fat, enzymes, hormones, and partial vitamins. Excessive or deficient amino acids will lead to the disorder of physiological function, even serious diseases, in which some amino acids act as the markers of cancers and diseases. Thus it is necessary to monitor the concentration of amino acids by ultrasensitive sensing.

Mazloum-Ardakani and co-workers built sensitive electrochemical sensor with glassy carbon electrode (GCE) as the substrate modified with GDs, Au NPs, and 4-(((4-mercaptophenyl)imino)methyl)benzene-1,2-diol (MIB) by self-assembly method [18]. After electrostatic adsorption of GDs on the GCE, Au NPs was immobilized on the modified electrode by electrodeposition. At last, MIB was grafted on the electrode surface to improve electrocatalytic activity and charge injection. GDs showed various electronic and optoelectronic properties, large surface area and high conductivity due to the quantum confinement and edge effects, which could enhance the electrocatalytic activity of the whole sensor. Au NPs with good biocompatibility, excellent conductivity, and catalytic properties also make them a promising electrocatalyst. The sensor was then used for the simultaneous determination of glutathione (GSH), tryptophan (Trp) and uric acid (UA), showing a linear range of GSH from 0.03–40.0 µm and 40.0–1300.0 µm with a detection limit of 9 nm. Another GDs and chitosan (CS) composite (GDs-CS) was used as the substrate of electrochemical sensor for the chiral recognition of tryptophan (Trp) enantiomers [19]. Here, CS provides a chiral microenvironment, and GDs can enhance the electrochemical response to improve the recognition efficiency as a mimic enzyme. In addition to amino acids, the electrochemical sensor based on CDs and GDs can also be used for the detection of nucleic acid. Wang *et al.* reported an electrochemical sensor with Ag nanoparticles (Ag NPs) and GDs modification for detection of guanine and adenine, which play a curial role in protein generation and genetic storage [20]. The sensor showed a superior electrocatalytic activity for guanine and adenine oxidation with rapid response and high sensitivity due to the excellent conductivity, abundant electrocatalytic active sites, and good electron transfer in synergistic effect of Ag NPs and GDs. The electrochemical response showed a good linear relationship with guanine concentration ranging from 0.015–430 µm with a detection limit of 0.01 µm, and with adenine ranging from 0.015–390 µm with a detection limit of 0.012 µm.

4) *Electrocatalyst for heavy metal ions detection*
The pollution of heavy metal elements is considered to be a serious problem in modern society, which can be accumulated in the organisms, leading to various diseases. After accumulation in the body, heavy metals cannot be biodegraded or metabolized, they can take intense interaction with proteins and enzymes to make them lose activity, and lead to chronic poisoning. Generally, the limit safety of heavy metal in water and food should be quite low to avoid potential health problem. Thus the sensing of heavy metal ions and related compounds is necessary, providing rapid, sensitive, and on-site detection. In heavy metal sensing techniques, the electrochemical technique shows great performance due to its rapid response, high sensitivity and selectivity. Carbon-based nanomaterials, including CDs and GDs, have been widely used for the detection of heavy metal owing to high conductivity, large surface area, high stability and biocompatibility. CDs and GDs as the substrate could result in enhanced electrochemical response for the redox reaction of heavy metals, similar to the enzyme catalysis amplification.

Ting *et al.* introduced GDs and Au NPs conjugation as the substrate of sensor for the sensitive detection of Hg^{2+} and Cu^{2+} [21]. As shown in Figure 4.2, Hg^{2+} and Cu^{2+} could be pre-concentrated onto the electrode due to the negatively charged carboxyl and hydroxyl groups of GDs in the form of R-COO-(Hg^{2+})-OOC-R and COO-(Cu^{2+})-OOC-R. Then the absorbed Hg^{2+} and Cu^{2+} are subsequently oxidized with enhanced response due to the synergistic integration between GDs and Au NPs. The linear relationship with Hg^{2+} ranged from 0.02–1.5 nm with a detection limit of

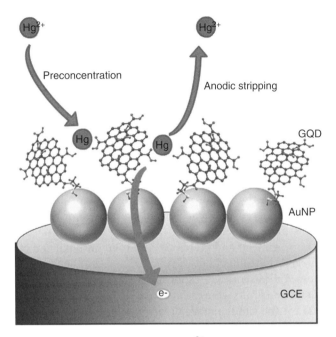

Figure 4.2 Schematic illustration of Hg^{2+} detection. *Source*: Ting 2015 [21]. Reproduced with permission of Elsevier.

0.02 nm, and that of Cu^{2+} from 500 pm to 1.5 nm with a detection limit of 50 pm. The simultaneous determination of Hg^{2+} and Cu^{2+} could be successfully completed by this sensor. Ou *et al.* reported another electrochemical sensing platform for the simultaneous determination of Zn^{2+}, Cd^{2+} and Pb^{2+} based on nanocomposites of chitosan (CS) and GDs conjugation hybrids (CS-GDs), which were prepared via the electrostatic attraction and H-bonding [22]. The CS-GDs could significantly enhance electrochemical performance of the sensor for the detection of heavy metals. The linear relationship ranged from 50–450 μg/l and the detection limit is 8.84 μg/l for Zn^{2+}, 1.9 μg/l for Cd^{2+}, 3.10 μg/l for Pb^{2+}.

4.2.2 Carriers for Probe Fabrication

CDs and GDs are also often used as the carriers labelled with various probes for the amplification of signal response. Similar as the substrate, CDs and GDs exhibit large surface area-to-volume ratio, exceeding physicochemical properties, and easily-functional property. In addition, CDs and GDs with quantum size are more suitable as the carriers for their superior dispersibility, biocompatibility, and electrostatic properties, which could high-effectively adsorb abundant probes such as organic molecules, proteins, DNA, enzyme, and nanomaterials. A selective and sensitive electrochemical sensor was developed with CDs as carriers for the determination of cerebral Cu^{2+} in rat brain micro-dialysates [23]. Methoxy functional (3-aminopropyl)trimethoxysilane (APTMS) was modified on the surface of glassy carbon electrode (GCE) via carbon–nitrogen linkage by electrochemical scanning method. Then CDs with carboxyl groups were attached to APTMS through silanization interaction as carriers of the whole biosensor. The N-(2-aminoethyl)-N,N',N'-tris-(pyridine-2-yl-methyl)ethane-1,2-diamine (AE-TPEA) was immobilized on the surface of CDs through amide interaction to specific recognize Cu^{2+}. Here, the large surface area and modification with carboxyl and hydroxyl group on CDs successfully realized the modification of GCE and immobilization of TPEA, turning into the central part of biosensor. Differential pulse anodic stripping voltammetry (DPASV) was employed to detect the adsorbed Cu^{2+} concentration, from 1–60 μm as a linear range relationship, and the detection limit was calculated to be ~100 nm. Figure 4.3 showed another electrochemical sensor with CDs as carriers for the simultaneous detection of ascorbic acid (AA), uric acid (UA), dopamine (DA), and acetaminophen (AC) [24]. CDs were introduced as the template to synthesize Au NPs/carbon dots nanocomposite (Au/C NC), on which thiol functional ferrocene derivative (Fc-SH) could be modified as the signal probes for the electrocatalysis of redox reaction. Here, CDs exhibited superior physicochemical properties for the synthesis of nanocomposites. What is more, signal probes could be immobilized on the graphene sheets via interaction between CDs and graphene, and then immobilized on the GCE. Au/C NC played a crucial bridge-effect between Fc-SH and graphene sheet. This electrochemical sensor showed ultrasensitive response to AA, UA, DA, and AC with excellent linear range, the detection limits were estimated to be 1.00, 0.12, 0.05 and 0.10 mm. The sensor was applied for the simultaneous determination of four components in serum and urine samples, with recovery of 94.0–104.0%.

In Hu's report, a specific and sensitive electrochemical sensor was developed for quantitative determination of miRNA [25]. Through the hybridization reaction among

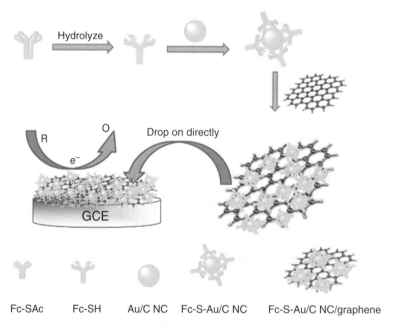

Fc-SAc   Fc-SH   Au/C NC   Fc-S-Au/C NC   Fc-S-Au/C NC/graphene

Figure 4.3 A quadruplet electrochemical platform for ultrasensitive and simultaneous detection of ascorbic acid, dopamine, uric acid and acetaminophen based on a ferrocene derivative functional Au NPs/carbon dots nanocomposite and graphene. *Source*: Yang 2016 [24]. Reproduced with permission of Elsevier. (*See color plate section for the color representation of this figure.*)

thiol-tethered probes (capture DNA), aminated indicator probe (NH_2-DNA), and target DNA (miRNA-155), the GDs modified with carboxyl and hydroxyl groups were captured as carriers on the surface of golden electrode via amide interaction. Since the large surface area, strong π–π conjugation and electrostatic adsorption of GDs, horseradish peroxidase (HRP) could be immobilized on the sensor by noncovalent assembly, promoting an electrochemical enzyme sensor. An increased electrochemical current signal was obtained after HRP catalyzed the hydrogen peroxide (H_2O_2)-mediated oxidation of 3,3′,5,5′-tetramethylbenzidine (TMB), in direct proportion to the target DNA concentration. The detection range of miRNA-155 was from 1 fm to 100 pm, and the detection limit was 1 fm. GDs with good biocompatibility, large surface area, and remarkable conductivity could effectively introduce abundant HRP, providing higher electrochemical signals.

4.2.3 Signal Probes for Electrochemical Performance

Unique photo-physical properties make CDs and GDs significant distinction from other carbon-based nanomaterials. Their strong photoluminescence (PL) properties and up-converted PL behavior, making CDs and GDs estimated to be promising in photovoltaic devices, biosening and imaging. Different from traditional organic fluorescent materials and inorganic quantum dots, CDs and GDs exhibit low cost, low toxicity, superior stability, multiple functionalization, and attractive biocompatibility, showing a potential tendency to substitute traditional QDs or fluorescent dyes as PL probes.

Electrochemiluminescence (ECL) technique, one branch of electrochemistry, has already gained attention due to the low cost, high sensitivity, and low background noise. One of the key parts of ECL sensing is the probe with strong and stable PL property, which could be stimulated to light from electrochemically generated reagents. Traditional ECL probes include organic dyes and quantum dots, which are high toxic and unstable. Instead of these, CDs and GDs are successfully introduced as new environmental-friendly ECL probes to develop high-performance ECL immunosensors. The work of ECL sensing based on the CDs and GDs will be introduced later in the second part subsequently.

In addition to the ECL sensor with CDs or GDs as signal probes based on their PL properties, CDs and GDs could also be used as signal probes in other electrochemical sensing. As shown in Figure 4.4, Liu *et al.* reported an electrochemical sensor for the detection of DNA with peroxidase-like magnetic $ZnFe_2O_4$–GDs ($ZnFe_2O_4$/GDs) nanohybrid as a mimic enzymatic probe [26]. Capture probe ssDNA (S1) was first immobilized on a graphene sheets (GS) and Pd nanowires (NWs) modified glassy carbon electrode (GCE), then target ssDNA (S2) was captured by S1 through hybridization reaction. S2 could further capture complementary ssDNA (S3) modified $ZnFe_2O_4$/GDs signal probes by hybridization reaction. In this work, GDs exhibited intrinsic peroxidase-like activities, as large surface area, distinct crystal sizes and aromatic structure, periphery carboxylic groups. $ZnFe_2O_4$/GDs composites provided highly-efficient enhanced catalytic performance, due to conjugation with GDs, while pure $ZnFe_2O_4$ only possessed relatively low peroxidase-like catalytic activity as a mimic enzyme. This sensor exhibited a wide linear relationship with target DNA ranging from 0.1 fm to 5 nm with a detection limit of 0.062 fm.

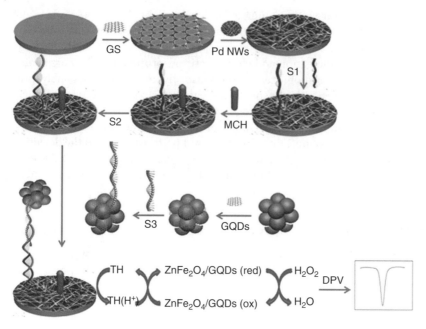

Figure 4.4 Graphene–palladium nanowires based electrochemical sensor using $ZnFe_2O_4$–graphene quantum dots as an effective peroxidase mimic. *Source*: Liu 2014 [26]. Reproduced with permission of Elsevier. (*See color plate section for the color representation of this figure.*)

4.2.4 Metal Ions Sensing

Metal ion detection is an important and basic field in electrochemical sensing. Punrat *et al.* reported an electrochemical sensor for Cr (VI) ion determination with stopped-flow analysis coupled with voltammetric technique [27]. In this ion sensor, a polyaniline/GDs modified screen-printed carbon electrode was used. The GDs were synthesized from citric acid in a botton-up manner, and were mixed with aniline monomer in an optimized ratio. This mixture was further injected into an electrochemical flow cell for the electro-polymerization of the aniline monomer. The obtained polyaniline/GDs modified screen-printed carbon electrode was then applied in the sensing of Cr(VI) ion and could be used continuously with a high throughput of more than 90 samples per hour. An ion sensor with 'green' electrochemical strategy synthesized GDs was reported by Ananthanarayanan *et al.* [28]. They electrochemically exfoliated GDs from three-dimensional graphene grown by chemical vapor deposition (CVD), and used this kind of GDs for ferric ion (Fe^{3+}) detection. They attempted different protocols for electro-chemical exfoliation and found that by applying a constant voltage of 5 V for 100 s they could get high yield of GDs and preserve the 3D structure of graphene. Their GDs dispersion appeared blue fluorescence under UV (365 nm), which could selectively quenched by Fe^{3+} ion with no obvious effect from Mg^{2+}, Fe^{2+}, Zn^{2+}, Co^{2+}, Ni^{2+}, Cd^{2+}, and K^+ ions. They used AFM measurements to confirm the aggregation of GDs induced by Fe^{3+} ion.

4.2.5 Small Molecule Sensing

Certain small molecules are vital to biological systems. The determination of these kinds of molecules are significant in bioanalysis areas. Blood glucose is one of the most important indexes for human health. Many technologies are pursued to develop advanced glucose sensors, electrochemical method is one of them. In most of the electrochemical biosensors for glucose, enzyme immobilization is a tricky step. Razmi *et al.* explored the enzyme absorption and electron transfer promotion ability of graphene dots and fabricated a biosensor for glucose detection [29]. At the first stage, they simply coated GDs, which were synthesized by hydrothermal method, on carbon ceramic electrode (CCE) surface through by-the-drop casting method. Then they modified glucose oxidase (GOx) onto GDs. Finally, they examined the performance of this GOx–GQD/CCE based glucose biosensor. Electrochemical impedance spectroscopy (EIS) is a useful electrochemical technique offering detailed information about the charge transfer resistance (Rct) changes of the electrode interface. Their study of EIS on GDs showed that in comparison with graphene, the conductivity of GDs was much lower. They believed that the size tuneable properties of GDs like energy gap (Eg) which would increase with the decrease of size, could be estimated from Rct, with an initial equation $Eg \propto K \cdot Rct$. After assembling GOx, the Rct of the electrode interface increased dramatically. GDs as the substrate could effectively immobilize the GOx and adsorb the Glu molecules, which could improve the electrocatalytic efficiency. The high surface area, good biocompatibility, hydrophilic edges of GDs and the hydrophobic interactions between GDs plane and GOx, led to strong absorption of enzyme on the surface of GDs. Thus, they found graphene dots as excellent substrate for enzyme immobilization. Furthermore, they investigated the performance of this glucose biosensor with electrochemistry method. The enhanced performance of the biosensor was attributed in the remarkable conductivity, large surface area, and good biocompatibility of GDs. It is worth mentioning that, GDs could enhance the electrocatalytic activity of GOx as a mimic enzyme due to the synergistic effect with GOx. The results showed a very fast electron transfer process and good sensitivity and accuracy in glucose determination.

Dopamine is another significant molecule for biological systems. It is a neurotransmitter in the mammalian central and peripheral nervous systems regulating a series of neuronal functions [30]. A GDs and nafion composites based electrochemical sensor was built for the detection of dopamine by Pang *et al.* [31]. Their GDs were synthesized by a hydrothermal approach from graphene sheets. The carboxyl groups and negative charge of GDs provided good stability and enabled interaction with amine functional groups of dopamine. The interaction and electron transfer between GDs and dopamine could be further strengthened by π–π stacking force. This GDs-nafion based sensor showed good responses in the range of 5 nm to 100 μm. Other satisfactory dopamine sensor based on GDs was further reported [16]. Huang and co-workers built a simple, sensitive and reliable DA biosensor modified with the composites of CDs and chitosan (CS) [32]. The CDs–CS composites modified GCE electrode exhibited an enhanced the redox response of DA by differential pulse voltammetry (DPV) with a linear range from 0.1–30 μm. The detection limit was 11.2 nm (S/N = 3). Jiang *et al.* prepared a type of N-doped carbon dots (NCDs) from diethanolamine (DEA) via a simple microwave-assisted technique [33]. NCDs could attract the electroactive DA due to abundant hydroxyl and amino groups on the surface. The NCDs modified electrode exhibited a superior sensitive electrochemical response toward DA in a linear range from 50 nm to 8 μm with a detection limit of 1.2 nm. The sensor has high sensitivity and selectivity.

L-cysteine, which is an amino acid in natural proteins, plays crucial roles such as cancer indicator, free radical scavenger, antioxidant and antitoxin in biological systems [34]. Wang and coworkers built a polypyrrole (PPy) and GDs/Prussian Blue (PB) nanocomposites based electrochemical sensor for the determination of L-cysteine [35]. They grafted the PPy/GDs@PB nanocomposites on a graphite felt (GF) substrate (PPy/GDs@ PB/GF), wherein GDs played an important role in promoting the synthesis process of PB. The PPy film was electro-polymerized to improve the electrochemical stability of the nanocomposites modified electrode. GDs are considered as the excellent electron donors and acceptor to be a promising electrode substrate. GDs functionalized with abundant oxygen-containing groups can effectively enhance the hydrophilia and biocompatibility. PB is a classical mixed-valence transition hexacyano- metallorganics with excellent electrochemical and electrocatalytic properties. There is evidence to prove that carbon nanomaterials could promote the electrocatalytic activity of PB. PPy was used to improve the stability of substrate. This PPy/GDs@PB/GF nanocomposites modified electrode exhibited excellent activities in the electrocatalytic oxidation and electrochemical detection of L-cysteine. In this work, the electrochemical sensor based on GDs@PB showed an excellent activity for the electrocatalytic oxidation of L-cysteine, simple, fast, and sensitive. The linear relationship ranged from 0.2–50 μm and 50–1000 μm, with detection limit of 0.15 μm. In another report, a GDs-chitosan nanocomposite film was prepared by electrodeposition and successfully used in the electrochemical chiral recognition of tryptophan enantiomers, wherein GDs could improve the recognition efficiency and amplify the electrochemical signals [19].

Hydrogen peroxide (H_2O_2) is one of the most common enzymatic product of many kinds of biological processes. Thus, the detection of H_2O_2 is of great importance for biological analysis study. Li and co-workers reported a free-enzymatic electrochemical sensor based on the composites of CDs and octahedral cuprous oxide (Cu_2O) as the substrate and electrocatalyst for the determination of glucose and H_2O_2 [36]. CDs acted as a peroxidase-like catalyst of glucose and H_2O_2, and Cu_2O with high-index facets showed a high catalytic activity. Combined CDs and Cu_2O could further enhance the electrocatalytic performance of glucose oxidation and H_2O_2 reduction with high sensitivity and stability. The linear relationship to glucose ranged from 0.02–4.3 mm, with a detection limit of 8.4 μm, and the linear relationship to H_2O_2 ranged from 5 μm to 5.3 mm, with a detection limit of 2.8 μm. Zhang *et al.* reported an electrochemical sensor with covalently assembled GDs/Au electrode for living cell hydrogen peroxide detection [37]. They first developed a simple route to prepare periphery carboxylic groups enriched graphene quantum dots through a photo-Fenton reaction [38]. Their synthetic method provided a more intact aromatic basal plane structure in GDs than the micrometer-sized GO sheets. It has been reported that GO could show peroxidase like activity that could catalyze the reaction of certain compounds such as hydroquinone, tetramethylbenzidine (TMB), and others in the presence of hydrogen peroxide [39]. While, their research showed that GDs exhibited higher peroxidase activity than the micrometer-sized GO. Upon this finding, they assembled the GDs with cysteamine as cross-linker on Au electrode and demonstrated the properly preserved catalytic property of the GDs. Their electrochemical sensor showed great performance and stability in H_2O_2 detection with a wide linear detection range and a low detection limit. A

H_2O_2 sensor was fabricated based on electrochemical and fluorescent properties of GDs [40]. They first fabricate a PVA/GQD nanofibrous membrane of GDs via electrospinning GDs with polyvinyl alcohol (PVA). This nanofibrous membrane had a three-dimensional structure with a high specific surface area which was excellent for electrolytes adsorption and reactants diffusion. Then with fluorescent and electrochemical sensing methods, this PVA/GQD nanofibrous membrane was applied in the dual-purpose sensor for the determination of H_2O_2 and glucose. Recently, an enzyme-free H_2O_2 sensor was built using a chitosan-GDs/silver nanocube nanocomposite (Chit-GDs/AgNCs) modified electrode [41]. This Chit-GDs/AgNCs/GE displayed good electrocatalytic activity toward H_2O_2 reduction and gave a rapid response and wide linear range during electrochemical sensing. According to certain studies, the structure defects of GDs could be manipulated by doping heteroatoms into the π conjugated system [42, 43]. For example, in the nitrogen-doped GDs (N-GDs) the chemically bonded N atoms drastically changed the electronic characteristics of GDs and offered more active sites in the structure, leading to certain unexpected properties [44].

Certain small molecules are highly toxic for environment and human species. Etoposide (ETO) is a potential anticancer drug for the treatment of various disease such as small-cell lung cancer and other solid tumours. It is necessary to monitor the ETO concentration in the cancer therapy for its important effect of treatment. Nguyen *et al.* demonstrated a CDs based electrochemical sensor for the detection of etoposide (ETO) with enhanced electrochemical performance [45]. The response of the sensor exhibited a good linear relationship with ETO concentration from 0.02–10 µm with a detection limit of 5 nm . Then, the sensor was used for ETO investigation in real samples during the treatment of prostate cancer cell line PC3. The result showed that Trinitrotoluene is one of the most hazardous chemicals in our environment. Cai and coworkers reported a N-GDs based electrochemical sensor for trinitrotoluene [46]. Their N-GDs were synthesized by oxidative ultrasonication of graphene oxide at low temperature and then reduced and nitrogen doped by hydrazine. This N-GDs could act as catalyst allowing electrochemical detection of 2,4,6-trinitrotoluene. The linear scan voltammogram (LSV) results showed a linear response range from 1–400 ppb with a detection limit of 0.2 ppb. Bisphenol A (BPA) is an important endocrine disrupter which can interfer the endogenous gonadal steroid hormones and induce abnormal differentiation of reproductive organs [47]. An electrochemical sensor with molecularly imprinted polypyrrole/GDs (MIPPy/GDs) composite was designed for the detection of bisphenol A in water samples [48]. The good electroconductibility, adsorbability and water solubility of GDs make them excellent sensing materials in electrochemical platforms. The MIPPy/GDs composite layer was prepared onto a glassy carbon electrode surface by the electropolymerization of pyrrole with BPA as a template. This composite layer could specifically recognize BPA with the imprinted sites and lead to the decrease of the diffusion of $K_3[Fe(CN)_6]$ at the MIPPy/GDs electrode surface and the decrease of peak currents in cyclic voltammetry (CV) and differential pulse voltammetry (DPV) measurements. Other nanocomposites with GDs for the detection of small molecules keep being developed recently. A nanocomposite consisted of molybdenum disulphide (MoS_2) and GDs was used in an electrochemical laccase biosensor for the determination of total polyphenolic content [49]. This MoS_2-GDs nanocomposite was proved to be suitable support for laccase immobilisation and have interesting electrochemical properties. The developed laccase biosensor showed efficiently response to caffeic acid in the range of 0.38–100 µm and was successfully applied in the detection of total polyphenolic content from red wine samples. Arvand and coworkers reported a nanocomposite consisting of GDs, magnetic nanoparticles and carboxylated multiwalled carbon nanotubes for the electrode surface modification in an electrochemical sensing platform for L-DOPA determination in sunflower seed, sesame seed, pumpkin seed and fava bean seed [50]. Zhou *et al.* developed a gold/proline-functionalized GDs nanostructure for electrochemical detection of p-acetamidophenol. The GDs and gold nanoparticles could achieve significant synergy to improve the sensing sensitivity [51]. Similar GDs and gold nanoparticles nanocomposite was also reported in the electrochemical sensing of quercetin in biological samples [52].

4.2.6 Protein Sensing

Sensitive and effective detection of specific biomolecules such as protein and DNA is critical in clinical medical diagnosis. Avian leukosis viruses are one of the most common kind of avian retroviruses associated with neoplastic diseases [53]. As shown in Figure 4.5,

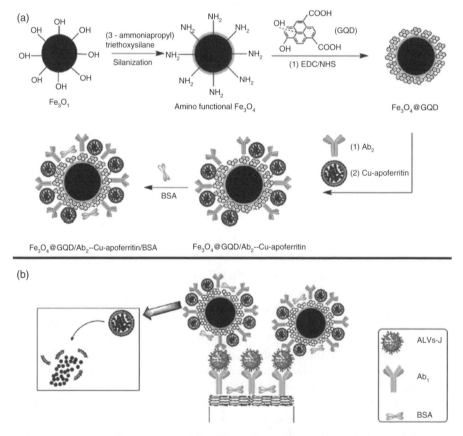

Figure 4.5 Electrochemical immunosensor with graphene quantum dots and apoferritin-encapsulated Cu nanoparticles double-assisted signal amplification for detection of avian leukosis virus subgroup J. *Source*: Wang 2013 [54]. Reproduced with permission of Elsevier. (*See color plate section for the color representation of this figure.*)

Wang and coworkers developed a novel sandwich electrochemical immunoassay based on GDs, Fe_3O_4 nanospheres and apoferritin-encapsulated Cu (Cu-apoferritin) nanoparticles for ultrasensitive detection of avian leukosis virus subgroup J (ALVs-J) [54]. In this immunoassay, GDs were used for the immobilization of both primary ALVs-J antibodies (Ab1) and secondary ALVs-J antibodies (Ab2). They first synthesized the $Fe_3O_4@$ GDs hybrid as a substrate to conjugate Ab2 and Cu-apoferritin nanoparticles which worked as electroactive probes. Apoferritin is a kind of protein with a spherical protein shell which is capable of accommodating around 4500 atoms [55]. Then taking into consideration of the huge surface area and functional groups of GDs, they used GDs in the immunosensor platform for capturing numerous Ab1. Thus, the performance of this immunosensor was greatly amplified by the two strategies, (1) the introduction of a large amount of electroactive probes by the protein cage of apoferritin and (2) the immobilization of high content of antibodies and Cu-apoferritin by GDs. After the sandwich-type assembly, Cu was released and detected by differential pulse voltammetry (DPV). This electrochemical immunosensor displayed excellent analytical performance in the detection of ALVs-J, with a detection limit of 115 TCID50/ml (S/N = 3).

4.2.7 DNA/RNA Sensing

Zhao *et al.* designed a graphene dots based electrochemical aptasensor platform to detect target DNA or target protein [56]. In this system, GDs were readily modified onto the surface of pyrolytic graphite (PG) electrode, and the probe single-stranded DNA (ssDNA-1) could be easily immobilized basing on the strong $\pi-\pi$ stacking interaction between the nucleobases of ssDNA and graphene dots. Thanks to the excellent conductivity of carbon materials, this platform showed very fine electrochemical response. While, the probe ssDNA could inhibit the electron transfer between the electrode and the electrochemical active species $[Fe(CN)_6]^{3-/4-}$. In the presence of target molecule, which in this case was ssDNA-2 or thrombin, the structure of the probe ssDNA would be altered due to the interaction between the target molecules and probe ssDNA. Thus, the immobilization of the probe ssDNA on the GDs modified electrode would be removed and the electrochemical response would be consequently changed, leading to the detection of target molecules. The sequence of the probe ssDNA could be designed for different kinds of target molecules, showing great application potential in electrochemical biosensing fields. Another electrochemical biosensor for miRNA-155 detection mentioned before, GDs and horseradish peroxidase (HRP) enzyme catalytic amplification were used to improve the sensing performance [25]. While, in the presence of target RNA, the double-stranded structure was constructed and the modified HRP would effectively catalyze hydrogen peroxide mediated 3,3′,5,5′-tetramethylbenzidine (TMB) oxidation with an increased electrochemical current signal.

4.3 Electrochemiluminescence Sensors

Electrochemiluminescence (also called electrogenerated chemiluminescence, ECL) is a powerful and convenient technique in electrochemical sensing fields. It has promising application potential and has been developed rapidly. The remarkable advantages of ECL technology such as excellent sensitivity and reproducibility, fast response speed, wide response range, simple operation procedures and low-cost devices make it highly attractive for a variety of applications in sensing areas. In particular, ECL technology allows relatively accurate control on the potential, position and time of the light-emitting reactions, and has no requirement for external light source. ECL process is a kind of chemiluminescence triggered by electrochemical approach, which involves light emission from electrogenerated species undergoing highly energetic electron-transfer and forming excited states [57]. A series of quantum dots have been reported to show great ECL properties [58, 59]. The anodic and cathodic ECL behaviors of CDs and GDs were also reported [60, 61]. CDs and GDs are successfully introduced as new environmental-friendly ECL probes to develop high-performance ECL immunosensors. Generally, there are two dominant pathways to produce ECL process, the annihilation pathway and the coreactant pathway [62]. The ECL reactions of CDs and GDs mostly follow the coreactant pathway, using hydrogen peroxide or peroxydisulfate as coreactants [61, 63].

Based on the great ECL properities of GDs and CDs, different kinds of ECL sensors were built for a variety of analytes detection applications. Li and co-workers first observed the ECL emissions from their as-prepared GDs and designed a GDs based ECL sensor for Cd^{2+} ion [63]. Cd^{2+} ion is known as a kind of highly toxic metal ion and a carcinogen for mammals, which can accumulate in many kinds of organs of mammals like kidney, lung, liver and spleen. Using the specific quenching effect of Cd^{2+} ion on the ECL emission of GDs, they built an ECL sensing platform to accurately determinate Cd^{2+} ion. First, they prepared a kind of greenish-yellow luminescent GDs (g-GDs and b-GDs) via cleaving graphene oxide under acid conditions with the assistance of microwave irradiation. Their samples of analyte contained various metal ions including Ni^{2+}, Pb^{2+}, Cu^{2+}, Co^{2+}, Fe^{2+} and Cd^{2+}, with almost 92% quenching of GDs ECL from Cd^{2+} ion while no obvious quenching effect from other ions. Furthermore, they found that after adding certain amount of ethylenediaminetetraacetic acid (EDTA) which is a strong metal ion chelator, the quenched GDs ECL emission could be almost completely recovered. They concluded that the functional groups such as hydroxyl and carboxyl groups on GDs could act as coordination groups for certain metal ions and induce the aggregation of GDs and the quenching of ECL. In the presence of Cd^{2+} ion, the CVs of GDs showed a new strong peak at about −1.5 V, which might originate from the Cd^{2+}-GDs complex. Furthermore, Dong *et al.* reported a GDs/L-cysteine coreactant ECL system for sensing lead ions (Pb^{2+}) [64]. This GDs/L-cysteine system was found to exhibit strong cathodic ECL signal which was mainly dependent on some key factors including the reduction of GDs, the oxidation of L-cysteine and the presence of dissolved oxygen. They proposed that the possible ECL mechanism was related to the unstable intermediates produced during the oxidation of L-cysteine, and dissolved oxygen was an important factor to the formation of these intermediates. Experimental results showed that this GDs/L-cysteine system emitted very weak ECL signal in the absence of dissolved oxygen whereas gave a strong emission with the presence of dissolved oxygen. Then they applied this system for Pb^{2+} ion sensing basing on the quenching effect of Pb^{2+} ion on the ECL signal. However, they found that Pb^{2+} ion had nearly no quenching effect on the exited-state GDs, implying the quenching effect of Pb^{2+} ion was most likely through inhibiting the formation of the unstable intermediates during L-cysteine oxidation.

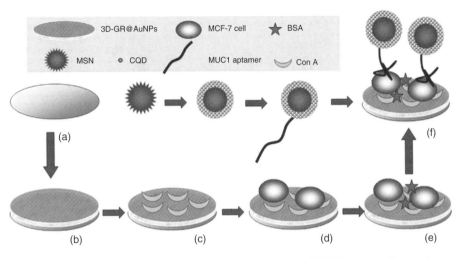

Figure 4.6 Aptamer-based electrochemiluminescent detection of MCF-7 cancer cells based on carbon quantum dots coated mesoporous silica nanoparticles. *Source*: Su 2014 [65]. Reproduced with permission of Elsevier. (*See color plate section for the color representation of this figure.*)

As shown in Figure 4.6, a novel ECL sensor based on CDs coated mesoporous silica nanoparticles (MSNs) was developed for the detection of MCF-7 cancer cells [65]. Three-dimensional macroporousAuNPs@graphene complex (3D-GR/Au NPs) was introduced as the substrate of the whole sensor, to immobilize concanavalin A (Con A), which then adsorbed MCF-7 cancer cells. CDs coated MSNs as ECL signal probes exhibited low cytotoxicity and good biocompatibility, which could specifically bind mucin1 on cancer cells after conjugation with mucin 1 aptamer. This ECL sensor showed a sensitive and stable performance for the determination of MCF-7 cancer cells ranging from 500 to 2×10^7 cells/ml. The detection limit is 230 cells/ml, much lower than other immunoassays toward cancer cells. Another report by Deng showed a sensitive ECL immunosensor for the detection of human carcinoembryonic antigen (CEA) with CDs modified Pt/Fe nanoparticles (Pt/Fe@CDs) as signal probes [66]. Pt nanoparticles dotted graphene–carbon nanotubes complex (Pt/GR–CNTs) was substrate used for immobilized the primary anti-CEA antibodies (Ab1), which later captured CEA and then captured target anti-CEA antibodies (Ab2) modified Pt/Fe@CDs probes through sandwich immunoassay. The Pt/Fe@CDs probes exhibited a good linear response range from 0.003–600 ng/ml of CEA with a detection limit of 0.8 pg/ml, much lower than other CEA electrochemical immunosensors. Yang *et al.* designed an ECL immunosensor with GDs coated porous PtPd nanochains as labels for tumor marker detection in Figure 4.7 [67]. In this sensing platform, gold-silver nanocomposite-functionalized graphene was used for electrode modification, which could increase the capacity of primary antibodies and improve the electronic transmission rate. Porous PtPd nanochains with good conductivity and large surface area were used to conjugate a large number of GDs ECL labels and second antibodies. With the great ECL property of GDs and the enhancement from gold-silver nanocomposite-functionalized graphene and porous PtPd nanochains, this ECL immunosensor exhibited good performance in biomolecules sensing with carbohydrate antigen 199 as a model analyte. A label-free ECL immunosensor based on aminated GDs and carboxyl GDs for prostate specific antigen

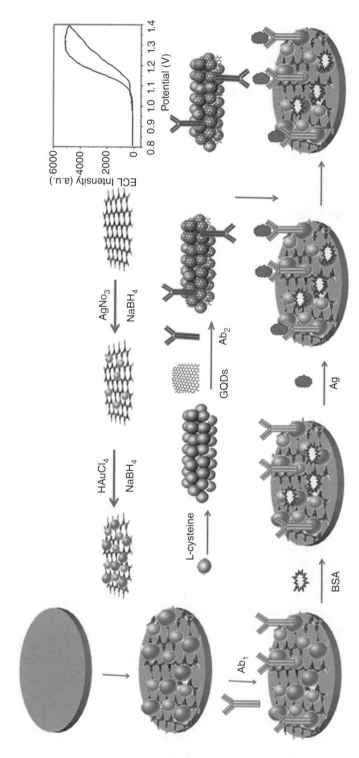

Figure 4.7 Gold–silver nanocomposite-functionalized graphene based electrochemiluminescence immunosensor using graphene quantum dots coated porous PtPd nanochains as labels. *Source:* Yang 2014 [67]. Reproduced with permission of Elsevier. (*See color plate section for the color representation of this figure.*)

(PSA) detection was reported by Wu *et al.* [68]. In this system, the Au/Ag-rGO composites were used as electrode material for largely loading aminated GDs, carboxyl GDs and antibodies of PSA. The GDs modified electrode showed high ECL intensity with $K_2S_2O_8$ as coreactant. Under the optimal conditions, ECL signal of this immunosensor decreased with the increase of PSA concentration in the range of 1 pg/ml to 10 ng/ml.

The ECL sensing based on CDs or GDs could be used for not only the detection of metal ions, proteins, antigens and cancer cells, but also for the detection of organic molecules and DNA. Liu reported an ultrasensitive ECL sensor for the detection of pentachlorophenol (PCP) based on graphene quantum dots–CdS nanocrystals (GDs–CdS NCs) as signal probes [69]. The presence of GDs improved the ECL intensity up to 5-folds than pure CdS NCs, because of the high fluorescent activity and outstanding electric conductivity of GDs. By this ECL sensor, PCP, as an effective inhibition of ECL response, could be detected with a linear range of 0.01–500 ng/ml and a low detection limit of 3 pg/ml. The possible detection mechanism was shown in Figure 4.8. On the electrode, CdS NCs could be reduced to generate CdS$^-$, on the other hand, CdS NCs also could get a hole from hydroxyl radical (OH$^•$) to form oxidized CdS$^+$. The recombination of CdS$^-$ and CdS$^+$ further generated excited states of CdS* with a light emission. At the presence of PCP, PCP could be oxidized into TCQ, leading to energy transfer process, which reduced the generation of CdS* and resulted in a decrease of ECL response.

Lou and co-workers reported a novel ECL sensor for the detection of DNA based on GDs with excellent ECL activity as the signal probes and site-specific cleavage of BamHI endonuclease [70]. In their research, they found that the difference between photoluminescence and ECL spectral peaks of GDs suggesting negligible defect existed on the surface of GDs for ECL signal generation. As for the assembly of GDs probes, they used

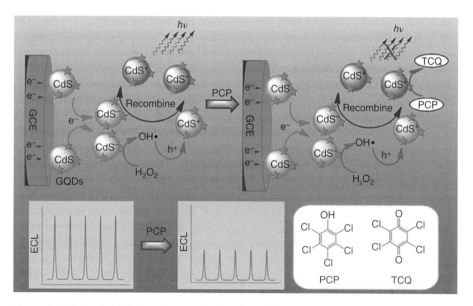

Figure 4.8 Illustrative ECL detection mechanism for PCP based on GQDs–CdS NCs/GCE. *Source:* Liu 2014 [69]. Reproduced with permission of the Royal Society of Chemistry. (*See color plate section for the color representation of this figure.*)

the bidentate chelation of the Dithiocarbamate DNA (DTC-DNA) which provided a strong affinity between the ligands and the gold surface by bidentate anchoring (S–Au–S bonds) as the capture probe. Then GDs were immobilized on the terminal of DTC-DNA by amide interaction as the signal probes. Hepatitis C virus-1b genotype complementary DNA (HCV-1b cDNA) as the target DNA was captured on the DTC-DNA by hybridization reaction. BamHI endonuclease could recognize and cleave the duplex symmetrical sequence, leading to the formation of double-stranded DNA (dsDNA) fragments and the release of GDs from the electrode surface. Thus, basing on the changes of ECL intensity before and after the cleavage of DNA hybrid, this signal-off ECL biosensor could be used for DNA detection with a linear range from 5 fm to 100 pm and a detection limit of 0.45 fM at signal-to-noise ratio of three. Another ECL biosensor used boron doped GDs to detect oncogene microRNA-20a [71]. The boron doped GDs (B-GDs) which had an atomic percentage of boron of 0.67–2.26% were synthesized by electrolytic exfoliation of B doped graphene rods. This B-GDs were proved to possess improved luminous performance and stability and decreased resistance. ECL sensing results showed excellent analytical property toward the detection of miRNA-20a in a linear range of 0.1–104 pm with high selectivity among similar bases strands.

ECL resonance energy transfer (ECL-RET) is an emerging technology which resembles fluorescence (or Förster) resonance energy transfer (FRET) in many aspects. Resonance energy transfer is a phenomenon in which energy is transferred from a luminescent donor to a proximal acceptor via nonradiative dipole-dipole interactions [72]. Compared with conventional FRET, the most important advantage of the ECL-RET method is that no excitation light source is required, thus avoiding the problems of light scattering, high background noise, auto-fluorescence and direct acceptor excitation [72, 73]. Another advantage is that ECL-RET can happen at a relative long distance in comparison with FRET [74]. Furthrmore, in comparison with bioluminescence resonance energy transfer (BRET) [75–77] and chemiluminescence resonance energy transfer (CRET) [78–80], the ECL-RET technique is much more versatile since it initiates the energy-transfer process by electrochemical potential rather than chemical or biological stimulation [72]. Though ECL-RET are attracting increasing attention in sensing applications, the field of ECL-RET exploration is still in its infancy [81, 82]. Because perfect energy overlapped donor/acceptor pairs are crucial to optimal ECL-RET efficiency, while in early works concerning ECL-RET, the donors or acceptors are mostly nonadjustable, which makes it difficult to find a suitable donor/acceptor pair [58]. Under these circumstances, graphene dots with spectra tunable property have become especially appealing potential donor and acceptor. Lu and coworkers designed an ECL-RET biosensing platform for DNA damage detection with GDs as ECL emitters and gold nanoparticles as accepters [83]. Their approach for the preparation of GDs combined thermal reduction and cleaving in an oxidation procedure with UV-light irradiation reduction, which greatly shortened the reaction time and reduced the use of chemical reagents. And they firstly observed and reported the ECL-RET effect between GDs and Au NPs. Upon this finding, they linked Au NPs with a probe of single-stranded DNA (cp53 ssDNA) to get the probe Au NPs-ssDNA, then quenched the ECL signal of GDs via non-covalent binding of this probe to GDs. After the Au NPs-ssDNA probe hybridized with target p53 DNA, the non-covalent interaction between the Au NPs and GDs was disturbed, leading to the recovery of the ECL signal. This novel ECL-RET DNA biosensor showed a detection limit of 13 nm towards the model target p53 ssDNA, and could be used for DNA damage detection basing on the different bonding ability between damaged target DNA and normal DNA. Another example of GDs based ECL-RET biosensor was reported by Wen *et al.* [84]. They developed a double-quenching of GDs ECL strategy for protein kinase A (PKA) detection. In this system, the ECL intensity of GDs was significantly quenched by the G-quadruplex–hemin DNAzyme catalyzed coreactant consuming reaction and the ECL-RET between GDs and Au NPs. G-quadruplex–hemin DNAzyme, composed of a single-stranded guanine-rich nucleic acid and hemin, is an interesting DNAzyme with HRP-like activity to catalyze the reduction of hydrogen peroxide which is one of the most common coreactant in ECL systems [85]. They first assemblied the GDs functionalized substrate peptide onto the electrode, forming a strong and stable ECL emission source in the presence of coreactant hydrogen peroxide. After phosphorylating the substrate peptide by PKA using ATP as the co-substrate, the phosphorylated linker DNA and G-quadruplex–hemin DNAzyme functionalized Au NPs were attached onto the surface via Zr^{4+}-mediated reaction between the phosphorylated linker DNA and the phosphorylated peptide. Then, the G-quadruplex–hemin DNAzyme catalyzed the reduction of coreactant hydrogen peroxide and the proximal Au NPs quenched the ECL of GDs due to the ECL-RET effect. Thus, the ECL signal decreased greatly with the double-quenching effect. This platform showed excellent analytical performance in the highly sensitive detection of PKA activity, indicating promising potential in protein kinase-related biochemical research. Recently, Liang and coworkers designed a graphene materials based ECL-RET biosensor with GDs as emitters and graphene oxide as acceptors for monitoring and detecting casein kinase II (CK2) activity [86]. The biosensor was built by covalent modifying GDs onto chitosan film coated electrode and further functionalized with peptides by amide reaction. In the presence of adenosine 5'-triphosphate and model protein CK2, the peptides would be phosphorylated and capture the anti-phosphoserine antibody conjugated graphene nanocomposites. Due to the ECL-RET effect between GDs and graphene nanocomposites, ECL signal was quenched positively correlated with CK2 activity. Thus, this platform could be used for sensitive kinase activity detection and quantitative kinase inhibitor screening.

4.4 Photoelectrochemical Sensing

The process of photoelectrochemistry (abbreviated PEC) is just the reverse of ECL, it refers to the photon-electricity conversion resulted from the charge transfer during the oxidation–reduction reaction caused by photoactive species upon illumination [87]. Like ECL and other well developed analytical techniques, PEC is also an evolutionary development of the electrochemical method. It has many same advantages as the ECL technique such as high sensitivity, simple instrumentation and low cost. While, it also has unique characteristics. For example, in a typical PEC detection, light is essential for the excitation of the photoactive species and the detection readout is transduced electrical signal. It has been reported that the size-dependent bandgap nature of carbon-based materials including CDs and GDs was particularly interesting for photoelectrochemical applications [88]. As superior electron receptors with abundant photo-physical properties, CDs and GDs are gradually introduced into photoelectric field instead of traditional semiconductor quantum dots (QDs), also for their environmental and biological friendship, low-cost and low toxicity. CDs and GDs exhibit good conductivity, effective photocarrier generation and separation combining with metal

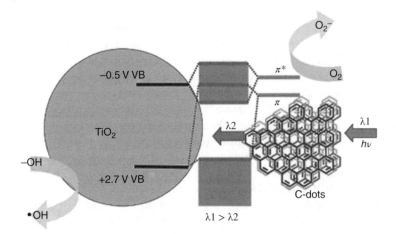

Figure 4.9 Proposed formation of dyadic structure with charge transfer-type orbital hybridizations at the surface of TiO$_2$ and C-dots. *Source*: Ming 2012 [89]. Reproduced with permission of Royal Society of Chemistry.

oxides, leading to charge transfer from photoactive layer to electron transport layer and reduce of charge recombination rate. In this line, CDs and GDs have been considered as potential photoelectic materials in photodetector photovoltaics, photocatalyst and photoelectrochemical sensing fields. Based on the photoelectrochemical properties of CDs and GDs, a series of PEC sensors were built.

Ming *et al.* presented a facile synthesis method of CDs with a high crystalline nature, excellent dispersibility, and remarkable photoluminescence properties [89]. Then, CDs were introduced to prepare TiO$_2$/CDs nanohybrids. The as-obtained TiO$_2$/CDs nanohybrids exhibited higher photoelectrical activities for the interaction between CDs and TiO$_2$. Figure 4.9 showed the possible photoelectric mechanism of the TiO$_2$/CDs. The π states of CDs and conduction band states of TiO$_2$ generated the electronic coupling, broadening absorption band of TiO$_2$ to visible light, reducing the original band gap. Since that, it is easier to stimulate electron transition to conduction band. The CDs with up-conversion property could transfer long wavelength irradiation light into shorter wavelength emission light, promoting the efficiency of photon utilization. The O$_2$ absorbed on the surface of CDs, the generation of photoreactive species (e$^-$, O$_2^-$ and •OH), can accelerate the oxidation rate of methyl blue (MB). CDs also play a role of electron trap to regulate electron–hole recombination probability. GDs onto metal oxide electron transport layer is also full of prospects in photoelectric field.

Sudhagar reported a type of hybrid TiO$_2$-GDs heterostructure nanowires, which exhibited an enhanced photocurrent output and enhanced light harvesting efficiency, as promising architectures for photoelectrochemical applications [90]. The charge transfer mechanism in Figure 4.10 is similar as above. Under the light irradiation above 400 nm wavelength, the electron was excited to conduction band, generating electron–hole pairs. Photogenerated holes at the valence band of GDs were injected to the solution and integrated with electron donors, and photogenerated electrons were injected into TiO$_2$ and then the contact due to more negative conduction band of GDs compared to that of TiO$_2$. GDs sensitization may also contribute to light scattering,

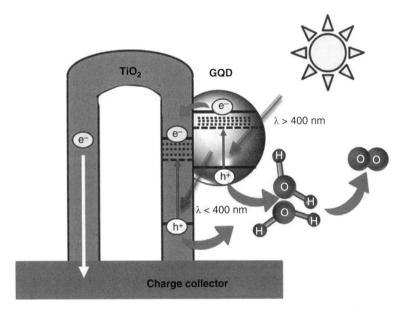

Figure 4.10 Photoexcitation charge carriers generated at GQDs/TiO$_2$ interfaces. *Source*: Sudhagar 2016 [90]. Reproduced with permission of Elsevier.

leading to the enhancement of photovoltaic performance. Shen and coworkers used 0D graphene quantum dots instead of 2D graphene sheets in their research [91]. Their results showed GDs exhibited new phenomenon because of quantum confinement and edge effects. They also found that the surface-passivation agents on GDs had great influence on the photoelectrochemical property of GDs. The photon-to-electron conversion capability of GDs-PEG and GDs were very different.

Yan et al. reported another type of GDs–TiO$_2$ nanocomposites, gaining enhanced photoelectrochemical signal under visible-light irradiation [92]. The mechanism of enhancement is quite different from that of previous report, since GDs here are majorly considered as excellent electrical conductivities, improving separation efficiency of electron and hole pairs. The photocurrent could be further sensitized by dopamine (DA), developing a novel photoelectrochemical sensor for sensitive determination of DA. The enhanced photocurrent showed a linear relationship with DA concentration from 0.02–105 µm with a detection limit of 6.7 nm. Liu et al. designed a label-free PEC aptasensor for chloramphenicol (CAP) determination using nitrogen-doped graphene quantum dots [93]. They synthesized the NGDs via a facile one-step hydrothermal method and these NGDs showed highly efficient photon-to-electricity conversion upon visible light irradiation. The UV–visible absorption spectra gave the evidence that nitrogen doping could obviously enhance the absorption capacity of GDs in visible light region, which promoted the PEC activity of GDs. On the other side, the π-conjugated structure of NGDs improved the immobilization of aptamers by π–π stacking interaction. With the presence of the CAP, it would be captured by aptamer and further reacted with photogenerated holes of NGDs to get an enhanced photocurrent signal. This aptamer/NGDs based photoelectrochemical sensor exhibited high sensitivity and selectivity with a linear PEC response in the range of 10–250 nm. Another more recent example is an enhanced photoelectrochemical cytosensor for fibroblast-like synovio-cyte cells (FLS cells) with visible light-activated GDs and g-C$_3$N$_4$ sensitized TiO$_2$ nanorods [94]. Preparing heterojunctions, which can energetically match conduction bands and valence bands of different materials, is the most frequently used strategy to improve the photocatalytic performance of photoactive materials [95, 96]. In this cytosensor, they used GDs and carboxylated g-C$_3$N$_4$ as the heterojunction sensitizers for TiO$_2$ nanorods, leading to a wide band gap of 3.2 eV. Then the nanohybridization of TiO$_2$ nanorods/carboxylated g-C$_3$N$_4$/ NGDs worked as the photon-to-electron genera-tor in this PEC cytosensor. After the immunoreaction between CD95 antibody and FLS cells, a decrease in photocurrent intensity occured due to the steric hindrance and thus the target cells could be determined. Similar synergistic effect of GDs and TiO$_2$ nano-particles was also applied in a PEC sensor for dopamine determination [92]. In this system, the GDs-TiO$_2$ nanocomposites were fabricated by a simple physical adsorption method. This nanocomposites showed enhanced PEC signal upon visible-light irradia-tion, with nearly 30-fold and 12-fold enhanced photocurrent than that of GDs and TiO$_2$, respectively. Moreover, the photocurrent of this nanocomposites was selectively sensi-tized by dopamine, providing a strategy for the determination of dopamine. Other pho-toelectrochemical active nanohybrids were also reported in the PEC sensors. Tian et aln developed a GDs-silicon nanowires nanocomposites as signal transduction and bio-compatible nano-scaffold for antibody immobilization in a PEC platform to detect microcystin-LR (MC-LR) in water samples [97]. The results showed that GDs distinctly improved the photoelectrochemical performance of silicon nanowires. The measure-ment mechanism was based on the specific recognition of MC-LR caused photocurrent decrease. This immunosensor possessed good sensitivity and was successfully applied in real water samples.

4.5 Conclusions

This chapter has discussed recent progress in the development of electrochemical sens-ing based on C-dots and G-dots and their typical applications. Though considerable achievements have made in the past decade, this filed is still at the early stage. Growing interest will be focused on the preparation of multifunctional C-dots and G-dots to improve the sensitivity and widen the detection range, either via advanced synthetic methods or composite materials with other functional nanomaterials.

References

1 Zheng, X.T., Ananthanarayanan, A., Luo, K.Q. Chen, P. Glowing graphene quantum dots and carbon dots: Properties, syntheses, and biological applications. Small, 2015, 11, 1620–1636.

2 Sun, H., Wu, L., Wei, W. Qu, X. Recent advances in graphene quantum dots for sensing. Materials Today, 2013, 16, 433–442.

3 Chen, M., Zhao, C., Chen, W., *et al*. Sensitive electrochemical immunoassay of metallothionein-3 based on K$_3$[Fe(CN)$_6$] as a redox-active signal and C-dots/Nafion film for antibody immobilization. Analyst, 2013, 138, 7341–7346.

4 Gao, Q., Han, J., Ma, Z. Polyamidoamine dendrimers-capped carbon dots/Au nanocrystal nanocomposites and its application for electrochemical immunosensor. Biosensors and Bioelectronics, 2013, 49, 323–328.

5 Ji, H., Zhou, F., Gu, J., *et al.* Nitrogen-doped carbon dots as a new substrate for sensitive glucose determination. Sensors 2016, 16, 630.

6 Muthurasu, A. Ganesh, V. Horseradish peroxidase enzyme immobilized graphene quantum dots as electrochemical biosensors. Applied Biochemistry and Biotechnology, 2014, 174, 945–959.

7 Dong, J., Hou, J., Jiang, J. Ai, S. Innovative approach for the electrochemical detection of non-electroactive organophosphorus pesticides using oxime as electroactive probe. Analytica Chimica Acta, 2015, 885, 92–97.

8 Sadhukhan, M., Bhowmik, T., Kundu, M. K. Barman, S. Facile synthesis of carbon quantum dots and thin graphene sheets for non-enzymatic sensing of hydrogen peroxide. RSC Advances, 2014, 4, 4998–5005.

9 Umrao, S., Jang, M. H., Oh, J.H., *et al.* Microwave bottom-up route for size-tunable and switchable photoluminescent graphene quantum dots using acetylacetone: New platform for enzyme-free detection of hydrogen peroxide. Carbon, 2015, 81, 514–524.

10 Jahanbakhshi, M., Habibi, B. A novel and facile synthesis of carbon quantum dots via salep hydrothermal treatment as the silver nanoparticles support: Application to electroanalytical determination of H_2O_2 in fetal bovine serum. Biosensors and Bioelectronics, 2016, 81, 143–150.

11 Yang, C., Hu, L.-W., Zhu, H.-Y., *et al.* rGO quantum dots/ZnO hybrid nanofibers fabricated using electrospun polymer templates and applications in drug screening involving an intracellular H_2O_2 sensor. Journal of Materials Chemistry B, 2015, 3, 2651–2659.

12 Ju, J., Chen, W. In situ growth of surfactant-free gold nanoparticles on nitrogen-doped graphene quantum dots for electrochemical detection of hydrogen peroxide in biological environments. Analytical Chemistry, 2015, 87, 1903–1910.

13 Liu, Y., Li, W., Li, J., Shen, H., Li, Y., Guo, Y. Graphene aerogel-supported and graphene quantum dots-modified γ-MnOOH nanotubes as a highly efficient electrocatalyst for oxygen reduction reaction. RSC Advances, 2016, 6, 43116–43126.

14 Doroodmand, M.M., Deylaminezhad, M. Electrochemical study on the intercalation properties of hydroxyl anion for the reversible conversion of graphene quantum dots into carbon dots. Journal of Electroanalytical Chemistry, 2015, 756, 161–170.

15 Jiang, G.H., Jiang, T.T., Zhou, H.J., *et al.* Preparation of N-doped carbon quantum dots for highly sensitive detection of dopamine by an electrochemical method. RSC Advances, 2015, 5, 9064–9068.

16 Li, Y.H., Jiang, Y.Y., Mo, T., *et al.* Highly selective dopamine sensor based on graphene quantum dots self-assembled monolayers modified electrode. Journal of Electroanalytical Chemistry, 2016, 767, 84–90.

17 Akhtar, N., El-Safty, S.A., Abdelsalam, M.E., Kawarada, H. One-pot fabrication of dendritic NiO@carbon–nitrogen dot electrodes for screening blood glucose level in diabetes. Advanced Healthcare Materials, 2015, 4, 2110–2119.

18 Mazloum-Ardakani, M., Aghaei, R., Abdollahi-Alibeik, M., Moaddeli.A. Fabrication of modified glassy carbon electrode using graphene quantum dot, gold nanoparticles and 4-(((4-mercaptophenyl)imino)methyl) benzene-1,2-diol by self-assembly method and

investigation of their electrocatalytic activities. Journal of Electroanalytical Chemistry, 2015, 738, 113–122.

19 Ou, J., Tao, Y., Xue, J., *et al.* Electrochemical enantiorecognition of tryptophan enantiomers based on graphene quantum dots–chitosan composite film. Electrochemistry Communications, 2015, 57, 5–9.

20 Wang, G., Shi, G., Chen, X., *et al.* A glassy carbon electrode modified with graphene quantum dots and silver nanoparticles for simultaneous determination of guanine and adenine. Microchimica Acta, 2015, 182, 315–322.

21 Ting, S.L., Ee, S.J., Ananthanarayanan, A., *et al.* Graphene quantum dots functionalized gold nanoparticles for sensitive electrochemical detection of heavy metal ions. Electrochimica Acta, 2015, 172, 7–11.

22 Ou, J., Tao, Y.X., Ma, J.F., Kong, Y. Well-dispersed chitosan-graphene quantum dots nanocomposites for electrochemical sensing platform. Journal of the Electrochemical Society, 2015, 162, H884–H889.

23 Shao, X. L., Gu, H., Wang, Z., *et al.* Highly selective electrochemical strategy for monitoring of cerebral Cu^{2+} based on a carbon dot-TPEA hybridized surface. Analytical Chemistry, 2013, 85, 418–425.

24 Yang, L., Huang, N., Lu, Q., *et al.* A quadruplet electrochemical platform for ultrasensitive and simultaneous detection of ascorbic acid, dopamine, uric acid and acetaminophen based on a ferrocene derivative functional Au NPs/carbon dots nanocomposite and graphene. Analytica Chimica Ata, 2016, 903, 69–80.

25 Hu, T., Zhang, L., Wen, W., *et al.* Enzyme catalytic amplification of miRNA-155 detection with graphene quantum dot-based electrochemical biosensor. Biosensors and Bioelectronics, 2016, 77, 451–456.

26 Liu, W.Y., Yang, H.M., Ma, C., *et al.* Graphene-palladium nanowires based electrochemical sensor using $ZnFe_2O_4$-graphene quantum dots as an effective peroxidase mimic. Analytica Chimica Acta, 2014, 852, 181–188.

27 Punrat, E., Maksuk, C., Chuanuwatanakul, S., *et al.* Polyaniline/graphene quantum dot-modified screen-printed carbon electrode for the rapid determination of Cr(VI) using stopped-flow analysis coupled with voltammetric technique. Talanta, 2016, 150, 198–205.

28 Ananthanarayanan, A., Wang, X.W., Routh, P., *et al.* Facile synthesis of graphene quantum dots from 3D graphene and their application for Fe^{3+} sensing. Advanced Functional Materials, 2014, 24, 3021–3026.

29 Razmi, H., Mohammad-Rezaei, R. Graphene quantum dots as a new substrate for immobilization and direct electrochemistry of glucose oxidase: Application to sensitive glucose determination. Biosensors and Bioelectronics, 2013, 41, 498–504.

30 Zhang, A., Neumeyer, J.L., Baldessarini, R.J. Recent progress in development of dopamine receptor subtype-selective sgents: Potential therapeutics for neurological and psychiatric disorders. Chemical Reviews, 2007, 107, 274–302.

31 Pang, P.F., Yan, F.Q., Li, H.Z., *et al.* Graphene quantum dots and nafion composite as an ultrasensitive electrochemical sensor for the detection of dopamine. Analytical Methods, 2016, 8, 4912–4918.

32 Huang, Q., Hu, S., Zhang, H., *et al.* Carbon dots and chitosan composite film based biosensor for the sensitive and selective determination of dopamine. Analyst, 2013, 138, 5417–5423.

33 Jiang, Y., Wang, B., Meng, F., *et al.* Microwave-assisted preparation of N-doped carbon dots as a biosensor for electrochemical dopamine detection. Journal of Colloid and Interface Science, 2015, 452, 199–202.

34 Hou, C., Fan, S., Lang, Q., Liu, A. Biofuel cell based self-powered sensing platform for L-Cysteine detection. Analytical Chemistry, 2015, 87, 3382–3387.

35 Wang, L., Tricard, S., Yue, P., *et al.* Polypyrrole and graphene quantum dots @ Prussian Blue hybrid film on graphite felt electrodes: Application for amperometric determination of L-cysteine. Biosensors and Bioelectronics, 2016, 77, 1112–1118.

36 Li, Y., Zhong, Y., Zhang, Y., *et al.* Carbon quantum dots/octahedral Cu_2O nanocomposites for non-enzymatic glucose and hydrogen peroxide amperometric sensor. Sensors and Actuators B: Chemical, 2015, 206, 735–743.

37 Zhang, Y., Wu, C., Zhou, X., *et al.* Graphene quantum dots/gold electrode and its application in living cell H_2O_2 detection. Nanoscale, 2013, 5, 1816–1819.

38 Zhou, X., Zhang, Y., Wang, C., *et al.* Photo-fenton reaction of graphene oxide: A new strategy to prepare graphene quantum dots for DNA cleavage. ACS Nano, 2012, 6, 6592–6599.

39 Liu, M., Zhao, H., Chen, S., *et al.* Interface engineering catalytic graphene for smart colorimetric biosensing. ACS Nano., 2012, 6, 3142–3151.

40 Zhang, P.P., Zhao, X.N., Ji, Y.C., *et al.* Electrospinning graphene quantum dots into a nanofibrous membrane for dual-purpose fluorescent and electrochemical biosensors. Journal of Materials Chemistry B, 2015, 3, 2487–2496.

41 Jiang, Y.Y., Li, Y.H., Li, Y.C., Li S.X. A sensitive enzyme-free hydrogen peroxide sensor based on a chitosan-graphene quantum dot/silver nanocube nanocomposite modified electrode. Analytical Methods, 2016, 8, 2448–2455.

42 Li, Y., Zhao, Y., Cheng, H., *et al.* Nitrogen-doped graphene quantum dots with oxygen-rich functional groups. Journal of the American Chemical Society, 2012, 134, 15–18.

43 Dey, S., Govindaraj, A., Biswas, K., Rao, C.N. Luminescence properties of boron and nitrogen doped graphene quantum dots prepared from arc-discharge-generated doped graphene samples. Chemical Physics Letters, 2014, 595–596, 203–208.

44 Jin, S.H., Kim, D.H., Jun, G.H., *et al.* Tuning the photoluminescence of graphene quantum dots through the charge transfer effect of functional groups. ACS Nano, 2013, 7, 1239–1245.

45 Nguyen, H.V., Richtera, L., Moulick, A., *et al.* Electrochemical sensing of etoposide using carbon quantum dot modified glassy carbon electrode. Analyst, 2016, 141, 2665–2675.

46 Cai, Z.W., Li, F.M., Wu, P., *et al.* Synthesis of nitrogen-doped graphene quantum dots at low temperature for electrochemical sensing trinitrotoluene. Analytical Chemistry, 2015, 87, 11803–11811.

47 Calafat, A.M., Kuklenyik, Z., Reidy, J.A., *et al.* Urinary concentrations of bisphenol A and 4-nonylphenol in a human reference population, Environmental Health Perspectives, 2005, 113, 391–395.

48 Tan, F., Cong, L., Li, X., *et al.* An electrochemical sensor based on molecularly imprinted polypyrrole/graphene quantum dots composite for detection of bisphenol A in water samples. Sensors and Actuators B: Chemical, 2016, 233, 599–606.

49 Vasilescu, I., Eremia, S.A., Kusko, M., *et al.* An electrochemical sensor based on molecularly imprinted polypyrrole/graphene quantum dots composite for detection of bisphenol A in water samples. Biosensors and Bioelectronics, 2016, 75, 232–237.

50 Arvand, M., Abbasnejad, S., Ghodsi, N. Graphene quantum dots decorated with Fe_3O_4 nanoparticles/functionalized multiwalled carbon nanotubes as a new sensing platform for electrochemical determination of L-DOPA in agricultural products. Analytical Methods, 2016, 8, 5861–5868.

51 Zhou, X.Y., Li, R.Y., Li, Z.J., *et al.* Ultrafast synthesis of gold/proline-functionalized graphene quantum dots and its use for ultrasensitive electrochemical detection of p-acetamidophenol. RSC Advances, 2016, 6, 42751–42755.

52 Li, J. J., Qu, J. J., Yang, R., *et al.* A Sensitive and selective electrochemical sensor based on graphene quantum dot/gold nanoparticle nanocomposite modified electrode for the determination of quercetin in biological samples. Electroanalysis, 2016, 28, 1322–1330.

53 Zhou, G., Cai, W., Liu, X., *et al.* A duplex real-time reverse transcription polymerase chain reaction for the detection and quantitation of avian leukosis virus subgroups A and B. Journal of Virological Methods, 2011, 173, 275–279.

54 Wang, X., Chen, L., Su, X., Ai, S. Electrochemical immunosensor with graphene quantum dots and apoferritin-encapsulated Cu nanoparticles double-assisted signal amplification for detection of avian leukosis virus subgroup J. Biosensors and Bioelectronics, 2013, 47, 171–177.

55 Liu, G., Lin, Y. Nanomaterial labels in electrochemical immunosensors and immunoassays. Talanta, 2007, 74, 308–317.

56 Zhao, J., Chen, G., Zhu, L., Li, G. Graphene quantum dots-based platform for the fabrication of electrochemical biosensors. Electrochemistry Communications, 2011, 13, 31–33.

57 Hu, L., Xu, G. Applications and trends in electrochemiluminescence. Chemical Society Reviews, 2010, 39, 3275–3304.

58 Li, L., Chen, Y., Lu, Q., *et al.* Electrochemiluminescence energy transfer-promoted ultrasensitive immunoassay using near-infrared-emitting CdSeTe/CdS/ZnS quantum dots and gold nanorods. Scientific Reports, 2013, 3, 1529.

59 Ji, J., He, L., Shen, Y. *et al.* High-efficient energy funneling based on electrochemiluminescence resonance energy transfer in graded-gap quantum dots bilayers for immunoassay. Analytical Chemistry, 2014, 86, 3284–3290.

60 Dong, Y., Dai, R., Dong, T., *et al.* Photoluminescence, chemiluminescence and anodic electrochemiluminescence of hydrazide-modified graphene quantum dots. Nanoscale, 2014, 6, 11240–11245.

61 Lu, J., Yan, M., Ge, L., *et al.* Electrochemiluminescence of blue-luminescent graphene quantum dots and its application in ultrasensitive aptasensor for adenosine triphosphate detection. Biosensors and Bioelectronics, 2013, 47, 271–277.

62 Miao, W. Electrogenerated chemiluminescence and its biorelated applications. Chemical Reviews, 2008, 108, 2506–2553.

63 Li, L.-L., Ji, J., Fei, R., *et al.* A facile microwave avenue to electrochemiluminescent two-color graphene quantum dots. Advanced Functional Materials, 2012, 22, 2971–2979.

64 Dong, Y., Tian, W., Ren, S., *et al.* Graphene quantum dots/L-Cysteine coreactant electrochemiluminescence system and its application in sensing lead(II) ions. ACS Applied Materials and Interfaces, 2014, 6, 1646–1651.

65 Su, M., Liu, H., Ge, L., *et al.* Aptamer-based electrochemiluminescent detection of MCF-7 cancer cells based on carbon quantum dots coated mesoporous silica nanoparticles. Electrochimica Acta, 2014, 146, 262–269.

66 Deng, W., Liu, F., Ge, S., *et al.* A dual amplification strategy for ultrasensitive electrochemiluminescence immunoassay based on a Pt nanoparticles dotted graphene-carbon nanotubes composite and carbon dots functionalized mesoporous Pt/Fe. Analyst, 2014, 139, 1713–1720.

67 Yang, H.M., Liu, W.Y., Ma, C., *et al.* Gold-silver nanocomposite functionalized graphene based electrochemiluminescence immunosensor using graphene quantum dots coated porous PtPd nanochains as labels. Electrochimica Acta, 2014, 123, 470–476.

68 Wu, D., Liu, Y., Wang, Y., *et al.* Label-free electrochemiluminescent immunosensor for detection of prostate specific antigen based on aminated graphene quantum dots and carboxyl graphene quantum dots. Scientific Reports, 2016, 6, 20511.

69 Liu, Q., Wang, K., Huan, J., *et al.* Graphene quantum dots enhanced electrochemiluminescence of cadmium sulfide nanocrystals for ultrasensitive determination of pentachlorophenol. Analyst, 2014, 139, 2912–2918.

70 Lou, J., Liu, S.S., Tu, W.W., Dai, Z.H. Graphene quantums dots combined with endonuclease cleavage and bidentate chelation for highly sensitive electrochemiluminescent DNA biosensing. Analytical Chemistry, 2015, 87, 1145–1151.

71 Zhang, T.T., Zhao, H.M., Fan, G.F., *et al.* Electrolytic exfoliation synthesis of boron doped graphene quantum dots: a new luminescent material for electrochemiluminescence detection of oncogene microRNA-20a. Electrochimica Acta, 2016, 190, 1150–1158.

72 Hu, T., Liu, X., Liu, S., *et al.* Toward understanding of transfer mechanism between electrochemiluminescent dyes and luminescent quantum dots. Analytical Chemistry, 2014, 86, 3939–3946.

73 Li, M., Li, J., Sun, L., *et al.* Measuring interactions and conformational changes of DNA molecules using electrochemiluminescence resonance energy transfer in the conjugates consisting of luminol, DNA and quantum dot. Electrochimica Acta, 2012, 80, 171–179.

74 Deng, S., Ju, H. Electrogenerated chemiluminescence of nanomaterials for bioanalysis. Analyst, 2013, 138, 43–61.

75 Angers, S., Salahpour, A., Joly, E., *et al.* M. Detection of β2-adrenergic receptor dimerization in living cells using bioluminescence resonance energy transfer (BRET). Proceedings of the National Academy of Sciences of the United States of America, 2000, 97, 3684–3689.

76 Pfleger, K.D., Eidne, K.A. Illuminating insights into protein-protein interactions using bioluminescence resonance energy transfer (BRET). Nature Methods, 2006, 3, 165–174.

77 Branchini, B.R., Rosenberg, J.C., Ablamsky, D.M., *et al.* Sequential bioluminescence resonance energy transfer–fluorescence resonance energy transfer-based ratiometric protease assays with fusion proteins of firefly luciferase and red fluorescent protein. Analytical Biochemistry, 2011, 414, 239–245.

78 Freeman, R., Liu, X., Willner, I. Chemiluminescent and chemiluminescence resonance energy transfer (CRET) detection of DNA, metal ions, and aptamer–substrate complexes using hemin/G-quadruplexes and CdSe/ZnS quantum dots. Journal of the American Chemical Society, 2011, 133, 11597–11604.

79 Huang, X., Li, L., Qian, H., *et al.* A resonance energy transfer between chemiluminescent donors and luminescent quantum-dots as acceptors (CRET). Angewandte Chemie, 2006, 118, 5264–5267.

80 He, Y., Huang, G., Cui, H. Quenching the chemiluminescence of acridinium ester by graphene oxide for label-free and homogeneous DNA detection. ACS Applied Materials and Interfaces, 2013, 5, 11336–11340.

81 Wu, M.-S., He, L.-J., Xu, J.-J., Chen, H.-Y. RuSi@Ru(bpy)$_3^{2+}$/Au@Ag$_2$S nanoparticles electrochemiluminescence resonance energy transfer system for sensitive DNA detection. Analytical Chemistry, 2014, 86, 4559–4565.

82 Qi, W., Wu, D., Zhao, J., *et al.* Electrochemiluminescence qesonance energy transfer based on Ru(phen)$_3^{2+}$-doped silica nanoparticles and its application in 'turn-on' detection of ozone. Analytical Chemistry, 2013, 85, 3207–3212.

83 Lu, Q., Wei, W., Zhou, Z., *et al.* Electrochemiluminescence resonance energy transfer between graphene quantum dots and gold nanoparticles for DNA damage detection. Analyst, 2014, 139, 2404–2410.

84 Liu, J., He, X., Wang, K., *et al.* highly sensitive electrochemiluminescence assay for protein kinase based on double-quenching of graphene quantum dots by G-quadruplex–hemin and gold nanoparticles. Biosensors and Bioelectronics, 2015, 70, 54–60.

85 Zhou, H., Zhang, Y.-Y., Liu, J., *et al.* Efficient quenching of electrochemiluminescence from K-doped graphene-CdS:Eu NCs by G-quadruplex-hemin and target recycling-assisted amplification for ultrasensitive DNA biosensing. Chemical Communications, 2013, 49, 2246–2248.

86 Liang, R.-P., Qiu, W.-B., Zhao, H.-F., *et al.* Electrochemiluminescence resonance energy transfer between graphene quantum dots and graphene oxide for sensitive protein kinase activity and inhibitor sensing. Analytica Chimica Acta, 2016, 904, 58–64.

87 Zhao, W.W., Xu, J.J., Chen, H.Y. Photoelectrochemical DNA biosensors. Chemical Reviews, 2014, 114, 7421–7441.

88 Son, Y.-W., Cohen, M.L., Louie, S.G. Energy gaps in graphene nanoribbons. Physical Review Letters, 2006, 97, 216803.

89 Ming, H., Ma, Z., Liu, Y., *et al.* Large scale electrochemical synthesis of high quality carbon nanodots and their photocatalytic property. Dalton Transactions, 2012, 41, 9526–9531.

90 Sudhagar, P., Herraiz-Cardona, I., Park, H., *et al.* Exploring graphene quantum dots/TiO$_2$ interface in photoelectrochemical reactions: Solar to fuel conversion. Electrochimica Acta, 2016, 187, 249–255.

91 Shen, J., Zhu, Y., Yang, X., *et al.* One-pot hydrothermal synthesis of graphene quantum dots surface-passivated by polyethylene glycol and their photoelectric conversion under near-infrared light. New Journal of Chemistry, 2012, 36, 97–101.

92 Yan, Y., Liu, Q., Du, X., *et al.* Visible light photoelectrochemical sensor for ultrasensitive determination of dopamine based on synergistic effect of graphene quantum dots and TiO$_2$ nanoparticles. Analytica Chimica Acta, 2015, 853, 258–264.

93 Liu, Y., Yan, K., Okoth, O.K., Zhang, J. A label-free photoelectrochemical aptasensor based on nitrogen-doped graphene quantum dots for chloramphenicol determination. Biosensors and Bioelectronics, 2015, 74, 1016–1021.

94 Pang, X., Zhang, Y., Liu, C., *et al.* Enhanced photoelectrochemical cytosensing of fibroblast-like synoviocyte cells based on visible light-activated ox-GQDs and carboxylated g-C$_3$N$_4$ sensitized TiO$_2$ nanorods. Journal of Materials Chemistry B, 2016, 4, 4612–4619.

95 Hou, Y., Wen, Z., Cui, S., *et al.* Constructing 2D porous graphitic C_3N_4 nanosheets/nitrogen-doped graphene/layered MoS_2 ternary nanojunction with enhanced photoelectrochemical activity. Advanced Materials, 2013, 25, 6291–6297.

96 Sun, L., Zhao, X., Jia, C.-J., *et al.* Enhanced visible-light photocatalytic activity of g-C_3N_4-$ZnWO_4$ by fabricating a heterojunction: investigation based on experimental and theoretical studies. Journal of Materials Chemistry, 2012, 22, 23428–23438.

97 Tian, J., Zhao, H., Quan, X., *et al.* Fabrication of graphene quantum dots/silicon nanowires nanohybrids for photoelectrochemical detection of microcystin-LR. Sensors and Actuators B: Chemical, 2014, 196, 532–538.

5

Electroanalytical Applications of Graphene

Edward P. Randviir and Craig E. Banks

Manchester Metropolitan University, UK

5.1 Introduction

The field of electrochemistry is more active than ever before, owing to the shift in social climate, where consumers are now demanding smaller, faster electronics, longer battery lifetimes, renewable energy, point-of-care medicinal treatments, electric cars, environmentally friendly fuelled cars, and many other devices that require a deep understanding of charge transfer reactions at interfaces. The fundamental understanding of charge transfer reactions can relate to several things: the diffusion characteristics of the active species in solution, the nature of the electrolyte, the chemical reactions involved in the mechanism, the electrochemical reactions involved in the mechanism, and perhaps most important of all, the surface that the charge transfer reactions occur upon. The latter has seen a shift in the past two decades away from traditional conductive, relatively chemically inert, and therefore expensive electrode materials such as platinum and gold, and moved towards cheaper materials and nanomaterials in an attempt to provide cheaper alternatives to traditionally expensive problems.

Out of the many carbon materials used throughout the past ten years, there is a current focus upon the properties of graphitic materials because they display some fascinating properties in different spatial dimensions (see below). This shift towards carbon nanomaterials has inspired this chapter, which will focus upon the electroanalytical properties of graphene, a common member of the graphitic materials family.

Graphene is an sp^2 hybridized hexagonal arrangement of carbon atoms, arranged in a characteristic honeycomb lattice structure. Each carbon bonds to three other carbons; two singly-bonded, and one doubly bonded (though resonant structures allow such bonds to interchange), and bonded to no other atoms, save for any lattice defects that are terminated by some form of oxygenated species, and the edges of graphene that are the same. This chapter will discuss the origins of graphene, the types of graphene available and their potential uses, and the electroanalytical properties of the many types of graphene available to the researcher today.

Nanocarbons for Electroanalysis, First Edition.
Edited by Sabine Szunerits, Rabah Boukherroub, Alison Downard and Jun-Jie Zhu.
© 2017 John Wiley & Sons Ltd. Published 2017 by John Wiley & Sons Ltd.

5.2 The Birth of Graphene

Out of the many significant discoveries made within the past two decades, it could be argued that graphene has captured the imagination of scientists and the public alike more so than any other has. This can be reflected in the academic field by the number of papers published with the keyword 'graphene' in its title since 2005 (Figure 5.1) [1].

The starting point for such a flurry of interest was undoubtedly when graphene was initially discovered[1] in the research papers reported by Geim and Novoselov in 2004/05 [2, 3]. The initial reports were widely celebrated by many because of the relative simplicity in the approach to graphene isolation, simply using sticky tape to peel it off a slab of graphite, coupled with the plethora of physical properties of graphene that opened up a seemingly endless number of potential research opportunities and societal benefits. Following the conception of graphene was a more in-depth investigation into the physical properties of the material, which yielded some surprising and potentially ground-breaking results. The electric field effect reported in graphene in 2005 [3] was quantified later in some publications, with electron mobilities reported as high as $200\,000\,\mathrm{cm^2\,V^{-1}120s^{-1}}$ [4], theoretically providing graphene with a huge advantage in the high speed consumer electronics industry. The high intrinsic strength and Young's modulus reported for grapheme [5] allowed the material to be flexible, and its electronic structure was found to be arranged in such a way that prevented almost anything from permeating the material [6]. Disseminating the full complement of properties is beyond the scope of this chapter, but a list of selected properties are provided in Table 5.1 for interested readers.

One property that is not listed in Table 5.1 is the explicit electrical anisotropy exhibited by graphene, something that has caused much debate in the field of electrochemistry, and is something that may be seen as a hindrance for the progress of graphene research. Electrical anisotropy is the property of having different electronic properties based upon the directional vector probed; in other words, graphene conducts to different

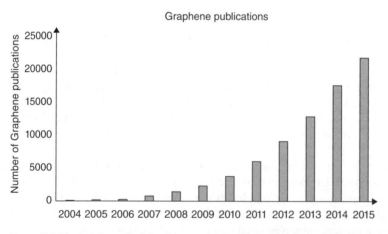

Figure 5.1 The number of publications containing the keyword *graphene* since 2004.

---

1 'Discovered' in this context alludes to the reporting of single layer graphene by Geim and Novoselov in 2005. This may be contentious in an academic context but for the purposes of this book chapter we assume graphene was discovered in 2005.

Table 5.1 A selected non-exhaustive list of properties of graphene, the combination of which has excited scientists across the globe.

Property	Value	Potential application	Ref
Electron mobility	$200\,000\,\mathrm{cm^2\,V^{-1}\,s^{-1}}$	High speed microelectronics	[4]
Young's modulus	$1\,\mathrm{TPa}$	Flexible nanomaterials	[5]
Intrinsic strength	$130\,\mathrm{GPa}$	Flexible nanomaterials	[5]
Optical transparency	97.7%	Visually transparent conductors	[55]
Thermal conductivity	$3800 - 5300\,\mathrm{WmK^{-1}}$	High temperature electronics	[56]

levels in different directions. As detailed in Figure 5.2, graphene is a 2D material that consists of two planes, termed the edge and basal plane.

If one considers graphene in three planes (x, y, z), the 2D sheet of graphene is an excellent conductor in the x and y planes, or from edge plane to edge plane. In the z direction graphene is actually a poor conductor, because like all electronic systems, current flows along the path of least resistance, so in the case of graphene current can easily flow between the sp^2 hybridized electronic orbitals of graphene in the x and y directions, but in the z direction there are no such orbitals. This prevents the flow of current in the z direction and therefore the charge transfer reactions on the basal plane are approximately three orders of magnitude slower than on the edge plane. This can be a significant problem for graphene electroanalysis if graphene sheets orientate themselves with the z direction facing the electrolyte.

One exception to this rule is the phenomenon observed when probing a graphene basal plane with a micro-droplet cell. Latest reports suggest there is unequivocal evidence that the basal plane of graphite is at least as active towards out-sphere redox probes as noble metal electrodes [7]. However, this phenomenon is yet to be replicated on a large scale. Indeed, the same group previously demonstrated that the basal plane is largely inactive on a macro scale, and the majority of electrochemical activity happens along step edges in mechanically exfoliated grapheme [8]. For the purposes of this chapter, we therefore assume that the basal plane of graphene is effectively electrochemically

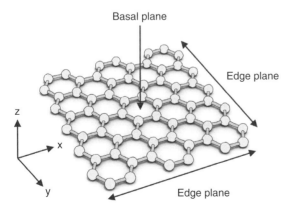

Figure 5.2 Illustration of the basal and edge planes of a single graphene sheet.

inert on the macro scale, and the edge plane of graphene is electrochemically active on the macro scale.

The wide-ranging properties that emerged from initial graphene research thus indicated a material that was unparalleled in its versatility, and has become a focal point for research ever since. However, before moving towards the electroanalytical applications on graphene in detail, one must understand that there are many barriers for graphene to becoming a success, the first of which is knowing the different types of graphene that are required for each application.

5.3   Types of Graphene

Graphene is undoubtedly unique in several ways, as alluded to in the introduction, but it also has many limitations that require researchers to exercise a level of control in order to maximise the desired properties from graphene. For example, graphene has been touted as a potential semiconductor for microelectronics, but its zero band gap [9] prevents its use as a semiconductor without manipulating graphene in some way to behave as a semiconductor. Therefore it is sometimes necessary to actually create 'defective' graphenes! Of course, the term 'defective' in this case doesn't actually mean that the material itself will be defective, but it implies that defects have to be introduced to pristine graphene sheets (physically or chemically) in order for the material to be of any use. However, creating a consistent 'defective' graphene is a significant challenge, and despite the several methods of graphene production, there are few that provide a consistent large-scale batch that is effective for its intended purpose. For interested readers there are several examples of creating tunable graphenes for semiconductors (see for example [10]). Despite this, one can roughly classify types of graphene from their synthetic method, and there is a remarkable relationship between the synthetic method, the quality of the graphene, and the application of that particular graphene, as summarised in Figure 5.3 [11].

For example, smart windows require optical transparency and the highest possible conductivity, therefore a pristine graphene would be more applicable, and would likely have to be fabricated through a Chemical Vapour Deposition method such as the method reported by Bae *et al.* [12]. Conversely, a semiconductor requires tunable electronic properties, which is a something that can only be afforded by wrinkling [13, 14] or doping [15] of graphene to create an energy gap between the conduction and valence bands of graphene. While such graphenes have been attempted as electroanalytical platforms, there have been some difficulties with them, as will be discussed later.

This leads onto solution-based graphenes that are specifically prepared for electroanalysis; an area that has been highly controversial and almost impossible to keep up to date with due to sheer volume of research papers reporting electrocatalytic effects of graphene. Solution-based graphenes can be created by many wet synthesis methods (Hummers [16], Staudenmaier [17], Hofmann [18] methods) that normally use strong acids and powerful oxidising agents to exfoliate graphene from graphite powders. This method carries advantages of being able to produce large volumes of graphitic oxide that can be reduced chemically, thermally, or electrochemically [19] into so-called graphene, and the material produced often reduces the thermodynamic barriers for adiabatic reactions in electrochemistry, making it an attractive

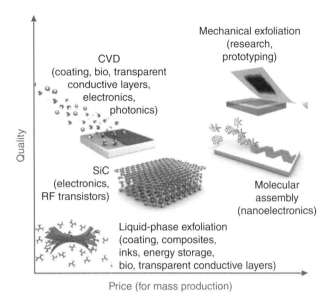

Figure 5.3 Comparison of the price of mass production compared to the quality of graphene. *Source*: Novoselov 2012 [11]. Reproduced with permission of Nature Publishing Group.

nanomaterial for electroanalysis. However, it is contentious to term such materials as graphene, because almost all solution-based methods firstly create graphite oxide or graphene oxide, two materials that behave considerably different to graphene in terms of electrochemistry [20]. Secondly, solution-based graphitic oxides contain a large amount of defects that alter the electronic nature of the material, and thirdly the material tends to flocculate in solution anyway so that the material is no longer single layer, but multi-layer [21]. Fourth and finally, graphitic oxides often contain metal ions such as manganese, and anions such as sulphate, both of which can affect the electrochemical performance of the material considerably [22]. Resultantly, the field of graphene electroanalysis can sometimes be a tricky field to navigate properly, yet there are some tips that may be useful for researchers. Often graphene oxides give many different properties because the methods used are not well understood, leave electroactive residues in the resulting graphene oxide, are not temperature controlled properly, or are not stored properly. Therefore, a useful check for the researcher is as follows:

1) Understand the *methods* used in research papers: modified Hummers methods are atisfactory, but the details need to be reported fully.
2) Identify the *quality* of chemicals used in the fabrication: reagent grade chemicals are not good enough as they will leave residues behind.
3) Assess the *procedures* in detail: check for temperature control, atmosphere, chemical and physical changes.
4) Scrutinise *storage*: are the samples concentrated enough to precipitate, what solvent are the samples stored within.
5) How are stored samples *checked* for homogeneity: sonication of samples is often a good way to ensure graphene oxides remain as graphene oxides in a solution.

If such procedures are followed, it is likely that a reproducible graphene oxide can result, and therefore the research from such graphene oxides can be trusted. If this is not the case, one should proceed with upmost caution.

In addition to the graphene types described above, free-standing graphene foams and graphene screen-printed electrodes (SPEs) have also been researched as electroanalytical platforms. Such graphenes are unique in their own right and give different properties, depending upon the needs of the researcher. The remainder of this chapter will explore the research, general findings, and pertinent examples of four different graphene types: chemical vapour deposition (CVD) graphene, free-standing 3D graphene foam, graphene SPEs, and solution-based graphene.

5.4 Electroanalytical Properties of Graphene

The previous sections have detailed an introduction to the start of graphene research and the developed methods for synthesizing graphene, the latter of which is an important point with respect to the relative applications of graphene. With this in mind, this section is split into four sections, each assessing the research conducted in the specific area and detailing the challenges faced for using each graphene as an electroanalytical platform. The chapter will conclude by offering an outlook on the future of graphene electroanalysis using the dearth of knowledge available and our best possible predictions based upon the challenges faced by each type of graphene.

5.4.1 Free-standing 3D Graphene Foam

The first of the four types of graphene discussed is free-standing 3D graphene foam (GF), a relatively new material that is probably slightly moronic in its name, given that graphene is meant to be a 2D material! However, such a material does indeed exist and is used in applications such as supports for supercapacitors [23], electrodes for batteries [24], and environmental clean-up [25], as well as electrochemical sensing [26].

The material was first reported in 2011 by Cao *et al.* who designed the fabrication method for apparent supercapacitor-based applications [23]. The innovation was required to explore the possibility of extracting the most from graphene's surface area in order for supercapacitors to undergo the required redox reactions that deliver pseudo-capacitance and thus a material that could deliver higher current loads. In order to design the 3D graphene network, researchers used an existing nickel foam structure as a base substrate in a CVD rig, instead of a flat nickel surface that would normally create single layer CVD grapheme [27]. The result was a foam structure coated in graphene, and therefore the term 'free-standing 3D graphene foam' was realised.

The advantage of GF is that it allows the user to exploit the higher surface area for electrochemical purposes if used as a working electrode. Such a property could improve current collectors on fuel cells or supercapacitors, and amplify the currents signals generated in voltammetric applications; therefore it could be a desirable material for electroanalysis. On the other hand its highly hydrophobic nature poses a significant problem for aqueous electrochemistry because the material requires liquid penetration for its properties to be meaningful, and the stress imposed upon the material by the hydrophobicity when in contact with water is often enough to damage the material. Therefore, it is necessary to decrease the surface tension of water using a surfactant or latex particle, otherwise experiments are limited to organic solvents and ionic liquids.

Table 5.1 lists several efforts by researchers to utilise GF as an electrode material in electroanalytical applications [26, 28–35]. Inspection of the column labelled 'Electrode' tells the story of GF electroanalysis quite simply: as a standalone material it suffers as an electrode; yet when decorated with another material such as CuO or Co_3O_4, the electrochemistry suddenly come to life. One of the major contributing factors to this is indeed the above-mentioned hydrophobicity of GF, as was pointed out in the research paper by Brownson *et al.* [21]. Research actually demonstrated a contact angle between GF and water of 120°, which is comparable with waterproof materials such as Teflon film. The disadvantages of this phenomenon are described previously. Researchers therefore choose to overcome this obstacle by decorating the GF with another material, often in nanoparticle form, to promote a more favourable wetting of the 3D network. However, it could also be argued that while graphene does exhibit excellent conductive properties across its lattice, the 'wire' points of graphene lie at its edges. Therefore, because the basal planes are more exposed in a GF than the edges, in reality a bare 3D graphene network might not be useful for electroanalysis at all, as the thermodynamic boundaries required to activate the electron transfer reactions are much higher along the basal plane of graphene than on the edge.

One route towards improving the electrochemical response of GF that has not yet been discussed is the process of manipulating the GF surface without introducing nanoparticles or other dopants. This has been the focus of some literature reports, however, and the results have been interesting. One such report, reported by Brownson *et al.* [28], details a procedure to introduce a pre-washing of the GF electrode with acetone, a process that allows aqueous penetration of the GF, and therefore allows the research to utilise the high surface area offered by the GF. Through careful manipulation of the GF, their paper managed to obtain a reasonable electroanalytical performance of bare GF towards uric acid, with a limit of detection (LoD) of 1 µM and a working linear range of 20–150 µM [28]. The procedure is quick, simple, and the value of information derived from it can be enormously beneficial. Perhaps the major drawbacks of such an approach are inherent of GF itself, rather than of the method proposed (see Table 5.2).

The material is difficult to use in electroanalytical experiments due to the unknown electrode area that has to be estimated using debateable calculations; and the connection of the GF electrode to the solution using crocodile clips, for example, is likely to be a stumbling block for proper implementation of such graphene in electroanalysis. The material requires a disposable GF electrode device to be designed with a consistent surface area if it is to be taken seriously in the future for electroanalytical applications. Otherwise its utility will likely be limited to electrodes for closed systems such as batteries or fuel cells.

5.4.2 Chemical Vapour Deposition and Pristine Graphene

In the early days of graphene research it was often infeasible for research groups to conduct effective graphene work using mechanical exfoliation, because while such methods do create the highest purity graphene, it takes far too long to create enough to conduct meaningful electrochemical studies upon. Resultantly, researchers sought

Table 5.2 Table listing some selected works using GF as an electrode material in electroanalytical applications.

Target species	Electrode	LoD / μM	Sensitivity / μA mM^{-1} cm^{-2}	Ref
Dopamine	GF	0.025	620	[26]
	3D N-doped graphene	0.0010	9870	[30]
	GF/Co$_3$O$_4$	0.025	3390	[29]
Glucose	GF/Mn$_3$O$_4$	10	360	[31]
	GF/Ni(OH)$_2$	0.34	2650	[33]
Ascorbic acid	GF/CuO	0.43	2060	[34]
H$_2$O$_2$	GF/thionine	0.080	170	[32]
	GF/PtRu nanoparticles	0.040	1020	[35]
Uric acid	GF	1.0	16300	[28]

methods to produce high quality graphene in larger quantities. CVD techniques therefore offered a different route towards highly pure, single-layered graphene that was suitable for fundamental electrochemistry investigations. Briefly, CVD graphene methods are a bottom-up approach to graphene construction, where graphene is grown upon a metal substrate (nickel or copper) using gases such as methane as the carbon source. The graphenes produced are about as close to pristine graphene as is currently possible, with some methods reporting the possibility of producing up to 30 square inches at a time [12]. Another method of graphene synthesis is ultrasonic exfoliation of highly pure graphite in ethanol [36]. In this method, pure pristine graphene flakes remain suspended in ethanol and can be drop-casted upon an electrode surface to create a pristine graphene electrode. Such a method allows for quicker experimentation, but the graphene may suffer from the 'coffee ring' effect (discussed later), and restack to form multilayers when dry. Nevertheless, electrochemical studies of this graphene type have been conducted and demonstrated that pristine graphene does not offer beneficial electron transfer kinetics over materials such as graphite [36, 37].

The general pristine nature of CVD graphene, that is, long range order of graphene sheets that are free from defects, impurities, and terminations, effectively means that when the electrolyte in an electrochemical investigation moves into contact with the graphene working electrodes, the majority of the surface area covered is the graphene basal plane. The effect of this is that the charge transfer reactions remain relatively inefficient, as opposed to if the graphene was oriented in a manner that its edge was exposed to the electrolyte. This means that in terms of electroanalysis, CVD and pristine graphenes have no quite lived up to their original hype.

Resultantly there are very few works to speak of in terms of electroanalysis of CVD or pristine graphene, save for the works by Brownson *et al.* [38]. In their work, CVD graphene was employed for the detection of NADH and uric acid, two common biologically relevant analytes that the field of electroanalysis focusses upon. The initial fundamental study revealed that the CVD graphene behaved in a similar fashion to edge plane pyrolytic graphite, with a good peak resolution and low oxidation potential. This translated into a beneficial sensitivity of 260 μA cm^{-2} mM^{-1} for NADH and a limit of detection of 7.21 μM, which is well within the required detection limits for this target species which appear in the blood at a maximum off 900 μM [39]. Similarly for uric acid, a sensitivity of $480\,\mu A\,cm^{-2}\,mM^{-1}$ was obtained with a limit of detection of 8.84 μM, which is analytically useful given that the healthy range for adult females is as low as 14 μM, and the highest healthy level for an adult male is 42 μM [40]. One thing that cannot be understated, however, is the role of graphitic islands in CVD graphene, and the possibility of metal ion mediated voltammetry due to the CVD graphene growth substrate. Further work by Brownson *et al.* concluded that graphitic islands dominate the electrochemical response of CVD graphene and warned researcher that nickel in particular may give false voltammetric profiles [41].

This section has explored CVD and pristine graphene and shared insights into the reasons why researcher prefer not to choose such substrates as choice electrode substrates for electroanalytical charge transfer reactions. It is the opinion of the author that pristine and CVD graphenes will prove to be far more beneficial for their conductivity for small consumer electronics and smart devices, rather than for electrochemical sensing platforms, resulting from the largely electrochemically inactive nature of the graphene sheet. Thus the key to using graphene in an electroanalytical sense lies within manipulating the graphene sheet to behave in a more favourable way.

5.4.3 Graphene Screen-printed Electrodes

One of the most favoured methods of electroanalysis is to implement Screen-Printed Electrodes (SPEs) for the detection of biologically relevant analytes. SPEs are carbon-based electrodes that use printed conductive carbon layers to facilitate charge transfer reactions on a small, flexible substrate that is normally made from polyester. SPEs are lightweight, disposable, mass-producible, reproducible, repeatable, and fundamentally, they could allow the user to scale voltammetric measurements down into a point-of-care device for environmental [42], forensic [43], and medicinal applications [44], provided the methodologies were rigorous. SPEs have been used by many institutions and it is therefore no surprise that, given the hype surrounded graphene, that researchers would attempt to design graphene screen-printed electrodes (GSPEs) in the hope that they would open the door to some unique properties.

However, designing and fabricating GSPEs was not quite as simple as it may sound. SPEs are generally constructed by pressing a carbon ink through a screen onto a polyester substrate (see Metters *et al.* for further information on SPE design [45]) and subsequently curing the substrate in an oven. This method leaves two options for graphene inoculation: first, graphene can be mixed in with the ink before the ink is pressed through the screen; or second, the graphene can be printed over the top of the working electrode on the first print. The first option carries that advantage of printing the graphene and ink at the same time, but the user has less control over the fate of the graphene after it is printed, meaning that graphene may be allowed to restack, or even attach to the carbon black particles in the carbon ink. This means that the working electrode would essentially be a carbon black and graphene mix, and would not be entirely representative of the fundamental graphene response. The second option carries the disadvantage of adding an extra printing step, so the electrodes take longer to make, however they give the researcher an element of control over the graphene, as an extra printed later would allow graphene to be completely in contact with the electrolyte during experimentation.

When approaching GSPE literature, one must be conscious of a differentiation between GSPEs and graphene modified SPEs, two things that are completely different in nature. GSPEs infer that the graphene is either incorporated into the ink, or printed upon the working electrode separately. In many examples, graphene-based composite materials are drop-casted upon the SPE and labelled as GSPEs, something that we believe to be factually incorrect. The method described is simply the drop-casted method of graphene electrode preparation, something that is covered in the final section of this chapter. For the purposes of this chapter, we will focus only upon graphenes that are printed. The most prominent GSPE research papers were reported independently by Ping *et al.* [46], and Randviir *et al.* [47], while there have also been reports of the use of commercially available GSPEs that will be discussed too [48].

Possibly the first report of a GSPE was the work by Ping *et al.*, who presented their GSPE as the first of its kind. Their work used graphene that was fabricated through a Hummers method for graphitic oxide, which was then chemically reduced to graphene using ammonia and hydrazine. The chemically reduced graphene was then incorporated into the carbon ink used in the fabrication of the SPE [46]. While this method was the first of its kind, there were several key parts identified that made the resulting electroanalysis debateable. First, the SEM images presented in their work demonstrated the existence of graphitic islands, meaning that either the graphene had restacked to form graphite, or the surface of the SPE was simply a carbon black/graphene mix that was not truly representative of the fundamental properties of graphene. Indeed, the successful electroanalytical performance of the GSPE was a clear indication that graphitic edges were dominating the electrochemical responses. This is even before one accounts for the Hummers-type graphene used, which is well known to create highly defective graphenes once reduced. The most positive aspect of this work was that the resultant GSPE had the ability to differentiate the electrochemical signals of ascorbic acid, uric acid, and dopamine under lab conditions, in linear ranges from as low as $0.5\,\mu M$ up to $4.5\,mM$. Even more encouraging, the lab-based efforts took a step away from lab conditions and focussed upon real samples such as urine, with a reasonable level of success, and good level of analytical reproducibility with measurements yielding under 4% for most experiments.

The above experiments were conducted at a time when there was a limited understanding of Hummers-type graphene, which most researchers in the field now know to be careful when using. Of the few other examples of GSPEs, the report by Randviir *et al.* is probably the most rigorous investigation of in-house GSPEs, which were fabricated using two different methods [47]. The first method incorporates a graphene that is produced from a split plasma process, producing highly pure graphene with few basal plane defects (GSPE1). In this ink, a small amount of carbon black is mixed in to improve the conductivity of the ink. The second method uses an ink fabricated from a solution-based method, and incorporates polymeric binders into the ink, which are known to inhibit charge transfer reactions at the electrode/solution interface (GSPE2) [49]. The two methods also display radically divergent electroanalytical properties as a result of their construction. Raman spectroscopy of the GSPEs indicate that GSPE1 yields a more graphite-like SPE that one might expect to offer favourable electrochemical responses. On the other hand, the Raman spectroscopy of GSPE2 shows a more graphene-like characteristic fingerprint. This observation translates in electrochemical terms to a faster rate of charge transfer for GSPE1 than GSPE2. Electroanalytical experiments indeed demonstrated that GSPE1 allowed a sensitivity towards ascorbic acid of

$11.1\,\mu A\,cm^{-2}\,mM^{-1}$, while GSPE2 was ten times less $(1.02\,\mu A\,cm^{-2}\,mM^{-1}$. But this was not the full story because the preparation method for graphene for GSPE2 incorporated impurities into the structure, some of which had the ability to catalyse electrochemical reactions at significantly high potentials. For example, the sensitivity of uric acid was actually higher for GSPE2 than GSPE1, which exhibited 600 and $570\,\mu A\,cm^{-2}\,mM^{-1}$, respectively. Therefore, though two methods were suggested, they had to be chosen and implemented very carefully for future use.

Commercially available GSPEs have also been implemented by researchers in the context of medicinal technology research. One specific example of this was reported by Teixeira *et al.*, who used such GSPEs from DropSens (Spain) to determine the concentration of pregnancy markers in urine [48]. This example takes advantage of GSPEs that are fabricated at the commercial end, who modify the working electrode with a graphene solution. Inspection of their product datasheet would suggest that the graphene they use is either fabricated using a Hummers-type method or commercially obtained (containing surfactants), both of which can give rise to false voltammetric signals that are often attributed to graphene structures. Nevertheless, we refer to this example because it is assumed that the company provides a consistent electrode. The researchers modify their GSPE with polyaniline (PANI), which is a common method for preserving the conductive element of graphene, while avoiding the introduction of defects upon the surface. The electrode was also hybridized with an antigen to the target species, allowing a selective recognition protocol, while using bovine serum albumin to block non-specific binding sites. Electrochemical impedance spectroscopy is the most common electroanalytical technique to sue in such scenarios because the antibody-antigen interaction is essentially a bio-recognition event that blocks electrochemical reactions once the binding takes place. The resultant response to pregnancy markers yields a linear response from $0.001-50\,\mu M$, which expands the linear range of existing technologies by over two orders of magnitude.

The field of GSPEs has seen very limited activity for the field of electroanalysis, but the procedures reported are ones that hold some promise for future point-of-care devices in medicinal applications. The major stumbling blocks to GSPEs are understanding the true nature of the SPE once graphene has been incorporated, something that has only really been attempted by Randviir *et al.* Without a true rigorous experimentation with the electrode using standard redox probes, it is inevitable that false positives may arise during experimentation; this was especially apparent for GSPE2 in the work by Randviir *et al.* [47]. Resultantly, the field requires a considerable amount of work before any true GSPEs can have a real impact on point-of-care devices.

5.4.4 Solution-based Graphene

The final type of graphene discussed in this chapter is solution-based graphene.[2] Solution-based graphenes are particularly useful for electroanalytical purposes because they can be produced in large amounts, and therefore many experiments can be completed in a short period of time. Furthermore, the relative level of defects in solution-based graphene can be beneficial for electroanalysis because the defects give rise to

---

2   The term 'solution-based graphene', while factually inaccurate, is a typical description for this type of material. In fact it might be termed as 'graphene suspensions' given that graphene doesn't dissolve in any common suspending solvent such as water, ethanol, methanol, or isopropanol.

electrocatalytic effects in some cases. Solution-based graphenes are often prepared using powerful oxidising agents as discussed previously, and are stored at known concentrations in the solution phase, before being applied to an electrode.

The main method of solution-based graphene electrode assembly is the drop-casting method, whereby the user will pipette a known volume (and therefore, mass) of graphene upon an inverted electrode surface, before leaving to evaporate in air. This method requires careful transfer of the graphene solution, clean electrode substrates, fast-evaporating solvents, and protection from the atmosphere. But even if these things are controlled, it isn't always easy to create a uniform graphene coverage across a surface, due to the phenomenon known as the 'coffee ring effect' (Figure 5.4). Therefore, in addition to the common poor quality graphene used in solution-based applications, the method of electrode assembly is fundamental to allowing a restacking of the graphene sheets, and therefore the responses often become akin to graphite, much like the graphitic islands argument presented for CVD graphene [41, 50]. Nevertheless, if the researcher takes such factors into consideration, there is still scope to create electroanalytical sensing platforms that are viable. Readers are referred to our list provided previously to identify plausibility of solution-based graphene research.

Resultantly, one of the major problems faced is identifying a solution-based graphene production method that can be adequately controlled and reproduced. This is something that is not as easy as it may seem, and despite several years of solution-based graphene production, there are still discrepancies in electrode production methods that make it difficult to identify the better works. One such method of electrode fabrication that deserves attention is the work by Zhou *et al.*, who report a unique method for the controlled production of solution-based graphene electrodes. The graphenes produced are typical Hummers type graphenes, however the method employed avoids the 'coffee ring effect' described in Figure 5.4. The electrodes are fabricated using a spray-coating method through a patterned template, on to an underlying substrate in a similar manner to a screen printed electrode. The differences are that the resulting electrodes are thin film electrodes rather than thick film electrodes, and the working electrode is far more homogeneous than in the case of screen printed electrodes. Further information on the method is provided for interested readers in the paper by Zhou *et al.* [51]. This method does require a high quality solution-based graphene for it to be effective, however, which is

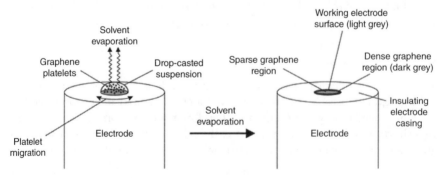

Figure 5.4 Schematic diagram of the 'graphene coffee ring effect' that is apparent when applying a graphene suspension upon an electrode surface. *Source*: Randviir 2014, http://pubs.rsc.org/is/content/articlehtml/2014/cp/c3cp55435j. Used under CC BY 3.0 https://creativecommons.org/licenses/by/3.0/.

something that is not so simple in solution-based graphenes. The method of choice in the previous example (and perhaps the most promising method for cleaning solution-based graphenes) is the method proposed by Gao *et al.*, who utilise hydrazine and thermal treatments to remove oxygen-containing functionalities from graphene sheets. [52]. The method is described as a de-epoxidation, where bridging epoxy groups are targeted using hydrazine, which then acts as a leaving group to reduce epoxides to allylic carbons.

Table 5.3 therefore lists some of the best electroanalytical applications of solution-based graphene. One common theme of the listed methods is the chemical reduction of

Table 5.3 Table listing some selected works using solution-based graphene as an electrode material in electroanalytical applications.

Target species	Electrode	Graphene type	LoD / μM	Sensitivity / μA mM^{-1} cm^{-2}	Ref
Ascorbic acid	N-doped graphene on GC	Hummers – chemically reduced with hydrazine	3.5	93.1	[57]
			50	83.5[1]	[58]
	Graphene-chitosan composite	Hummers – chemically reduced with hydrazine			
Uric acid	N-doped graphene on GC	Hummers – chemically reduced with hydrazine	0.57	132	[57]
			2.0	2030[1]	[58]
	Graphene-chitosan composite	Hummers – chemically reduced with hydrazine			
Dopamine	N-doped graphene on GC	Hummers – chemically reduced with hydrazine	0.25	205	[57]
			1.0	3360[1]	[58]
	Graphene-chitosan composite	Hummers – chemically reduced with hydrazine			
Glucose	Graphene with gold nanoparticles and chitosan	Hummers – assembled on cationic substrate as per Kovtyukhova *et al.* [59]	180	99.5	[60]
Lead	Graphene and Nafion®	Hummers – chemically reduced with hydrazine	97.0×10^{-6}	4.59×10^{-3}	[61]
Cadmium	Graphene and Nafion®	Hummers – chemically reduced with hydrazine	180×10^{-6}	10.1×10^{-2}	[61]
	Graphene and Nafion®	Hummers – chemically reduced with hydrazine	45.0×10^{-6}	Not reported	[62]
TNT	Graphene and porphyrin	Hummers – uncleaned	0.50 ppb	1.75 μA ppb^{-1} cm^{-2}	[63]
Glucose	Sulfonated graphene, Nafion® and gold nanoparticles on glassy carbon	Hummers – chemically reduced with hydrazine	5.0	67.2[1)]	[64]
	Nickel nanoparticle and graphene composite on glassy carbon	Hummers – uncleaned	0.10	891	[65]

1) Calculated from electrode diameter of 3mm and reading from the reported concentration versus current graphs.

graphene oxide using hydrazine and ammonia as described above, a method that reportedly creates a purer solution-based graphene that is free from problematic metal ions that may remain after preparation of Hummers-type graphene. Even so, care still has to be taken due to hydrazine's voltammetric activity. One example of the concerns surrounding solution-based graphene can be highlighted in the well-referenced work by Cui and Zhang, who employ a graphene/gold nanoparticle composite for the detection of epinephrine [53]. The method reports a limit of detection as low as $0.007\,\mu M$, a linear range panning two orders of magnitude, and a sensitivity of $25\,500\,\mu A\,mM^{-1}\,cm^{-2}$![1]

While the initial figures appear to be promising, they should be approached with caution. For example, the graphene itself is likely to be fabricated using a Hummers method without any cleaning, and therefore the graphene used is actually graphene oxide. Secondly, the graphene is commercially obtained in a period where methods of graphene production were poorly understood. During this time, surfactants were often used in commercially available graphenes to exfoliate the graphene sheets from graphite. Therefore, the electrochemical response observed may not be for epinephrine or graphene, but be due to the surfactants inherent in the graphene. Indeed, inspection of literature reports arguing this case demonstrate that the common surfactants used allow very similar peak potentials to the case discussed [54].

Despite the drawbacks, there are still some very good electroanalytical methods that can be employed for electroanalytical sensing applications. The field of graphene electroanalysis requires a more robust graphene fabrication method for the design and production of sensors that are repeatable and sustainable. Many methods in the literature report electrode designs that are extremely complex, particularly in the medical diagnostic field. This creates a conceptual issue with graphene-based electroanalysis because researchers pursue electrodes that are selective to certain targets, but the problem is that they commonly employ in excess of five components, which limits the commercial applicability of the sensors as they are fundamentally difficult to design without construction by hand. It is therefore suggested that the future of solution-based graphene electroanalysis requires some form of standardised method to be used in order to create a level of confidence with graphene literature such that researchers can study papers without the worries of inadequate production methods looming over them.

5.5 Future Outlook for Graphene Electroanalysis

The future of graphene as an electroanalytical platform most likely lies not within its charge transfer properties at an interface, but rather with its conducting capabilities and its ability to act as a conduit between an electrode material and a charge transfer mediating material, such as polyaniline. There are numerous examples in the academic ether that report graphene as an electronic linker, more like an molecular wire than as an electron transfer mediator itself, and it has consequently taken a firm seat at the top table of biosensor design. Many of such biosensors incorporate nanoparticles that are favoured for their high surface area, charge transfer activity, and binding ability to the edge and defect sites on a graphene sheet. However, as the material itself, graphene is often poorer than many other materials, and the graphenes that are reportedly useful may be hampered by inconsistent production methods and false positive electrochemistry.

The major positive aspects of current graphene electroanalysis is that the field has well and truly moved away from the idea that graphene itself is a charge transfer mediator, and also methods seem to have been developed for solution-based graphenes that are allowing more plausible investigations. Consequently, with another few years of research, several graphene-based electroanalytical platforms could start to emerge, albeit using other materials such as nanoparticles as their sensing aspects. Finally, there are several things that researchers need to consider with graphene research in future experiments that are commonly missed in several literature reports:

1) The effects changing the mass of graphene upon electrochemical signals;
2) The exploration of several underlying support materials instead of just glassy carbon;
3) Full physicochemical characterisation of graphenes and graphene modified electrodes prior to electroanalytical investigation;
4) Proper control methods using graphite, other graphenes, and potential interferents that may give rise to satellite peaks (eg MnO_2, Ni^{2+}, sodium dodecyl sulphate, sodium cholate hydrate);
5) The selection of graphene production method and a full assessment of its relative advantages and disadvantages.

References

1 Web of Knowledge http://wok.mimas.ac.uk/ (accessed 5 September 2016).
2 Novoselov, K.S.; Geim, A.K.; Morozov, S.V.; *et al.* Electric field effect in atomically thin carbon films. Science, 2004, 306 (5696), 666–669.
3 Novoselov, K.S.; Jiang, D.; Schedin, F.; *et al.* Two-dimensional atomic crystals. Proceedings of the National Academy of Sciences of the United States of America 2005, 102 (30), 10451–10453.
4 Bolotin, K. I.; Sikes, K. J.; Jiang, Z.; *et al.* Ultrahigh electron mobility in suspended graphene. Solid State Communications 2008, 146 (9–10), 351–355.
5 Lee, C.; Wei, X.; Kysar, J. W.; Hone, J. Measurement of the elastic properties and intrinsic strength of monolayer graphene. Science 2008, 321 (5887), 385–388.
6 Hu, S.; Lozada-Hidalgo, M.; Wang, F. C.; *et al.* Proton transport through one-atom-thick crystals. Nature 2014, 516 (7530), 227–230.
7 Unwin, P.R.; Güell, A.G.; Zhang, G. Nanoscale electrochemistry of sp^2 carbon materials: From graphite and graphene to carbon nanotubes. Accounts of Chemical Research 2016, 49 (9), 2041–2048.
8 Güell, A.G.; Cuharuc, A.S.; Kim, Y.-R.; *et al.* Redox-dependent spatially resolved electrochemistry at graphene and graphite step edges. ACS Nano 2015, 9 (4), 3558–3571.
9 Meric, I.; Han, M.Y.; Young, A. F.; *et al.* Current saturation in zero-bandgap, top-gated graphene field-effect transistors. Nature Nano 2008, 3 (11), 654–659.
10 Castro, E.V.; Novoselov, K.S.; Morozov, S.V.; *et al.* Biased bilayer graphene: Semiconductor with a gap tunable by the electric field effect. Physical Review Letters 2007, 99 (21), 216802.
11 Novoselov, K.S.; Falko, V. I.; Colombo, L.; *et al.* A roadmap for graphene. Nature 2012, 490 (7419), 192–200.

12 Bae, S.; Kim, H.; Lee, Y.; *et al.* Roll-to-roll production of 30-inch graphene films for transparent electrodes. Nature Nano 2010, 5 (8), 574–578.

13 Deng, S.; Berry, V., Wrinkled, rippled and crumpled graphene: An overview of formation mechanism, electronic properties, and applications. Materials Today 2016, 19 (4), 197–212.

14 Lim, H.; Jung, J.; Ruoff, R. S.; Kim, Y. Structurally driven one-dimensional electron confinement in sub-5-nm graphene nanowrinkles. Nature Communications 2015, 6, 8601 (1 – 6).

15 Rani, P.; Jindal, V.K. Designing band gap of graphene by B and N dopant atoms. RSC Advances 2013, 3 (3), 802–812.

16 Hummers, W.S.; Offeman, R. E. Preparation of graphitic oxide. Journal of the American Chemical Society 1958, 80 (6), 1339–1339.

17 Staudenmaier, L. Verfahren zur Darstellung der Graphitsäure. Berichte der Deutschen Chemischen Gesellschaft 1898, 31 (2), 1481–1487.

18 Hofmann, U.; König, E. Untersuchungen über Graphitoxyd. Zeitschrift für Anorganische und Allgemeine Chemie 1937, 234 (4), 311–336.

19 Dreyer, D.R.; Park, S.; Bielawski, C.W.; Ruoff, R.S. The chemistry of graphene oxide. Chemical Society Reviews 2010, 39 (1), 228–240.

20 Brownson, D.A.; Lacombe, A.C.; Gomez-Mingot, M.; Banks, C.E. Graphene oxide gives rise to unique and intriguing voltammetry. RSC Advances 2012, 2 (2), 665–668.

21 Brownson, D.A.; Figueiredo-Filho, L.C.; Ji, X.; *et al.* Freestanding three-dimensional graphene foam gives rise to beneficial electrochemical signatures within non-aqueous media. Journal of Materials Chemistry A 2013, 1 (19), 5962–5972.

22 Smith, J.P.; Foster, C.W.; Metters, J.P.; *et al.* Metallic impurities in graphene screen-printed electrodes can influence their electrochemical properties. Electroanalysis 2014, 26 (11), 2429–2433.

23 Cao, X.; Shi, Y.; Shi, W.; *et al.* Preparation of novel 3D graphene networks for supercapacitor applications. Small 2011, 7 (22), 3163–3168.

24 Wei, W.; Yang, S.; Zhou, H.; *et al.* 3D graphene foams cross-linked with pre-encapsulated fe3o4 nanospheres for enhanced lithium storage. Advanced Materials 2013, 25 (21), 2909–2914.

25 Dong, X.; Chen, J.; Ma, Y.; *et al.* Superhydrophobic and superoleophilic hybrid foam of graphene and carbon nanotube for selective removal of oils or organic solvents from the surface of water. Chemical Communications 2012, 48 (86), 10660–10662.

26 Dong, X.; Wang, X.; Wang, L.; *et al.* 3D graphene foam as a monolithic and macroporous carbon electrode for electrochemical sensing. ACS Applied Materials & Interfaces 2012, 4 (6), 3129–3133.

27 Kim, K.S.; Zhao, Y.; Jang, H.; *et al.* Large-scale pattern growth of graphene films for stretchable transparent electrodes. Nature 2009, 457 (7230), 706–710.

28 Figueiredo-Filho, L. C. S.; Brownson, D. A. C.; Fatibello-Filho, O.; Banks, C. E. Electroanalytical performance of a freestanding three-dimensional graphene foam electrode. Electroanalysis 2014, 26 (1), 93–102.

29 Dong, X.-C.; Xu, H.; Wang, X.-W.; *et al.* 3D graphene–cobalt oxide electrode for high-performance supercapacitor and enzymeless glucose detection. ACS Nano 2012, 6 (4), 3206–3213.

30 Feng, X.; Zhang, Y.; Zhou, J.; *et al.* Three-dimensional nitrogen-doped graphene as an ultrasensitive electrochemical sensor for the detection of dopamine. Nanoscale 2015, 7 (6), 2427–2432.

31 Si, P.; Dong, X.-C.; Chen, P.; Kim, D.-H. A hierarchically structured composite of Mn3O4/3D graphene foam for flexible nonenzymatic biosensors. Journal of Materials Chemistry B 2013, 1 (1), 110–115.

32 Xi, F.; Zhao, D.; Wang, X.; Chen, P. Non-enzymatic detection of hydrogen peroxide using a functionalized three-dimensional graphene electrode. Electrochemistry Communications 2013, 26, 81–84.

33 Zhan, B.; Liu, C.; Chen, H.; *et al.* Free-standing electrochemical electrode based on Ni(OH)2/3D graphene foam for nonenzymatic glucose detection. Nanoscale 2014, 6 (13), 7424–7429.

34 Ma, Y.; Zhao, M.; Cai, B.; Wang, W.; Ye, Z.; Huang, J., 3D graphene foams decorated by CuO nanoflowers for ultrasensitive ascorbic acid detection. Biosensors and Bioelectronics 2014, 59, 384–388.

35 Kung, C.-C.; Lin, P.-Y.; Buse, F. J.; *et al.* Preparation and characterization of three dimensional graphene foam supported platinum–ruthenium bimetallic nanocatalysts for hydrogen peroxide based electrochemical biosensors. Biosensors and Bioelectronics 2014, 52, 1–7.

36 Brownson, D.A.; Munro, L.J.; Kampouris, D.K.; Banks, C.E. Electrochemistry of graphene: not such a beneficial electrode material? RSC Advances 2011, 1 (6), 978–988.

37 Randviir, E.P.; Banks, C.E., The oxygen reduction reaction at graphene modified electrodes. Electroanalysis 2014, 26 (1), 76–83.

38 Brownson, D.A.; Gorbachev, R.V.; Haigh, S.J.; Banks, C.E. CVD graphene vs. highly ordered pyrolytic graphite for use in electroanalytical sensing. Analyst 2012, 137 (4), 833–839.

39 Uppal, A.; Ghosh, N.; Datta, A.; Gupta, P. K. Fluorimetric estimation of the concentration of NADH from human blood samples. Biotechnology and Applied Biochemistry 2005, 41 (1), 43–47.

40 Chemocare Hyperuricemia. http://chemocare.com/chemotherapy/side-effects/hyperuricemia-high-uric-acid.aspx (accessed 15 September 2016).

41 Brownson, D.A.; Banks, C.E. CVD graphene electrochemistry: the role of graphitic islands. Physical Chemistry Chemical Physics 2011, 13 (35), 15825–15828.

42 Šljukić, B.; Malakhova, N. A.; Brainina, K. Z.; *et al.* Screen printed electrodes and screen printed modified electrodes benefit from insonation. Electroanalysis 2006, 18 (9), 928–930.

43 Smith, J. P.; Metters, J. P.; Irving, C.; *et al.* Forensic electrochemistry: The electroanalytical sensing of synthetic cathinone-derivatives and their accompanying adulterants in 'legal high' products. Analyst 2014, 139 (2), 389–400.

44 Randviir, E.P.; Kampouris, D.K.; Banks, C.E. An improved electrochemical creatinine detection method via a Jaffe-based procedure. Analyst 2013, 138 (21), 6565–6572.

45 Metters, J.P.; Kadara, R.O.; Banks, C.E., New directions in screen printed electroanalytical sensors: an overview of recent developments. Analyst 2011, 136 (6), 1067–1076.

46 Ping, J.; Wu, J.; Wang, Y.; Ying, Y. Simultaneous determination of ascorbic acid, dopamine and uric acid using high-performance screen-printed graphene electrode. Biosensors and Bioelectronics 2012, 34 (1), 70–76.

47 Randviir, E.P.; Brownson, D.A.; Metters, J.P.; *et al.* The fabrication, characterisation and electrochemical investigation of screen-printed graphene electrodes. Physical Chemistry Chemical Physics 2014, 16 (10), 4598–4611.

48 Teixeira, S.; Conlan, R.S.; Guy, O.J.; Sales, M.G. Label-free human chorionic gonadotropin detection at picogram levels using oriented antibodies bound to graphene screen-printed electrodes. Journal of Materials Chemistry B 2014, 2 (13), 1852–1865.

49 Choudry, N.A.; Kampouris, D.K.; Kadara, R. O.; Banks, C.E. Disposable highly ordered pyrolytic graphite-like electrodes: Tailoring the electrochemical reactivity of screen printed electrodes. Electrochemistry Communications 2010, 12 (1), 6–9.

50 Brownson, D. A. C.; Varey, S. A.; Hussain, F.; *et al.* Electrochemical properties of CVD grown pristine graphene: Monolayer- vs. quasi-graphene. Nanoscale 2014, 6 (3), 1607–1621.

51 Zhou, M.; Wang, Y.; Zhai, Y.; *et al.* Controlled synthesis of large-area and patterned electrochemically reduced graphene oxide films. Chemistry: A European Journal 2009, 15 (25), 6116–6120.

52 Gao, X.; Jang, J.; Nagase, S. Hydrazine and Thermal reduction of graphene oxide: Reaction mechanisms, product structures, and reaction design. Journal of Physical Chemistry C 2010, 114 (2), 832–842.

53 Cui, F.; Zhang, X. Electrochemical sensor for epinephrine based on a glassy carbon electrode modified with graphene/gold nanocomposites. Journal of Electroanalytical Chemistry 2012, 669, 35–41.

54 Brownson, D.A.; Metters, J.P.; Kampouris, D.K.; Banks, C.E. Graphene electrochemistry: Surfactants inherent to graphene can dramatically effect electrochemical processes. Electroanalysis 2011, 23 (4), 894–899.

55 Nair, R.R.; Blake, P.; Grigorenko, A.N.; *et al.* Fine structure constant defines visual transparency of graphene. Science 2008, 320 (5881), 1308–1308.

56 Balandin, A.A.; Ghosh, S.; Bao, W.; *et al.* Superior thermal conductivity of single-layer graphene. Nano Letters 2008, 8 (3), 902–907.

57 Sheng, Z.-H.; Zheng, X.-Q.; Xu, J.-Y.; *et al.* Electrochemical sensor based on nitrogen doped graphene: Simultaneous determination of ascorbic acid, dopamine and uric acid. Biosensors and Bioelectronics 2012, 34 (1), 125–131.

58 Han, D.; Han, T.; Shan, C.; *et al.* Simultaneous determination of ascorbic acid, dopamine and uric acid with chitosan-graphene modified electrode. Electroanalysis 2010, 22 (17–18), 2001–2008.

59 Kovtyukhova, N.I.; Ollivier, P.J.; Martin, B.R.; *et al.* Layer-by-layer assembly of ultrathin composite films from micron-sized graphite oxide sheets and polycations. Chemistry of Materials 1999, 11 (3), 771–778.

60 Shan, C.; Yang, H.; Han, D.; *et al.* Graphene/AuNPs/chitosan nanocomposites film for glucose biosensing. Biosensors and Bioelectronics 2010, 25 (5), 1070–1074.

61 Li, J.; Guo, S.; Zhai, Y.; Wang, E. High-sensitivity determination of lead and cadmium based on the Nafion-graphene composite film. Analytica Chimica Acta 2009, 649 (2), 196–201.

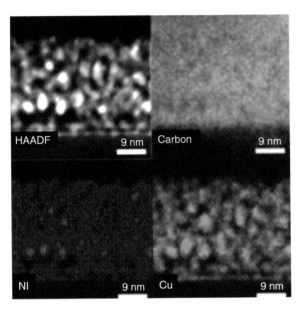

Figure 1.15 Nanostructural observation and elemental mapping obtained by HAADF-STEM and STEM-EDS measurements.

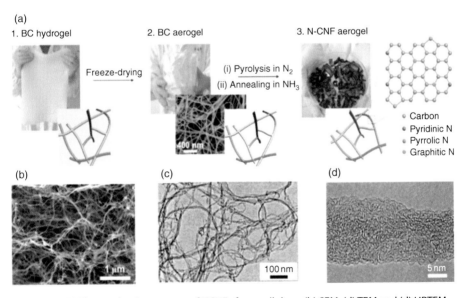

Figure 2.4 (a) The production process of NCNFs from cellulose; (b) SEM, (d) TEM and (d) HRTEM images of NCNFs. *Source:* Liang 2015 [32]. Reproduced with permission of Elsevier.

Nanocarbons for Electroanalysis, First Edition.
Edited by Sabine Szunerits, Rabah Boukherroub, Alison Downard and Jun-Jie Zhu.
© 2017 John Wiley & Sons Ltd. Published 2017 by John Wiley & Sons Ltd.

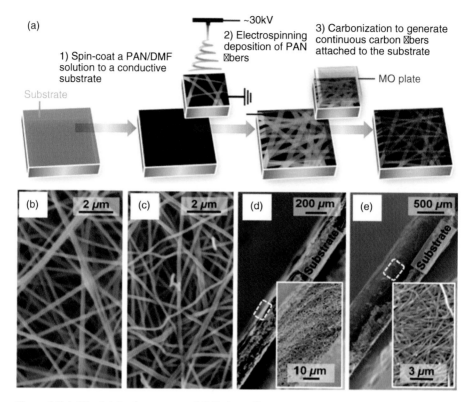

Figure 2.5 (a) The fabrication process of CNFs-based biosensor; SEM images of (b) PAN nanofibers and (c) CNFs; Cross-sectional SEM images of biosensors fabricated for different deposition times: (d) 12 h, (e) 68 h. *Source:* Mao 2014 [44]. Reproduced with permission of the American Chemical Society.

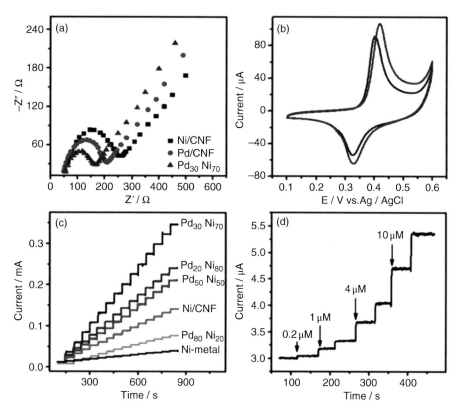

Figure 2.7 (a) Nyquist plots of Ni/CNFs, Pd/CNFs and Pd30Ni70/CNFs; (b) CVs of Pd30Ni70/CNFs in the presence (red) and absence (black) of glucose; (c) *I-t* curves of different electrodes with the addition of glucose; (d) *I-t* curve of Pd30Ni70/CNFs with the addition of varying concentrations of glucose. *Source:* Guo 2014 [60]. Reproduced with permission of the American Chemical Society.

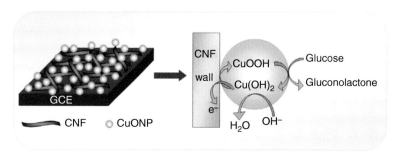

Figure 2.8 The mechanism of CuO/CNFs-based glucose biosensor. *Source:* Zhang 2013 [62]. Reproduced with permission of Elsevier.

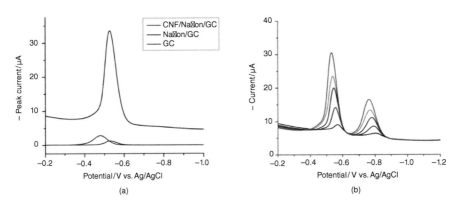

Figure 2.9 (a) Stripping voltammograms of CNF/Nafion/GCE, Nafion/GCE, and GCE in 0.1 M acetate buffer containing 0.5 µM Pb^{2+}; (b) stripping voltammograms of CNF/Nafion/GCE in 0.1 M acetate buffer with different concentration of Pb^{2+} and Cd^{2+}. *Source:* Zhao 2015 [69]. Reproduced with permission of the American Chemical Society.

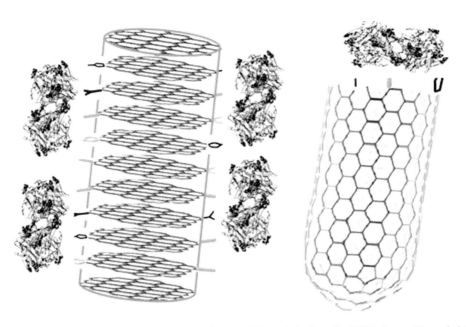

Figure 2.10 Immobilization of glucose oxidase on CNFs and single-walled CNTs. *Source:* Vamvakaki 2006 [10]. Reproduced with permission of the American Chemical Society.

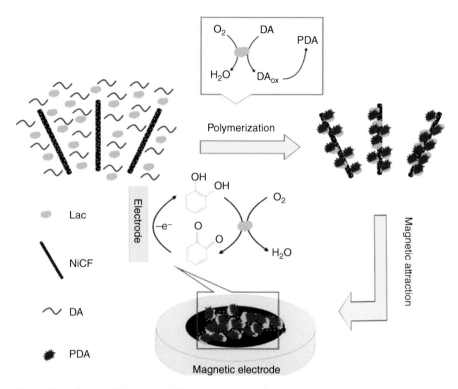

Figure 2.11 Scheme illustration of the construction of PDA/Lac/Ni/CNFs-based biosensors. *Source:* Li 2014 [87]. Reproduced with permission of the Royal Society of Chemistry.

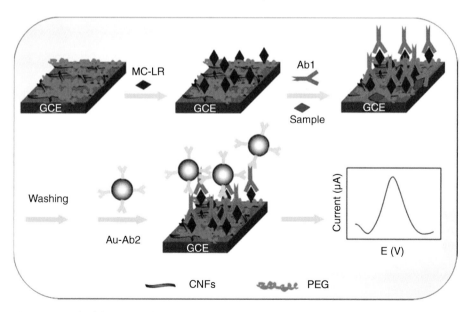

Figure 2.12 The fabrication of MC-LR immunosensor and its application to detect microcystin-LR. *Source:* Zhang 2016 [97]. Reproduced with permission of Elsevier.

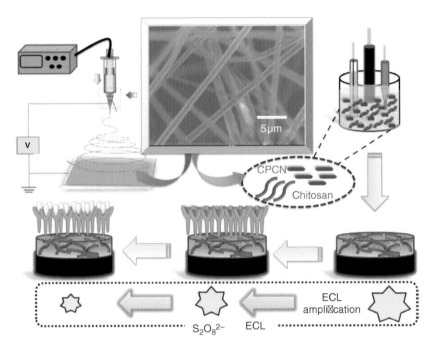

Figure 2.13 The fabrication process of ECL immunosensors based on CNTs@PNFs. *Source:* Dai 2014 [104]. Reproduced with permission of Elsevier.

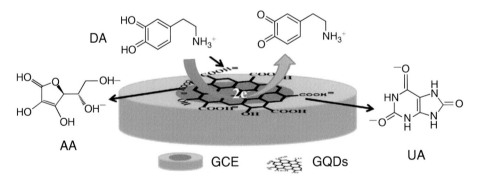

Figure 4.1 The schematic illustration of the reaction mechanism of DA on the GQDs–NHCH$_2$CH$_2$NH/GCE. *Source*: Li 2016 [16]. Reproduced with permission of Elsevier.

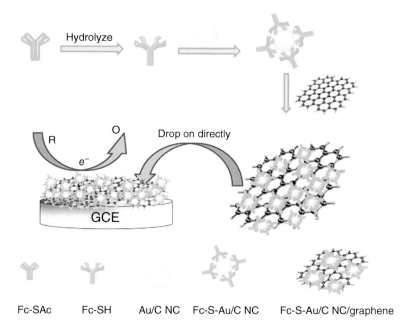

Fc-SAc   Fc-SH   Au/C NC   Fc-S-Au/C NC   Fc-S-Au/C NC/graphene

Figure 4.3 A quadruplet electrochemical platform for ultrasensitive and simultaneous detection of ascorbic acid, dopamine, uric acid and acetaminophen based on a ferrocene derivative functional Au NPs/carbon dots nanocomposite and graphene. *Source*: Yang 2016 [24]. Reproduced with permission of Elsevier.

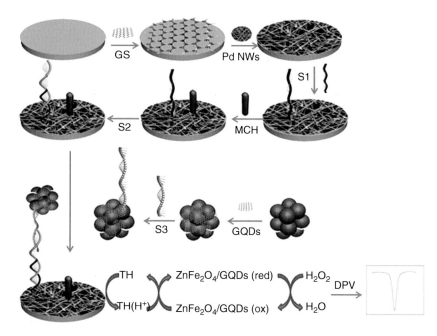

Figure 4.4 Graphene–palladium nanowires based electrochemical sensor using $ZnFe_2O_4$–graphene quantum dots as an effective peroxidase mimic. *Source*: Liu 2014 [26]. Reproduced with permission of Elsevier.

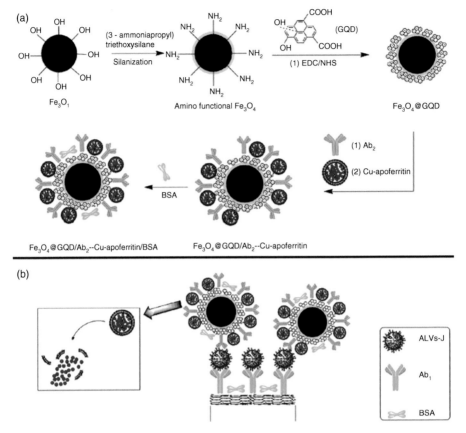

Figure 4.5 Electrochemical immunosensor with graphene quantum dots and apoferritin-encapsulated Cu nanoparticles double-assisted signal amplification for detection of avian leukosis virus subgroup J. *Source*: Wang 2013 [54]. Reproduced with permission of Elsevier.

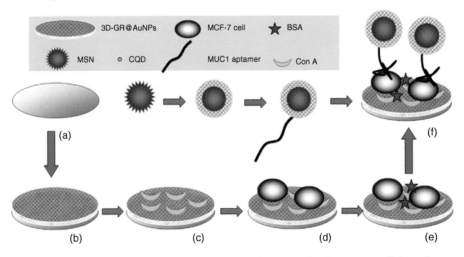

Figure 4.6 Aptamer-based electrochemiluminescent detection of MCF-7 cancer cells based on carbon quantum dots coated mesoporous silica nanoparticles. *Source*: Su 2014 [65]. Reproduced with permission of Elsevier.

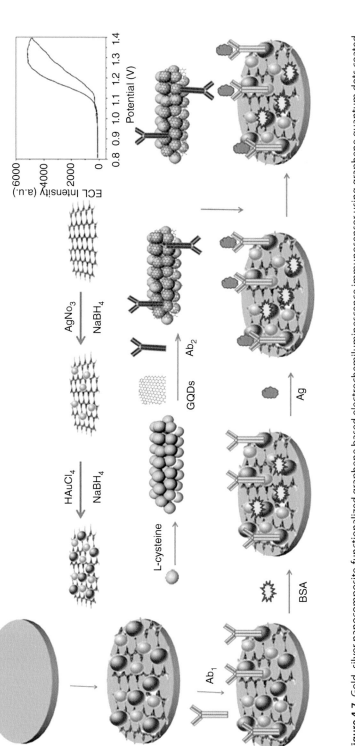

Figure 4.7 Gold–silver nanocomposite-functionalized graphene based electrochemiluminescence immunosensor using graphene quantum dots coated porous PtPd nanochains as labels. *Source:* Yang 2014 [67]. Reproduced with permission of Elsevier.

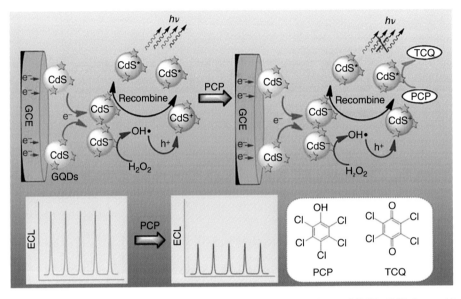

Figure 4.8 Illustrative ECL detection mechanism for PCP based on GQDs–CdS NCs/GCE. *Source*: Liu 2014 [69]. Reproduced with permission of the Royal Society of Chemistry.

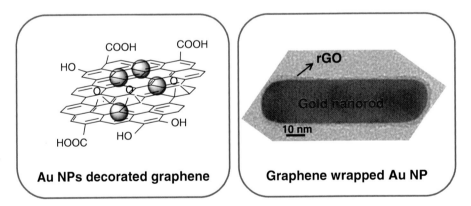

Figure 6.2 Two approaches for the integration of Au NPs with graphene: Au NPs *formed in situ* or *ex situ* on graphene nanosheets and graphene wrapped gold nanostructures (TEM image of a gold nanorod coated with a few layers of rGO). *Source*: Turcheniuk 2015 [18]. Reproduced with permission of the Royal Society of Chemistry.

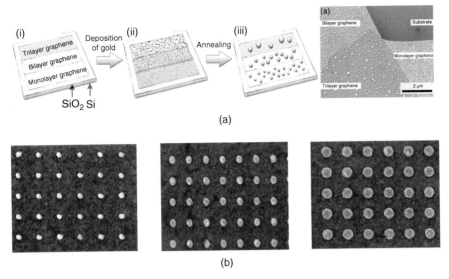

(a)

(b)

Figure 6.3 Decoration of CVD graphene with Au NPs: (a) Influence of the number of graphene layers on the morphology of the deposited nanostructures after annealing at 1260°C in vacuum for 30 s. *Source*: Zhou 2010 [22]. Reproduced with permission of the American Chemical Society. (b) SEM images of graphene decorated with Au NPs by e-beam lithography. The particles are 50 nm in height and 80 nm (i), 110 nm (ii), and 140 nm (iii) in diameter [20].

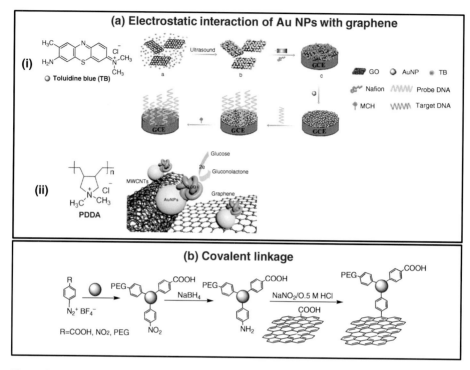

Figure 6.4 *Ex situ* approaches for the integration of Au NSs onto graphene nanosheets: (a) Electrostatic interactions: (i) Fabrication of a DNA sensor based on toluidine blue (TB) as molecular linkage between rGO and Au NPs *via* electrostatic interactions of the positively charged TB with negatively charged rGO. *Source*: Peng 2015 [23]. Reproduced with permission of Elsevier. (ii) Integration of positively charged PDDA capped Au NPs onto negatively charged rGO. Further integration of GODx resulted in the construcrtion of a glucose sensor. *Source*: Yu 2014 [24]. Reproduced with permission of Elsevier. (b) Covalent linkage between COOH groups of graphene and NH$_2$-functions on Au NPs. *Source*: Liu 2016 [26]. Reproduced with permission of Elsevier.

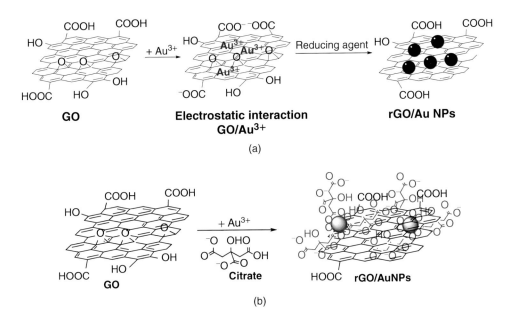

Figure 6.5 *In-situ* strategies for the formation of Au NPs loaded rGO matrixes: (a) electrostatic integration and reduction of GO/Au³⁺; (b) One-pot reduction and synthesis of rGO/Au NPs using citrate as reduction agent and capping agent of Au NPs.

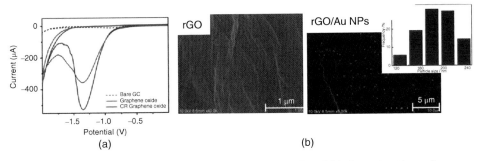

Figure 6.6 (a) Electrochemical activity of GO and chemically reduced GO determined by cyclic voltammetry (50 mM PBS, pH 7.4.) The bare glassy carbon (GC) electrode (dashed black line) is added for comparison. *Source*: Chng 2011 [58]. Reproduced with permission of John Wiley and Sons. (b) SEM images of electrochemically formed rGO on ITO and after electrodeposition of Au NPs. *Source*: Yang 2012 [49]. Reproduced with permission of the Royal Society of Chemistry.

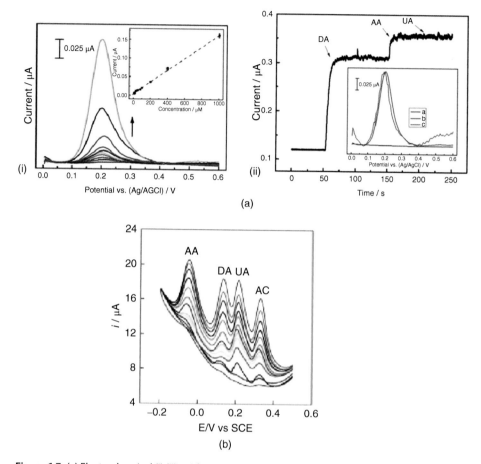

Figure 6.7 (a) Electrochemical (i) (ii) oxidation/reduction process of dopamine (DA); (i) DPV of dopamine (0-1000 μM) on ITO-rGO/Au NPs electrode; (ii) Interference study on ITO-rGO/Au NPs sensor for dopamine operated in amperometry mode at −0.3 V. *Source*: Yang 2012 [49]. Reproduced with permission of the Royal Society of Chemistry. (b) DPV profile of a mixture of AA, DA, UA, AC at ferrocene thiolate stabilized Fe_3O_4@Au NPs integrated on chitosan-modified rGO. *Source*: Liu 2013 [70]. Reproduced with permission of Elsevier.

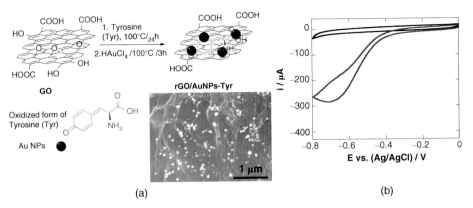

Figure 6.9 (a) Schematic illustration of the formation of rGO/Au NPs-Tyr nanocomposites together with an SEM image; (b) Cyclic voltammograms of rGO/Au NPs/Tyr/GC in N_2-saturated 0.1 M PBS solution (pH 7.4) in the absence (black) and presence (blue) of 10 mM H_2O_2; scan rate: 50 mV s^{-1} [91].

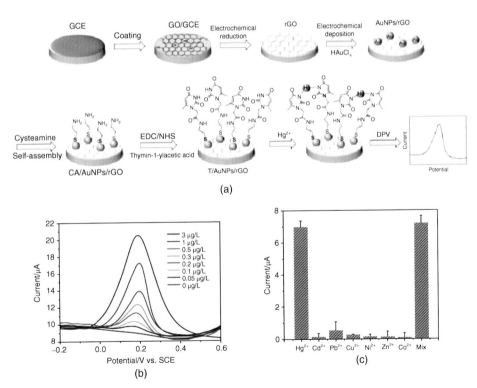

Figure 6.10 (a) Formation of thymidine functionalized rGO/Au NPs electrodes for Hg^{2+} detection; (b) DPV curves for dissolution of captured Hg^{2+}; (c) Electrochemical response to different metal ions. *Source*: Wang 2016 [48]. Reproduced with permission of Elsevier.

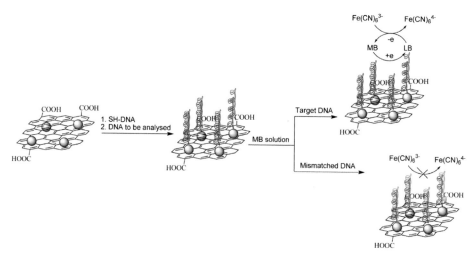

Figure 6.11 Schematic illustration of an electrochemical DNA sensor based on rGO/Au NPs modified glassy carbon electrode further modified by thiolated DNA ligands and using methylene blue (MB) to catalyze ferrocyanide reduction forming leucomethylene blue (LB). *Source*: Wang 2012[51]. Reproduced with permission of Royal Society of Chemistry.

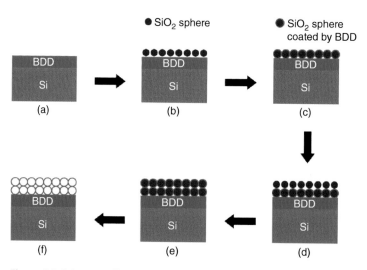

Figure 8.4 Schematic illustration showing the layer-by-layer growth technique for diamond foam electrodes showing: (a) boron-doped diamond (BDD) substrate growth on Si; (b) spin-coating of SiO_2 spheres on the substrate; (c) CVD diamond coating on SiO_2 templates; (d) spin-coating of the second SiO_2 layer, (e) the second CVD diamond coating on SiO_2 templates; (f) removal of SiO_2 templates. *Source*: Gao 2016 [47]. Reproduced with permission of the American Chemical Society.

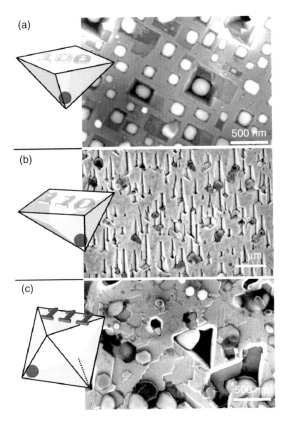

Figure 8.5 SEM images of observed pitting and channeling of (a): (100), (b): (110) and (c): (111) directed and etched single-crystal diamonds. The inset is a modeled octahedron which reflects symmetries of {111} oriented planes with indicated planes. Red circles indicate Ni. *Source*: Smirnov 2010 [55]. Reproduced with permission of AIP Publishing.

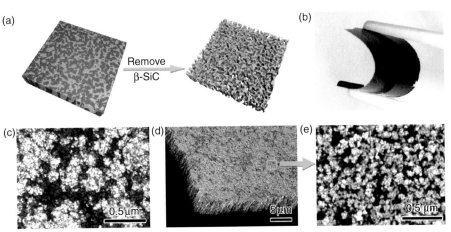

Figure 8.6 (a) Illustration of fabricating a diamond network from the composite film: the gray phase (β-SiC) is removed, leaving a yellow porous phase (diamond). (b) An optical photo of a flexible freestanding diamond network film. (c) SEM surface images of a nanocrystalline diamond/β-SiC composite film deposited with TMS/CH4 ratio of 1.5%. (d) Diamond network fabricated by etching the β-SiC phase from composite film shown in image c. (e) High-magnification SEM images of the surface of the film shown in image (d). *Source*: Zhuang 2015 [76]. Reproduced with permission of the American Chemical Society.

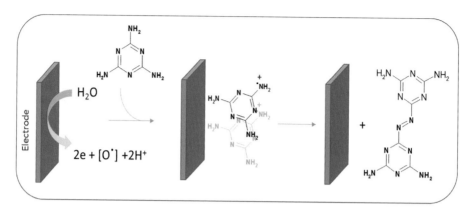

Figure 9.2 Representation for electrochemical preparation of g-C$_3$N$_4$ from melamine.

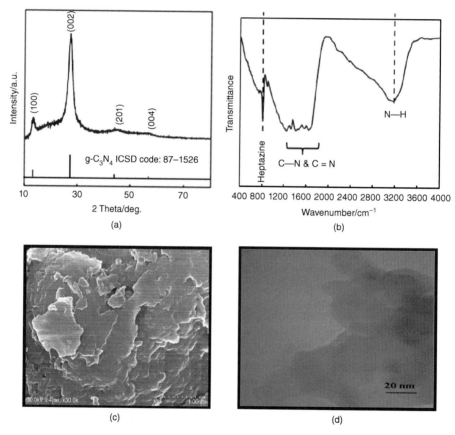

(a)

(b)

(c)

(d)

Figure 9.3 (a) XRD pattern of the g-C$_3$N$_4$ along with the standard pattern. (b) FTIR spectrum (c) SEM (d) TEM images of g-C$_3$N$_4$.

62 Li, J.; Guo, S.; Zhai, Y.; Wang, E. Nafion–graphene nanocomposite film as enhanced sensing platform for ultrasensitive determination of cadmium. Electrochemistry Communications 2009, 11 (5), 1085–1088.

63 Guo, C. X.; Lei, Y.; Li, C. M. Porphyrin functionalized graphene for sensitive electrochemical detection of ultratrace explosives. Electroanalysis 2011, 23 (4), 885–893.

64 Zhou, K.; Zhu, Y.; Yang, X.; Li, C. Electrocatalytic oxidation of glucose by the glucose oxidase immobilized in graphene-au-nafion biocomposite. Electroanalysis 2010, 22 (3), 259–264.

65 Zhang, Y.; Xiao, X.; Sun, Y.; *et al.* Electrochemical Deposition of nickel nanoparticles on reduced graphene oxide film for nonenzymatic glucose sensing. Electroanalysis 2013, 25 (4), 959–966.

6

Graphene/gold Nanoparticles for Electrochemical Sensing

Sabine Szunerits[1], Qian Wang[1,2], Alina Vasilescu[3], Musen Li[2] and Rabah Boukherroub[1]

[1] *Institute of Electronics, Microelectronics and Nanotechnology, Lille1 University, Villeneuve d'Ascq, France*
[2] *Key Laboratory for Liquid-solid Structural Evolution and Processing of Materials, Shandong University, Jinan, China*
[3] *International Center of Biodynamics, Bucharest, Romania*

6.1   Introduction

Different carbon materials are actually used in both analytical and industrial electrochemistry as they show wider potential window and are low cost materials in comparison to metal electrodes such as Au or Pt. Recently, graphene has captured great interest among physicists, chemists and materials scientists alike. It is also considered an ideal alternative material for the development of highly sensitive electrochemical based sensing platforms [1–5]. Graphene-based electrodes exhibit next to a large potential window, large 2-D electrical conductivity and fast heterogeneous electron transfer kinetics for various analytes; the involvement of additional 'chemical amplification effects' (e.g. in the case of aromatic structures such a dopamine interacting strongly with rGO via $\pi-\pi$ stacking interactions) has made these electrodes well adapted for electrochemical sensing (Figure 6.1).

These properties are not only offered by single-layer graphene, but also by multi-layer graphene, often termed stacked graphene platelets. Most importantly is that electrochemistry of graphene sheets is driven by its edges where heterogeneous electron transfer (HET) is fast. A defect-free basal plane is often electrochemically inert [6]. To increase the electron transfer rate on basal planes, different approaches can be undertaken. The incorporation of nanoparticles with graphene sheets to form graphene-nanoparticle hybrids is one of the most promising ways to tune the electrochemical properties of graphene-based sensors (Figure 6.1). Gold nanoparticles modified graphene nanosheets (G/Au NPs) are probably the most widely used sensing matrixes. The number of articles published since 2009 on the use of gold nanostructures decorated graphene for electrochemical sensing is steadily increasing. Au NPs of different shapes and sizes are readily accessible using solution-based techniques, have low toxicity and can be functionalized with sensing ligands. They are, therefore, well-suited for chemical and biological sensing applications. It has been argued that Au NSs could be used in fact in almost any domain related to sensing, ranging from electrochemical to optical, and fluorescence approaches. The exceptional optical properties of Au NSs,

Nanocarbons for Electroanalysis, First Edition.
Edited by Sabine Szunerits, Rabah Boukherroub, Alison Downard and Jun-Jie Zhu.
© 2017 John Wiley & Sons Ltd. Published 2017 by John Wiley & Sons Ltd.

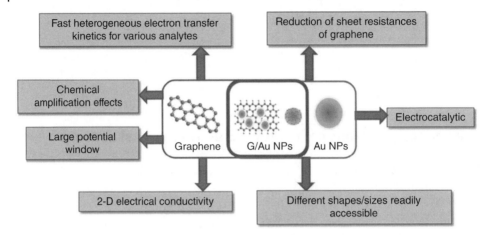

Figure 6.1 Synergetic effect of graphene and Au NPs.

including large optical field enhancements and their addressability *via* spectroscopic techniques have made them of particular interest as labels and for biosensing. The optical properties of Au NSs are dominated by the excitation of collective oscillations of the nanoparticles' conduction band electrons, called localized surface plasmon resonances (LSPR), by the incoming electromagnetic waves [7, 8]. As the position of the LSPR band is influenced by the refractive index of the surrounding sensing medium, such structures have been intensively used to detect molecular-binding events in the nano-environment of the particles [7, 9–11]. The other way to employ optical properties of Au NPs for sensing is to use discrete particle aggregation in response to biological elements, which results in a measurable color change. These colorimetric sensing systems exhibited relatively low detection limits with a high degree of sensitivity [12, 13]. The use of colloidal Au NPs as staining agents has become a standard procedure in biological microscopy and several reviews have discussed the use of Au NPs in a variety of diagnostic and therapeutic applications [12, 14–16].

For electronic- and electrochemical-based applications, the high electron mobility of graphene combined with the electrocatalytic properties of Au NPs and the possibility of reducing the sheet resistance of rGO-based devices have made G/Au NPs materials of high interest. The integration of Au NPs into graphene nanosheets allows in certain cases to efficiently improve the sensitivity and detection limit of the sensor due to enhancement of the electron transfer or through its catalytic reaction with certain analytes. This is the principal driving force for the growing research interest in graphene-metal nanohybrids. For the development of sensors, next to good electrical properties, the role of nanocomposites made of graphene and Au NPs also encompassed high loading and efficient immobilization of biorecognition elements with preservation of their activity and affinity.

Giving the growing interest in G/Au NPs hybrid structures, this chapter will survey the emerging applications of G/Au NPs scaffolds for electrochemical based sensing applications. Emphasis will be first placed on the currently available procedures for the embedding of Au NPs onto and into graphene-based nanosheets. In particular, we aim to provide a comprehensive review that covers the latest and most significant developments in this field and offer insights into future perspective. It is thus hoped that this chapter will inspire further interest from various disciplines interested in selective and sensitive electrochemical sensing platforms.

6.2 Interfacing Gold Nanoparticles with Graphene

There are currently various synthetic routes for the preparation of Au NPs decorated graphene nanocomposites [17]. Based upon the structural morphology of the final hybrid, these synthetic methods can be divided into two classes: Au NPs dispersed on graphene sheets and Au NPs wrapped by graphene and its derivatives (Figure 6.2). The main difference is the relative surface ratio between the Au NPs and the lateral dimensions of graphene. When the size of the Au NPs is in the range of a few nanometers, the particles are small and can be easily deposited onto the graphene nanosheets. When the Au NPs size becomes comparable with that of the 2D graphene sheets, wrapping around the particles occurs preferentially.

When it comes to electrochemical sensing applications, graphene wrapped Au NPs are rarely considered until now [19]. One of the only reports is that of Kim *et al.*, who used a substrate composed of 3D GO-encapsulated Au NPs to distinguish the differentiation stage of single neural stem cells using electrochemical and electrical techniques [19]. Furthermore, while deposition of Au NPs on the inert surface of CVD graphene can be performed using thermal evaporation, e-beam lithography or sputtering methods (Figure 6.3) [20–22], such approaches mostly result in higher costs for the fabrication of sensors and were not investigated until now as well.

However, a variety of *in-situ* and *ex-situ* approaches for the formation of Au NPs on GO and rGO matrixes have been considered as interfaces for further sensing applications. In the *in situ* approach, simultaneous chemical reduction of GO and Au salt is carried out in order to synthesize rGO/Au NPs composite materials while the *ex situ* approach involves the synthesis of Au NPs of desired size and shape and their transfer onto graphene, GO or rGO matrix.

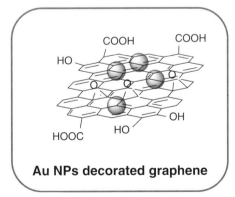 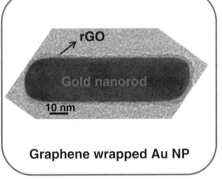

Au NPs decorated graphene   **Graphene wrapped Au NP**

Figure 6.2 Two approaches for the integration of Au NPs with graphene: Au NPs *formed in situ* or *ex situ* on graphene nanosheets and graphene wrapped gold nanostructures (TEM image of a gold nanorod coated with a few layers of rGO). *Source*: Turcheniuk 2015 [18]. Reproduced with permission of the Royal Society of Chemistry. (*See color plate section for the color representation of this figure.*)

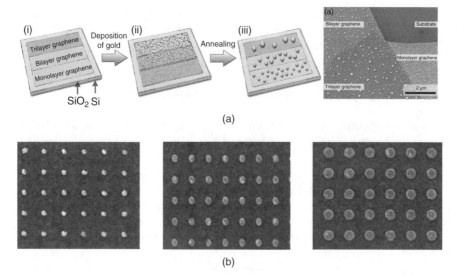

Figure 6.3 Decoration of CVD graphene with Au NPs: (a) Influence of the number of graphene layers on the morphology of the deposited nanostructures after annealing at 1260°C in vacuum for 30 s. *Source*: Zhou 2010 [22]. Reproduced with permission of the American Chemical Society. (b) SEM images of graphene decorated with Au NPs by e-beam lithography. The particles are 50 nm in height and 80 nm (i), 110 nm (ii), and 140 nm (iii) in diameter [20]. (*See color plate section for the color representation of this figure.*)

6.2.1 *Ex-situ* Au NPs Decoration of Graphene

In the *ex-situ* method, the Au NPs are synthesized separately from graphene nanosheets and are immobilized onto the graphene sheet through Au-NH and Au-S chemical bond formation with functional groups on graphene and its derivatives or by utilizing electrostatic, $\pi-\pi$ and/or van der Waals interactions for binding. The ex-*situ* method offers good control over size, shape and functionality of the Au NPs, being one of the main advantages of this approach. One of the limitations is that this approach is rather time consuming and involves several steps.

Toluidine Blue (TB) modified GO nanosheets were proposed by Peng *et al.* as an interesting matrix for linking Au NPs [23]. TB not only acts as an electrochemical active mediator due to its good electrochemical redox active properties, but also as a linker to connect Au NPs to rGO sheets due to the interaction between the amine groups in TB and Au NPs (Figure 6.4.a). Electrostatic interactions between positively charged poly(dialkyldimethylammonium chloride (PDDA) capped Au NPs and negatively charged rGO/MCNTs nanocomposites has been proposed by Yu *et al.* [24] (Figure 6.4.a). PDDA is not only an electronic conducting polymer but also a positively charged ionic matrix, used to cappe Au NPs. The positively charged Au NPs where then absorbed onto the negatively charged rGO/MWCNTs nanocomposites. Further integration of glucose oxidase (GOD) through electrostatic interaction with the Au NPs allowed to develop a glucose sensor.

The use of cysteine as linker molecule is in particular attractive as it contains both thiol and amino groups, linking to the carboxylic acid groups of graphene via its NH_2 group and to Au NRs *via* SH group, creating uniform distribution of Au NPs on the graphene interfaces.

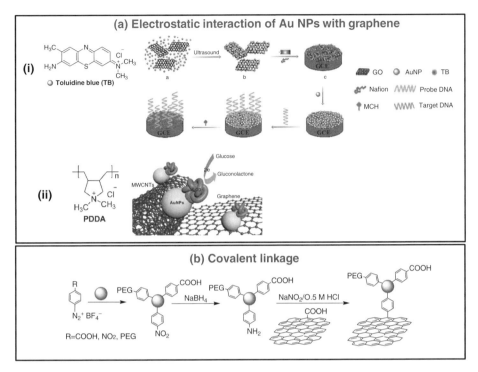

Figure 6.4 *Ex situ* approaches for the integration of Au NSs onto graphene nanosheets: (a) Electrostatic interactions: (i) Fabrication of a DNA sensor based on toluidine blue (TB) as molecular linkage between rGO and Au NPs *via* electrostatic interactions of the positively charged TB with negatively charged rGO. *Source*: Peng 2015 [23]. Reproduced with permission of Elsevier. (ii) Integration of positively charged PDDA capped Au NPs onto negatively charged rGO. Further integration of GODx resulted in the construcrtion of a glucose sensor. *Source*: Yu 2014 [24]. Reproduced with permission of Elsevier. (b) Covalent linkage between COOH groups of graphene and NH_2-functions on Au NPs. *Source*: Liu 2016 [26]. Reproduced with permission of Elsevier. (*See color plate section for the color representation of this figure.*)

The use of diazonium salts (e.g. 4-nitrobenzenediazonium tetrafluoroborate; 4-carboxyphenyl tetrafluoroborate) as molecular linkers to either graphene or Au NPs is another widely used strategy [25, 26]. For example, an electrochemical sensing interface composed of monolayered Au NPs chemically bound to rGO sheets using as molecular linker was proposed by Liu *et al.* lately [26]. Different diazonium salts were spontaneously anchored to Au NPs. Reduction of the 4-nitrobenzene groups attached to Au NPs to NH_2 groups was then used to anchor the Au NPs covalenlty to rGO *via* its carboxylic acid functions (Figure 6.4.b).

6.2.2 *In-situ* Au NPs Decoration of Graphene

While the *ex-situ* approach is well adapted to form controlled Au NSs loaded GO and rGO, most work focused on *in situ* approaches. The advantages of the *in situ* reduction is that generally there is no need for the use of capping agents or extra linker molecules, which can have a negative influence on the charge transfer characteristics. It is cost-effectuve and a one-pot synthesis. These advantages seem to override the limitation in

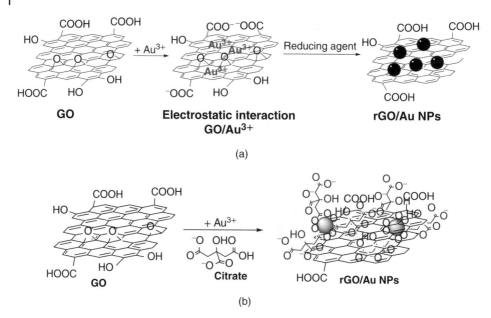

Figure 6.5 *In-situ* strategies for the formation of Au NPs loaded rGO matrixes: (a) electrostatic integration and reduction of GO/Au³⁺; (b) One-pot reduction and synthesis of rGO/Au NPs using citrate as reduction agent and capping agent of Au NPs. (*See color plate section for the color representation of this figure.*)

controlling size and morphology of the embedded Au NPs in the resulting composites. In general, a GO suspension is mixed with HAuCl₄ and the mixture is treated with a reducing agent (e. g. sodium borohydride, sodium citrate, hydrazine hydrate, ascorbic acid, glucose, etc.) resulting in the simultaneous reduction of Au^{3+} solution to Au^0 and GO to rGO (Figure 6.5) [27, 28]. Although the mechanism of reduction is not clearly established, involvements of three steps are hypothesized: the oxygen functionalities of GO provide the reactive nucleation sites. After reduction of Au^{3+} cations by the reducing agent such as citrate [29–31], sodium borohydride [32–34], ascorbic acid [35–37], tyrosine [38], poly(diallyldimethylammonium chloride) (PDDA) [39], sodium dodecyl sulfate(SDS) [40], ionic liquid [41], as well as sonolysis [42, 43], the growth of Au particles is initiated with the nanoparticles remaining attached to the GO sheet. The reduction agents simultaneously reduce GO to rGO, resulting in the formation of rGO/Au NPs nanocomposites. While so far it is impossible to distinguish the role of each oxygen containing group on the nucleation of Au NPs, oxygen-rich GO platforms are reported to promote Au NPs integration [34]. As a proof, thermally reduced GO, hydrazine reduced GO and conventional GO were used in a study to estimate the influence of level of reduction on the nucleation-growth [27]. Scanning electron microscopy (SEM) analysis of obtained graphene/Au NPs composites revealed the absence of Au NPs in case of hydrazine reduced GO, a homogeneous distribution of discrete Au NPs on the surface of GO and agglomerated Au NPs with a scarce distribution on the surface of thermally heated GO. Furthermore, Song *et al.* [44] reported that GO surfaces enriched with carboxylic acid groups did not facilitate Au seeding, while thiol-functionalized GO [44, 45] possesses high affinity to gold allocating the seeds in a highly ordered manner.

The use of sodium citrate as environmentally friendly reducing and stabilizing agent has become a widely used approach for the synthesis of rGO/Au NPs composites [31]. In this one pot synthesis, citrate ions act as a capping and reducing agent for the Au ions and GO. Au NPs of ~7 nm in size are uniformly distributed over the rGO nanosheets. A green *in situ* approach for the synthesis of aqueous stable rGO/Au NPs hybrids using PVP as a stabilizer and ascorbic acid as a reducing gent for both GO and Au ions was proposed by Iliut *et al.*, more recently [35]. The *in situ* growth of Au NPs on the surface of poly(diallyldimethylammonium chloride) (PDDA) modified GO using $NaBH_4$ as a reducing agent improved the dispersion of the Au NPs [46].

Next to the formation of rGO/Au NPs nanocomposites during reduction processes in mixed solution, the direct reduction of Au^{3+} on rGO was reported [47]. rGO thin films, prepared by GO reduction with hydrazine, were deposited by vacuum filtering a solution of rGO suspension on quartz and immersed in an aqueous solution of $HAuCl_4$. During the immersion process, Au NPs were produced by spontaneous reduction of Au^{3+} on the rGO sheets. The mechanism of reduction of Au^{3+} ions on rGO involves galvanic displacement and redox reaction by relative potential difference. It is believed that Au NPs deposition was promoted by electrons present on the negatively charged rGO. This was further supported by the negative zeta potential value (-39 mV) measured for the rGO aqueous suspension at pH 7, and the reduction potential of $+0.38$ V versus SHE estimated from ultraviolet photoelectron spectroscopy (UPS). This value is much lower than $+1.002$ V versus SHE corresponding to the reduction potential of $AuCl_4^-$, suggesting a spontaneous reduction of Au^{3+} ions on rGO.

6.2.3 Electrochemical Reduction

Electrochemical deposition is an efficient and green technique for the synthesis of rGO/Au NPs nanocomposites and has found to be rather appealing for the construction of graphene/Au NPs sensors [48–55]. Several strategies have been employed. Au NPs can be electrochemically deposited onto the surface of rGO under anodic scanning from 0.2 V to 1 V in 0.25 mM $HAuCl_4$ at 50 mV s^{-1}. Electrochemical reduction of GO into rGO following the suggested pH dependent mechanism:

$$GO + aH^+ + be^- \rightarrow rGO + cH_2 \text{ [56 – 57]}$$

followed by electrodeposition of Au NPs onto rGO is another possibility [52, 55]. Indeed, GO shows intrinsic electrochemical behavior as seen in Figure 6.6a, where the reduction wave is strongly dependent on the GO used [58, 59].

Yang *et al.* used this approach for the coating of ITO interfaces with rGO and sweeping the potential from 0.0 V to -1.5 V at a scan rate for 0.1 V s^{-1} for 100 cycles in deaerated 0.05 M PBS (pH 5) buffer [49]. Figure 6.6b displays a SEM image of the electrochemically reduced GO showing overlapped, corrugated and crumpled rGO nanosheets covereing the whole ITO interface. Au NPs could be formed on this rGO films by applying a constant potential of -0.2 V for 100 s in a $HAuCl_4$ (2.5 mM) solution (Figure 6.6b).

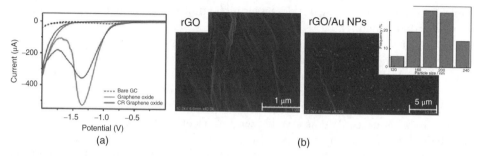

Figure 6.6 (a) Electrochemical activity of GO and chemically reduced GO determined by cyclic voltammetry (50 mM PBS, pH 7.4.) The bare glassy carbon (GC) electrode (dashed black line) is added for comparison. *Source*: Chng 2011 [58]. Reproduced with permission of John Wiley and Sons. (b) SEM images of electrochemically formed rGO on ITO and after electrodeposition of Au NPs. *Source*: Yang 2012 [49]. Reproduced with permission of the Royal Society of Chemistry. (*See color plate section for the color representation of this figure.*)

6.3 Electrochemical Sensors Based on Graphene/Au NPs Hybrids

Electrochemical sensors are by far the largest groups of sensors where G/Au NPs hybrids have been employed. Up to now a large variety of different analytes have been selectively sensed on graphene/Au NPs modified interfaces (Table 6.1) and the list is constantly increasing. These electrodes, when modified with the ligand of interest, have shown to be useful for the screening of DNA hybridization events, for the detection of specific proteins and biomarkers often based on graphene/Au NPs immunosensors, as well as for the sensing of small analytes such as dopamine, uric acid, glucose, serotonin, cholesterol, or different pesticides. In addition, these interfaces have shown their interest for the sensing of heavy metal ions. Sensing using graphene/Au NPs hybrids is probably the electrical interface with the most widely and widespread interest. This is based on the fact, that several analytes are more sensitively detected on these hybrid interfaces. For example, compared to a bare glassy carbon electrode (GCE), electrodes modified with poly(diallyldimethylammonium chloride) (PDDA) functional graphene/ Au NPs nanosheets, obtained by one-pot synthesis, provided both a 10^2 fold increase in the magnitude of the electrochemical signal for the detection of uric acid (UA) and a selectivity due to the excellent resolution of the DPV signals [60]. The reason was due to the presence of polycationic polyelectrolyte PDDA in the nanocomposite, concentrating UA at electrode surface by electrostatic interactions, and thus increasing the sensitivity and selectivity of detection. In the following, the sensing of different analytes using graphene/Au NPs electrodes will be discussed in more depth.

6.3.1 Detection of Neurotransmitters: Dopamine, Serotonin

Dopamine, chemically known as 3, 4-dihydroxy-L-phenylalanine, is one of the crucial catecholamine neurotransmitter widely distributed in mammalian brain tissues. It has a great influence on the central nervous, cardiovascular and endocrine system. Abnormal levels of dopamine can result in a variety of diseases such as Huntington's disease, Parkinson's disease or Schizophrenia. High dopamine levels can in addition

Table 6.1 Detection of several different organic and inorganic analytes.

Electrode	Analyte	Sensor characteristics	Ref.
rGO/Au NPs	Dopamine	LOD: 0.02 μM LR: 1.0-60 μM	[69]
G–PEI/AuNPs	Dopamine	LOD: 0.2 μM LR: 2.0-48 μM	[98]
rGO/Au NPs	Dopamine	LOD: 0.06 μM	[49]
Au NPs/β-cyclodextrin/ graphene	Dopamine Ascorbic acid Uric acid	LOD: 0.15 μM LR: 0.5-150 μM LOD: 10 μM LR: 0.03-2 mM LOD: 0.21 μM LR: 0.5-60 μM	[99]
rGO/Au NPs	Butyl-hydroxyanisole, tert-butylhydroquine	LOD: 41.9 ng mL^{-1} LR: 0.1-10 μg mL^{-1}	[55]
N-doped graphene aerogel/ Au NPs	Hydroquinone *o*-dihydrobenzene	LOD: 15 nM (HQ) LOD: 3.3 nM (DHB)	[71]
rGO/Au NPs	Hydroquinone resorcinol	LOD: 5.2 nM LOD: 2.3 resorcinol	[54]
rGO/Au NPs	Ascorbic acid	LOD: 100 nM LR: 0.11-0.6 mM	[100]
rGO/Au NPs-enzyme	Cholesterol	LOD: 0.05 μM LR: 0.05-0.35 mM S: 3.14 μA μM^{-1} cm^{-2}	[81]
PDDA-rGO/ AuNPs- aptamer	Ractopamine	LOD: 0.5 fM LR: 1 fM-10 μM	[101]
PDDA-rGO/AuNPs-enzyme	Paraoxon	LOD: 0.1 pM LR: 0.1 pM-5nM	[46]
rGO/AuNPs-antibody	17-β-estradiol	LOD: 0.1 fM LR: 1 fM-1 mM	[82]
Graphene/Au NPs	Diethylstilboestrol	LOD: 9.8 nM LR: 12 nM-12 μM	[102]
Graphene/Au Ps	Carbamazepine (antiepileptic drug)	LOD: 3.03 μM LR: 5 μM-10 mM	[103]
N-doped graphene/Au NPs	Chloramphenicol	LOD: 0.59 μM LR: 2 μM-80 μM	[83]
Graphene-DPB/ AuNPs-antibody	Aflatoxin B1	LOD: 1 fM LR: 3.2 fM–0.32 pM	[104]
Graphene/Au NPs-Anti-BPA	Bisphenol A	LOD: 5 nM LR: 0.01–10.0 μM	[105]

(Continued)

Table 6.1 (Continued)

Electrode	Analyte	Sensor characteristics	Ref.
graphene nanofibers/Au NPs	Bisphenol A	LOD: 35 nM LR: 80 nM-0.25 mM	[106]
carbon ionic liquid-rGO/Au NPS	Folic acid	LOD: 2.7 nM LR: 0.01-50 µM	[107]
rGO/Au NPs	Toxicant Sudan I	LOD: 1 nM LR: 0.01-70 µM	[108]
rGO/Au NPs	NADH	LOD: 1.13 nM LR: 50 nM-500 µM $S = 0.92\,\mu A\,\mu M^{-1}\,cm^{-2}$	[88]
rGO/Au NPs	NADH	LOD: 3.5 µM LR: 0.01-5 mM	[89]
rGO/Au NPs-chitosan-GODx	Glucose	LOD: 180 µM LR: 2-14 mM $S: 99.5\,\mu A\,\mu M^{-1}\,cm^{-2}$	[28]
MWCNT-rGO/Au NPs	Glucose	LOD: 0.8 µM LR: 5-175 µM $S: 29.72\,\mu A\,\mu M^{-1}\,cm^{-2}$	[24]
GO-Ag/Au NPs-mercaptophenylboronic acid	Glucose	LOD: 0.33 mM LR : 2-6 mM	[109]
GO-thionine-Au NPs	Glucose	LOD: 0.05 µM LR: 0.2-22 µM 0.2-13.4 mM	[78]
Au NPs/nitrogen-doped graphene	Glucose Dopamine	LOD: 12 µM LR: 0.04-16.1 mM LOD: 0.01 µM LR: 0.03-48 µM	[110]
Au NPs/polypyrrole/rGO	Glucose	LR: 0.2-1.2 mM $S: 123.8\,\mu A\,mM^{-1}\,cm^{-2}$	[111]
Au NPs/GO nanoribbon/carbon sheet	Glucose	LOD: 5 µM LR: 0.005-10 mM	[79]
N-doped graphene quantum dots/Au NPs	Hydrogen peroxide	LOD: 0.12 µM	[94]
Au NPs/rGO	Hydrogen peroxide	LOD: 0.1 µM LR: 0.02-10 mM	[112]
Hemin-graphene nanosheets/Au NPs	Hydrogen peroxide	LOD: 0.11 µM LR: 0.0003-1.8 mM	[113]
Au NPs/graphene-chitosan	Hydrogen peroxide	LOD: 1.6 µM LR: 0.005-35 mM	[114]

Table 6.1 (Continued)

Electrode	Analyte	Sensor characteristics	Ref.
rGO-Hemin-Au NPs	Hydrogen peroxide	LOD: 30 nM LR: 0.1-40 µM	[115]
hemoglobin/Au NPs/ZnO/graphene	Hydrogen peroxide	LOD: 0.8 µM LR: 6-1130 µM	[116]
rGO/Au NPs/poly(toluidine blue O)	Hydrogen peroxide	LOD: 0.2 µM LR: 0.002-1.077 mM	[93]
graphene sheets@CeO_2/Au NPs	Hydrogen peroxide	LOD: 0.26 µM LR: 0.001-10 mM	[117]
Au NPs/PDDA/rGO	Hydrogen peroxide	LOD: 0.44 µM LR: 0.5-500 µM	[90]
Au NPs/sulfonated graphene sheets	Hydrogen peroxide	LOD: 0.25 µM LR: 2.3-16 mM	[118]
rGO/Au NPs-aptamer	TNT	LOD: 3.6 pg mL^{-1} LR: 0.01-100 ng mL^{-1}	[86]
rGO/Au NPs (UV irradiated)	TNT	LR : 5-11.5 µg L^{-1}	[85]
Ionic liquid-rGO/Au NPs	Hg^{2+}	LR: 0.1-100 nM LOD: 0.03 nM	[96]
Au NPs/graphene	Cu^{2+}	LR: 5-100 nM LOD: 0.028 nM	[119]
Chitosan-rGO-/Au NPs- Hemoglobin	Nitrite	LOD: 0.01 µM LR: 0.05 – 1000 µM	[120]
SDS/BPG-graphene/Au NPs-Hb	Nitric oxide	LOD: 12 nM LR: 0.72-7.92 µM	[121]
Au NPs/rGO	Nitric oxide	LOD: 0.133 µM LR: up to 3.38 µM	[122]

BPA = Bisphenol A; BPG = basal plane graphite; DHB = o-dihydrobenzene; DPB = 2-(3, 4-dihydroxy phenyl) benzothiazole; GODx = glucose oxidase; Hb = Hemoglobin HQ = hydroquinone =; LR = linear range; LOD = limit of detection; S = sensitivity; PDDA = poly (dimethyl diallyl ammonium chloride).

lead to high blood pressure, which can cause health problems. Use of dopamine in animal feed has been banned in most countries and it is essential to prevent its illegal use. Rapid and accurate detection of dopamine at low costs has thus become of demand in clinical diagnostics as well as to control the safety of the meat food chain. The high electrochemical activity of dopamine makes its electrochemical detection rather appealing. Dopamine oxidizes in a two electron process with a transfer of two protons (Figure 6.7a). However, uric acid (UA) and ascorbic acid (AA) are coexisting with dopamine in the extracellular fluids of the central nervous system in mammals and can be oxidized at a potential close to that of dopamine. The use of graphene based electrodes has shown to enhance significantly the voltammetric selectivity towards dopamine [61–64]. The significant enhancement in k^0 using dopamine as redox agent and the

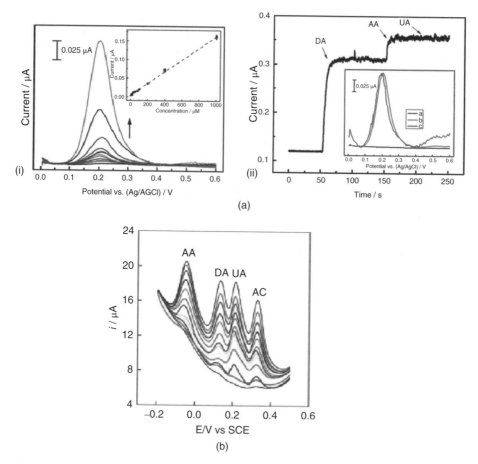

Figure 6.7 (a) Electrochemical (i) (ii) oxidation/reduction process of dopamine (DA); (i) DPV of dopamine (0-1000 μM) on ITO-rGO/Au NPs electrode; (ii) Interference study on ITO-rGO/Au NPs sensor for dopamine operated in amperometry mode at −0.3 V. *Source*: Yang 2012 [49]. Reproduced with permission of the Royal Society of Chemistry. (b) DPV profile of a mixture of AA, DA, UA, AC at ferrocene thiolate stabilized Fe$_3$O$_4$@Au NPs integrated on chitosan-modified rGO. *Source*: Liu 2013 [70]. Reproduced with permission of Elsevier. (*See color plate section for the color representation of this figure.*)

better electrochemical reversibility of dopamine on graphene based electrodes is due to the involvement of an additional 'chemical' amplification effect [65–68]. The aromatic structure of dopamine interacts strongly with rGO though π−π stacking interactions, being responsible for enhanced sensitivity as well as selectivity in electrochemical detection. Jiang and co-workers proposed a multilayer composed of layer-by-layer assembly of polysodium-4-styrenesulfonte (PSS) functionalized rGO and polyamidoamine dendrimer stabilized Au NPs as dopamine sensor with a detection limit of 0.02 μM and a learn rate from 1–60 μM [69]. Yang co-workers proposed ITO electrodes coated by rGO/Au NPs for the amperometric sensing of dopamine where dopamine can be detected with good selectivity over ascorbic acid and uric acid and a LOD of 60 nM (Figure 6.7b) [49]. Liu *et al.* showed the interest of ferrocene thiolate stabilized

Fe$_3$O$_4$@Au NPs integrated on chitosan-modified rGO for the sensing of dopamine next to ascorbic acid (AA), uric acid (UA) and acetaminophen (AC) (Figure 6.7c) [70]. The same principle was applied for the detection of hydroquinone or resorcinol, a benzene diol derivative [54, 71].

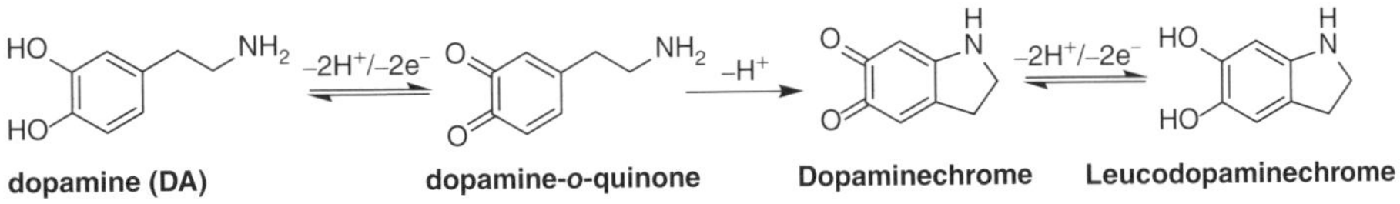

Scheme 6.1 Electrochemical oxidation/reduction process of dopamine (DA)

What has to be kept in mind is that mixing GO under basic condition with dopamine, results in the formation of poly(dopamine) capped rGO. Loading with Au NPs by electrochemical deposition allowed the sensing of cysteine through the formation of Au-S bonds and Michael addition products and a consequent decrease in redox current of Ru(NH$_3$)$_6$$^{3+}$ [72].

Another important neurotransmitter is serotonin, also known under the name 5-hydroxytryptamine. Serotonin is widely dispersed throughout the central nervous system, playing crucial roles in the regulation of mood, sleep and appetite. The analysis of serotonin level is of great value in the diagnostics of several diseases. Like dopamine it is electroactive and its presence can be analyzed by electrochemical methods. However, some factors limited the electrochemical detection of serotonin under physiological conditions. One is that the concentration of serotonin is too low to be detected in human fluidics; the other is that the redox potential of ascorbic acid, uric acid and dopamine overlap with that of serotonin. To resolve this problem, a rGO/polyaniline interface further modified with a serotonin imprinted polymer with embedded Au NPs was proposed as a sensing matrix. Differential pulse voltammetry in the presence of serotonin showed that the sensor was sensitive to 0.2–10 µM concentrations with a LOD of 11.7 nM [73].

6.3.2 Ractopamine

Ractopamine, 4-[3-[[2-Hydroxy-2-(4-hydroxyphenyl)ethyl]amino]butyl]phenol, is like dopamine, a hydroxyphenyl derivative and is a β-agonist, previously used as additive in animal feeds for decreasing deposition of fat and increasing protein accumulation in animal meat. Due to its toxicity for human health, ractopamine was banned as additive use in animal feed in many countries. A signal-off aptasensor for the β-agonist ractopamine was developed by modifying GCEs first with PDDA modified rGO followed by citrate-capped Au NPs [74]. Thiolated ractopamine aptamer was immobilized on the electrode and ractopamine detection was accomplished by DPV, with [Fe(CN)$_6$]$^{4-/3-}$ as redox probe. The decrease in oxidation current intensity recorded by DPV correlated with the concentration of ractopamine in the samples. The sensor design including the Au NPs-PDDA modified rGO composite facilitated the achievement of significantly better analytical characteristics compared to other sensors for ractopamine, i.e. a detection limit of 1.0×10^{-12} M and a wide linear range from 1.0×10^{-12} to 1.0×10^{-8} M. The electrochemical aptasensor was used to determine ractopamine in spiked swine urine samples as well as in real contaminated samples. The results provided by the aptasensor were in agreement to those obtained in parallel by standard UPLC-MS/MS method and the recovery factors in spiked samples were 92.6–103.0% with an RSD of 3.7–5.5% (n = 6), proving the accuracy of the measurements performed with the aptasensor.

6.3.3 Glucose

Glucose sensing is another common application of G/Au NPs hybrids. The synergetic effect of rGO matrix decorated with Au NPs on the electrochemical performance of a glucose sensor was demonstrated by several authors [24, 28]. Shan showed that rGO-Au NPs/chitosan composite electrodes modified with glucose oxidase exhibit ampero-metic responses to glucose from 2–10 mM at –0.2 V with a LOD of 180 µM [28]. The glucose sensor proposed by Yu *et al.* is based on glucose oxidase (GODx) modified rGO/MWCNTs, where PDDA-capped Au NPs were integrated. The new hierarchical nanostructures exhibit a large surface area and a more favorable microenvironment for electron transfer, resulting in a LOD for glucose of 4.8 µM and a sensitivity of $29.72\,\mathrm{mA\,M^{-1}\,cm^{-2}}$ [24]. GO modified with Au-Palladium (1:1) bimetallic nanoparticles and further modification with glucose oxidase formed a biosensor for glucose with a LOD of 6.9 µM [50]. The electrocatalytic process is expressed according to:

$$\mathrm{GODx\left(FADH_2\right) + O_2 \rightarrow GODx\left(FAD\right) + H_2O_2} \tag{6.1}$$

$$\mathrm{glucose + GODx\left(FAD\right) \rightarrow gluconolactone + GODx\left(FADH_2\right)} \tag{6.2}$$

In this EC catalytic process, the oxidized FAD is regenerated at the surface of the electrode and enhances the reductive peak current of FAD in the presence of oxygen. When glucose is added, the reduction peak decreases due to the consumption of GODx (FAD). Detection of the variation of the reduction current allows determining the glucose concentration (Figure 6.8).

In addition to glucose oxidase G/Au NPs composites, the excellent catalytic ability of gold has made the composite also of importance for non-enzymatic glucose sensing. Studies on the direct glucose oxidation mechanism on gold-based electrodes showed that the formation of a hydroxyl adsorption (OH_{ads}) layer on the gold surface leads to the dehydrogenation of glucose [75–77], as indicated in eqn (6.3) and (6.4). Kong *et al.* took advantage of thionine functionalized GO with embedded Au nanoparticles (NPs) for the enzyme-free detection of glucose with a low detection limit of 50 nM [78]. Ismail *et al.* investigated the enhanced catalytic of Au NPs by using graphene oxide nanorib-bons (GO NRs) as supporting material for non-enzymatic glucose sensor. The Au NPs decorated GO NRs was found significantly superior in catalytic ability towards glucose oxidation than the unsupported bare gold electrode, with a greatly enhanced current density ($\approx$200%) [79].

$$\mathrm{Au + [OH^-]_{ads} \rightarrow Au[OH]_{ads}} \tag{6.3}$$

$$\mathrm{Au[OH]_{ads} + glucose \rightarrow Au + gluconolactone} \tag{6.4}$$

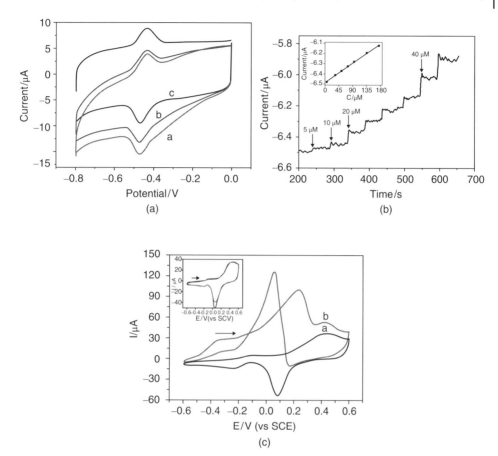

Figure 6.8 (a) CV of graphene/MWCNTs/Au NPs-GODx in air saturated PBS (trace a), air-saturated PBS + glucose (2 mM) (trace b), and N_2-saturated PBS (trace c) at a scan rate of 50 mV s^{-1}; (b) Amperometic response at –0.45 V upon successive addition of glucose in 0.1 M PBS. *Source*: Yu 2014 [24]. Reproduced with permission of Elsevier. (c) CVs of GO-thionine-Au NPs without (trace a) and with (trace b) 5 mM glucose in 0.1 M NaOH. *Source*: Kong 2012 [78]. Reproduced with permission of Elsevier.

6.3.4 Detection of Steroids: Cholesterol, Estradiol

The estimation of cholesterol level in serum has become an important step for the early diagnostic of many life-threatening diseases [80]. Tiwari and co-workers proposed a sodium dodecylbenzenesulfonate modified graphene/Au NPs matrix further bioconjugated with cholesterol oxidase and cholesterol esterase enzymes for the sensitive sensing of cholesterol. The anionic surfactant was used to disperse rGO and makes its surface hydrophilic, enabling immobilization of enzymes with preservation of enzyme activity and stable deposition of hydrophilic gold particles. The biosensor exhibited a linear response to cholesterol from 0.05–0.35 mM with a sensitivity of 3.14 µA µM^{-1} cm^{-2} and a LOD of 0.05 µM [81].

An electrochemical immunosensor for estradiol was built by modifying indium tin oxide (ITO) electrodes with a composite of rGO/Au NPs, followed by immobilization of anti-estradiol antibody [82]. Estradiol is the most important female sex hormone and is involved in inducing breast cancer. Dharuman *et al.* [82]. accomplished detection of estradiol with the above biosensor with a detection limit of 0.1 fmol, without to the need of any signal amplification strategy. The authors attributed the excellent sensitivity of their biosensor to the sensor design, including the formation of discrete structures of vertical oriented rGO/Au NPs composite on the ITO, to which the antibody was also efficiently affixed.

6.3.5 Detection of Antibacterial Agents

Chloramphenicol is a potent broad spectrum antibacterial agent that has been widely used to treat food-producing animals. While relatively inexpensive, highly effective with good pharmacokinetics properties, chloramphenicol is also associated with numerous toxic, adverse and fatal side effects, especially bone marrow depression. The EU imposes currently the maximum permissible amount of chloramphenicol in food products to be below 0.3 μg/kg. Nitrogen-doped graphene loaded with Au NPs [83] and Au NPs decorated rGO [84] have been proposed as sensing matrixes to achieve this goal. The $-NO_2$ groups of chloramphenicol get catalytically reduced at $-0.68\,V$ to $-NHOH$ functions, which allowed the construction of an amperometric sensing platform with a LOD of 0.25 μM for chloramphenicol with a linear range up to 2.95 μM [84].

6.3.6 Detection of Explosives Such as 2, 4, 6-trinitrotoluene (TNT)

TNT (2, 4, 6-trinitrotoluene) is one of the most widely used explosives in the world commonly used for military and industrial purposes. Due to its contamination of the environment and the risk to human health as well as growing homeland security concerns, detection of TNT has attracted increasing attention. Wang *et al.* investigated the electrocatalytic activity of rGO-Au NPs nanocomposite electrodes for TNT and reported a LOD of 11.5 ng mL^{-1} [85]. Electrochemiluminescence approach (ECL) is considered as one of the most promising analytical approaches due to its high sensitivity and ease of automatization. To overcome some shortcomings of these sensors such as complicated labeling and purification procedures, an anti-TNT aptamer based rGO/Au NPs sensor was developed. The rGO was functionalized with a ruthenium(II) ligand. Au NPs directly quench the ECL emission of the Ru(II) complex due to the energy transfer from the luminophore to the Au NPs. In the presence of TNT, the aptamer-Au NPs aggregate partly due to the aptamer-target interaction, reducing the quenching effect and leading to ECL signal restoration and strong ECL signal in the range of 0.01–100 ng mL^{-1}. A LOD of 3.6 pg mL^{-1} with high selectivity towards TNT was achieved over other nitrobenzene and nitrotoluene derivatives [86].

6.3.7 Detection of NADH

Nicotinamide adenine dinucleotide (NAD$^+$), as well as its reduced form NADH, is a vital co-enzyme couple that plays a significant role in energy production and consumption within the cells of living organisms in more than 300 enzymatic reactions catalyzed by dehydrogenases. The direct oxidation of NADH at the surface of classical electrodes requires high overpotentials and is often followed by poisoning the electrode surface due to the accumulation of oxidation products that strongly adsorb onto the electrode surface [87]. A rGO/Au NPs modified electrode formed through an *in situ* electrochemical reduction of GO and Au^{3+} allowed for NADH detection between 50 nM–500 μM with a LOD of 1.13 nM [88]. Bala and co-workers reported a NADH sensing platform based on glassy carbon electrodes modified with poly(allylamine hydrochloride) stabilized, Au NPs decorated rGO [89]. The composite materials promoted electron transfer between NADH and the electrode surface at 0.51 V with high sensitivity and detection limit of 3.5 μM for NADH in amperometric measurements.

6.3.8 Detection of Hydrogen Peroxide

The concept of using rGO/Au NPs for the catalysis of electrochemical reactions has also been employed to detect hydrogen peroxide, H_2O_2 (Figure 6.9). Fang *et al.* took advantage of the cationic properties of poly(diallyldimethylammonium chloride) (PDDA) functionalized graphene for the preparation of rGO/Au NPs heterostructure with enhanced electrochemical catalytic ability towards hydrogen peroxide [94]. The sensor exhibited a detection limit of 0.44 μM over a linear range from 0.5 μM to 0.5 mM. We recently investigated the reducing properties of tyrosine for the *in situ* synthesis of rGO/Au NPs hybrids from GO and Au ions [91]. The resulting nanocomposite showed electrochemical activity for nonenzymatic detection of H_2O_2 with a detection limit of 20 μM over a wide linear range from 0.02–25 mM. The layer-by-layer technique was investigated for the fabrication of a 3D Au NPs-embedded porous graphene (AuEPG) thin film for non-enzymatic H_2O_2 detection [92]. The detection limit of the AuEPG film-modified electrode was estimated to be 0.1 μM with a linear range from 0.5 μM to 4.9 mM. Jiang and co-workers used Au NPs-graphene hybrid structures for the detection of H_2O_2 in living cells. The results indicated that a higher efflux of H_2O_2 is observed in tumor cells versus normal cells [93]. Surfactant-free Au NPs on nitrogen-doped graphene quantum dots were proposed by Ju and Chen for H_2O_2 sensing with a LOD of 0.12 μM and a sensitivity of $186.22\ \mu A\ mM^{-1}\ cm^{-2}$. Importantly, this biosensor has shown great potential application for detection of H_2O_2 level in human serum samples

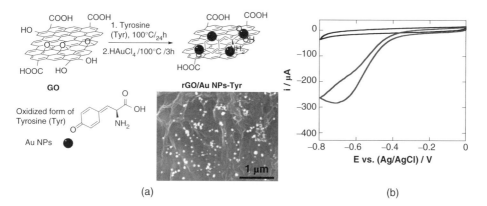

(a) (b)

Figure 6.9 (a) Schematic illustration of the formation of rGO/Au NPs-Tyr nanocomposites together with an SEM image; (b) Cyclic voltammograms of rGO/Au NPs/Tyr/GC in N_2-saturated 0.1 M PBS solution (pH 7.4) in the absence (black) and presence (blue) of 10 mM H_2O_2; scan rate: 50 mV s^{-1} [91]. (*See color plate section for the color representation of this figure.*)

and that released from human cervical cancer cells [94]. A two-dimensional rGO/Au NPs hybrid material where ultra small Au NPs (3 nm) were immobilized onto periodic mesoporous silica coated rGO was employed by Zhao *et al.* for the sensing of H_2O_2 with a LOD of 60 nM and a sensitivity of $39.2 \mu A \, mM^{-1} cm^{-2}$. The hybrid was found to be nontoxic and the sensor was used to detect a trace amount of H_2O_2 at a nM level released from living HeLa and HEpG2 cells [95].

6.3.9 Heavy Metal Ions

Heavy metal ions such as Hg^{2+} and Pb^{2+} impact the health of humans and the ecological environment. In the case of Hg^{2+} the safety value in drinking water is 1 μg/L defined by the World Health Organization. Au NPs have been widely used for the electrochemical detection of Hg^{2+} due to the high affinity of Au towards Hg which can enhance the effect of surface pre-concentrations. Ionic liquid functionalized rGO-Au NPs nanocomposites were proposed by Zhou for the accumulation of Hg^{2+} onto the electrode. After anodic stripping voltammetry, differential pulse voltammetry was employed for recording the signal of Hg^{2+} between 0.1–100 nM with a LOD of 0.03 nM [96].

More lately, special thymine-thymine (T-T) mismatches have been reported to show high selectivity for Hg^{2+} due to the formation of a T-Hg^{2+}-T complex [97]. Based on this principle, various DNA-based biosensors for Hg^{2+} were developed, but showed limited selectivity and lack of stability over time. To address these issues, Wang *et al.* proposed recently an alternative strategy for the electrochemical detection of Hg^{2+} based on thymine-1-acetic acid (T-COOH) modified rGO/Au NPs electrodes (Figure 6.10) [48]. The formed sensor displayed a LOD of 1.5 ng/L for Hg^{2+} and excellent selectivity.

6.3.10 Amino Acid and DNA Sensing

Amino acids are important fundamental units of proteins and vital participants in most of processes in living organisms. Many of them are also closely related to the development of diseases. The development of analytical approaches for these molecules has consequently been strongly persued. Early diagnosis is of importance for improving the changes of survival of a patient with cancer. One of the most common surfaces used to study DNA interactions are gold and Au NPs. Thiolated DNA can interact with Au NPs *via* the thiol groups and the DNA bases [123]. GO has emerged as a new material for interfacing with DNA. DNA interaction with GO occurs *via* aromatic stacking and hydrophobic interactions (Table 6.2). Mandler and co-workers showed that the surface of graphene/Au NPs electrodes can be easily modified by thiolated DNA through strong Au-S bonding [51]. When the target DNA completely hybridizes with the probe DNA, the electron is transferred from the electrode surface to methylene blue to catalyze ferrocyanide reduction (Figure 6.10). AT the same time the intercalated MB underwent electrochemical reduction generating leucomethylene blue (LB). MB can be regenerated trhough the chemical oxidation of LB by the solution-ophasee ferricyanide. A Toluidine blue modified rGO/Au NPs matrix was used by Peng *et al.* for the construction of a DNA biosensor [23]. Differential pulse voltammetry was employed to monitor the hybridization of DNA by measuring the changes in the peak current of Toluidine blue. Under optimal conditions, the decreased currents were proportional to the logarithm of the concentration of the target DNA in the range of $1 \times 10^{-11} - 1 \times 10^{-9} M$ with a limit of detection of $2.95 \times 10^{-12} M$ [23]. Ruiyi *et al.* demonstrated recently that

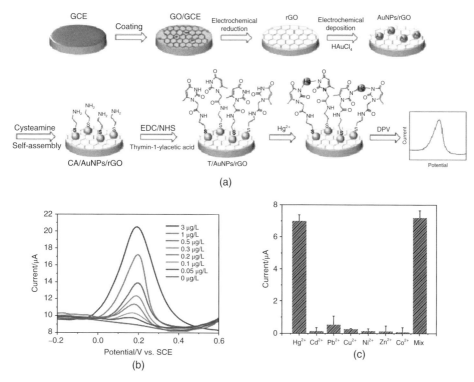

Figure 6.10 (a) Formation of thymidine functionalized rGO/Au NPs electrodes for Hg^{2+} detection; (b) DPV curves for dissolution of captured Hg^{2+}; (c) Electrochemical response to different metal ions. *Source:* Wang 2016 [48]. Reproduced with permission of Elsevier. (*See color plate section for the color representation of this figure.*)

N-doped graphene aerogels with embedded gold nanostars allow for the detection of double stranded DNA down to $3.9 \times 10^{-22}\,g\,mL^{-1}$ [141]. The analytical sensor was successfully applied for the electrochemical detection of circulating free DNA in human serum, which represents a big step further for the application of graphene-based sensors in molecular diagnostics.

L-Histidine is one of the naturally occurring amino acids and its determination in bodily fluids is of great importance for biological studies. A sensitive aptasensor for the detection of L-histidine based on the switching structures of the aptamer and rGO/Au NPs composite electrodes was reported by Liang *et al.* with a LOD of 0.1 pM and a linear rate from 0.1 pM–10 µM [124].

6.3.11 Detection of Model Protein Biomarkers

One of the first graphene-Au NPs hybrid sensors for the detection of proteins is the field-effect transistor (FET) proposed by Chen and co-workers in 2010 [125]. Thermally reduced GO sheets were decorated with 20 nm Au NPs, which were covalently conjugated to anti-immunoglobulin G (IgG) antibodies. By utilizing a method where the antibodies were conjugated to the Au NPs first, and then added to the rGO sheets, no direct modification to the rGO is needed, preserving the excellent electrical

Table 6.2 Protein sensing.

Electrode	Analyte	Sensor characteristics	Ref.
PDDA-rGO/Au NPs	Angiogenin	LOD: 0.064 pM LR: 0.1 pM–5 nM	[135]
rGO/Au NPs-antibody	CEA[1]	LOD: 0.01 ng mL^{-1} LR: 0.05–350 ng mL^{-1}	[126]
poly L-arginine-rGO/Au NPs- -antibody	CEA	LOD: 0.03ng mL^{-1} LR:0.5–200 ng mL^{-1}	[136]
Hemin-rGO/Au NPs-aptamer	CEA	LOD: :40 fg mL^{-1} LR: 0.0001–10 ng mL^{-1}	[130]
PEDOT-rGO/Au NPs	CEA	LOD: 0.1 pg mL^{-1} LR: 0.0004–40 ng mL^{-1}	[129]
MWCNTs-rGO/Au NPs-CeO[2]	CEA	LOD: 0.02 ng mL^{-1} LR: 0.05–100 ng mL^{-1}	[131]
rGO/Au NRs-antibody	Transferrin	LR: 0.0375–40 µg mL^{-1}	[137]
cyclodextrin-rGO/Au NPs-aptamer	Thrombin	LOD: 5.2 zeptoM LR: 16 zeptoM–8 aM	[133]
Orange II–rGO/Au NPs	Insulin	LOD: 6.0 aM LR: 10 aM–50 nM	[138]
Chitosan-rGO/Au NPS- antibody	Prolactin	LOD: 0.038 ng. mL^{-1} LR: 0.1–50 ng. mL^{-1}	[139]
rGO/Au NPs	L-histidine	LOD: 0.1pM LR: 10pM–10 µM	[124]
rGO/Au NPs-enzyme	L-lactate	LOD: 0.13 µM LR: 10µM–5 mM S: 154 µA mM^{-1} cm^2	[132]
Thionine-rGO/Au@Pd	Carbohydrate antigen 19-9	LOD: 006 U mL^{-1} LR : 0.015–150 U mL^{-1}	[128]
rGO/Au NPs-aptamer	ATP	LOD: 15 nM LR: 10 nM–4 mM	[140]
N-doped graphene aerogel/Au nanostar	cfDNA	LOD: 3.9 × 10^{-22} g mL^{-1}	[141]
rGO/Au NPs	DNA	LOD: 100 fM	[51]
PAN-rGO/-Au NPs	DNA	LOD: 2.11 pM LR: 10–1000 pM	[53]
rGO/Au NPs	Chinese Herbs	LOD: 11.7 fM LR: 100fM–10 nM	[142]
Toluidine blue–rGO/Au NPs	DNA	LOD: 2.95 pM LR: 0.1 pM–1 nM	[23]
rGO/Au NPs	cysteine	LOD: 0.1 nM	[72]

Table 6.2 (Continued)

Electrode	Analyte	Sensor characteristics	Ref.
PDDA-modified rGO/Au NPs	angiogenin	LOD: 0.064 pM LR: 0.1 pM–5 nM	[135]
rGO/Au NPs-anti-CEA	CEA	LOD: 0.01 ng mL^{-1} LR: 0.05–350 ng mL^{-1}	[126]
Graphene/Au-Fe$_3$O$_4$	prostate specific antigen	LOD: 5 ng mL^{-1} LR: 0.01–10 ng mL^{-1}	[143]
rGO/Au NRs-antibody	transferrin	LR : 0.0375–40 μg mL^{-1}	[137]
rGO-Au NPs	cells	LOD: 5.2×10^3 cells mL^{-1} LR: 1.6×10^4–1.6×10^7 cells mL^{-1}	[52]
rGO/Au NPs-antibodies	*E. coli* O157:H7	LOD: 1.5×10^2 cfu mL^{-1} LR: 1.5×10^2–1.5×10^7 cfu mL^{-1}	[144]
rGO/Au NPs	*M. tuberculosis*	LR : 1 aM^{-1} nM	[145]

1)   CEA: carcinoembryonic antigen; LOD: limit of detection; LR: linear range; PEDOT: Poly(3,4-ethylenedioxythiophene).

conductivity. The local geometric deformation upon binding of IgG to anti-IgG reduced the mobility of holes in rGO and consequently the conductivity of the rGO sheets, which resulted in a detection limit for IgG of 13 pM, with excellent selectivity when exposed to other protein mismatches such as IgM.

At the same time, the first reports of electrochemical based protein sensors emerged (Table 6.2). Zhong *et al.* designed a sandwhich-type immunosensor format to quantify cacinoembryonic antigen (CEA), a model tumor marker using nanogold-enwrapped graphene nanocomposites as labels [126]. The device consists of a glassy carbon electrode coated with Prussian blue, onto which Au NPs were electrochemically deposited and further modified with anti-CEA antibodies (Figure 6.11). After interaction with the CEA analyte, horseradish peroxidase (HRP) conjugated anti-CEA was used as a second antibody, which allowed the electrochemical conversion of hydrogen peroxide to water, being proportional to the amount of CEA. Due to the signal amplification strategy, a dynamic range of 0.05–350 ng mL^{-1} and a LOD of 0.01 ng mL CEA were achieved [126].

Three-dimensional macroporous rGO/Au NPs composite was proposed by Sun *et al.* for the sensitive detection of CEA in human serum [127]. In the proposed structures, the Au NPs were distributed not just on the surface, but also on the inside of graphene, which increased considerably the sensing area, resulting in capturing more primary antibodies as well as improving the electronic transmission rate. Using a sandwich-type immune assay, a detection range of CEA from 0.001–10 ng mL^{-1} with a low detection limit of 0.35 pg mL^{-1} were obtained. To this was added more recently a porous rGO/Au NPs nanocomposite platform with Au@Pd core/shell bimetallic functionalized rGO as signal enhancer with a sensitivity to CEA 19-9 of 0.006 U mL^{-1} [128]. A poly(3, 4-ethylenedioxythiophene)-doped rGO/Au NPs was used by Gao *et al.* The immunosensor

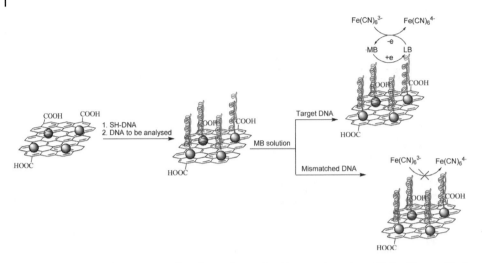

Figure 6.11 Schematic illustration of an electrochemical DNA sensor based on rGO/Au NPs modified glassy carbon electrode further modified by thiolated DNA ligands and using methylene blue (MB) to catalyze ferrocyanide reduction forming leucomethylene blue (LB). *Source*: Wang 2012 [51]. Reproduced with permission of Royal Society of Chemistry. (*See color plate section for the color representation of this figure.*)

displayed a LOD of $0.1\,pg\,mL^{-1}$ for CEA with a linear range from $0.0004–40\,ng\,mL^{-1}$ and could be used for the analysis of CEA in human samples [143]. Using hemin functionalized rGO/Au NPs nanocomposite and assembling a CEA specific aptamer to the matrix, a CEA sensor with a LOD of $40\,fg\,mL^{-1}$ was recently reported [130].

Pang *et al.* proposed lately an immunoassay strategy for the detection of Human IgG (HIgG) based on the decrease of electrochemiluminescence (ECL) of $S_2O_8^{2-}$ upon immune reaction [131]. The rGO/Au NPs electrode was prepared by first reduction of GO to rGO at the electrode surface followed by electrochemical reduction of $HAuCl_4$ to form gold nanostructures on rGO. Modification with anti-IgG resulted in an electrode interface with significantly increased ECL signal of $S_2O_8^{2-}$, which could be suppressed specifically upon immunoreaction with a linear response from $0.02–100\,ng\,mL^{-1}$ IgG and a detection limit of $1.3\,pg$ (Figure 6.12b).

However, the examples are not limited to CEA. An amperometric biosensor for L-lactate tumor biomarker (LOD $= 0.13\,\mu M$) was proposed by Bala and co-workers using L-lactate dehydrogenase (LDH) loaded rGO/Au NPs interfaces [132]. The LDH enzyme acted as a catalyst of L-lactate transformation in pyruvate, simultaneously with the reduction of the enzyme cofactor NAD^+ to NADH. Quantitative determination of L-lactate relied on the amperometric determination of NADH based on the direct correlation between the consumed L-lactate and the NADH formed in the enzymatic reaction. Inclusion of rGO-Au NPs nanocomposite in the biosensor enabled the achievement of NADH detection at lower overpotentials with a higher sensitivity compared to bare or to other modified electrodes.

Attomolar detection of thrombin could be achieved with an electrochemical aptasensor which included a composite of thio-β-cyclodextrin-functionalized rGO/Au NPs [133]. A thrombin aptamer labeled with the electroactive probe ferrocene was used for both specific recognition of thrombin and electrochemical detection by DPV. The

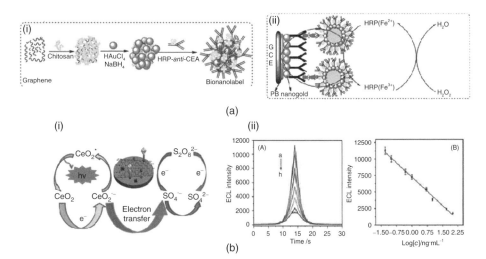

Figure 6.12 (a) Fabrication process of nanogold-wrapped graphene nanocomposite (i), its use as a label for immunosensor (ii) (Zhong 2010 [126]). Reproduced with permission of Elsevier; (b) (i) Schematic diagram of the electrochemiluminescent Immunosensor for the detection of carcinoembryonic antigen based on MWCNTS-rGO/Ag NPs-CeO$_2$ electrodes, (ii) ECL response for the sensor to different CEA concentrations together wirth calibration curve. *Source*: Pang 2015 [131]. Reproduced with permission of American Chemical Society.

aptamer was immobilized on electrode surface by host-guest affinity interaction between ferrocene and cyclodextrin. Integration of Au NPs within the SH-β-CD-Gr/Au NPs composite allowed improving the detection limit for thrombin by two orders of magnitude, compared to the case when the electrode was modified only with thio-β-cyclodextrin functionalized graphene. This dramatic improvement in analytical properties was credited to the good electrical properties of Au NPs, combined with their high surface area allowing quantitative immobilization of cyclodextrin and resulting, further on efficient immobilization of ferrocene-labeled aptamer by host-guest interactions. The thio-β-cyclodextrin acted not only as a linker for the labeled aptamer, but also as a dispersant for graphene and Au NPs; the SH-β-CD-Gr/Au NPs composite was obtained by a one-pot strategy.

A highly sensitive sandwich-type electrochemiluminescence (ECL) immunosensor was lately proposed for the quantitative determination of alpha fetoprotein (AFP) using gold nanoparticles decorated carbon black intercalated reduced graphene oxide (Au-rGO@ CB) as sensing platform and nanoporous silver (NPS) loaded Ru(bpy)$_3^{2+}$ as labels. Under optimal conditions, the designed immunosensor exhibited wider linear range from $0.0001-30\,ng\,mL^{-1}$ with a relative lower detection limit of $33\,fg\,mL$ for AFP detection [134].

6.4 Conclusion

In this chapter it became clear, that graphene/Au NPs matrixes have witnessed tremendous interest as interfaces for electrochemical sensing applications as these hybrid materials display extraordinary synergetic properties when combined rather than individually.

Due to its unique basal plane structure graphene can be loaded with nanometer to micrometer sized Au NPs, while the high surface area of graphene improves interfacial contact. The low cost of fabrication, together with robustness, rapidity, sensitivity and selectivity have made these electrodes of remarkable advantage over others. These composite materials have, in some instances, overcome the limitations of the present electroanalytical platforms such as limit of detection, specificity, catalytic properties… Some of the work is of high quality and, in addition of showing interesting sensing properties, it also contains detailed characterizations of the newly developed graphene-related nanomaterials. However, some articles lack such characterization, making it more difficult to appreciate the reported sensing characteristics. The importance of a well characterized sensing matrix cannot be underestimated. Most articles which use the term 'graphene' referred to a multi-layered structure prepared from graphite and thus to GO or rGO. The morphology and the electrochemical performance of such multi-layer graphene materials, prepared by top down approaches, differ significantly depending on their method of preparation and a detailed characterization of the graphene precursor before using it in electrochemistry is essential if results are to have any meaning.

The general interest of graphene based electrodes over other carbon materials such as carbon nanotubes is that rGO can be fabricated from graphite (which is inexpensive), while the opposite is true for CNTs, which are synthesized using NPs as templates from carbon-containing gases. Graphene nanocomposites can provide more uniform and greater electroactive site distribution and density in order to decrease overpotentials, compared to graphite, and larger surface area (even larger than that of SWCNTs) for immobilization of biomolecules, making it thus such a promising matrix for sensing. The possibility of obtaining doped and porous graphene matrixes is another advantage as heteroatoms such as nitrogen can provide electrocatalytic properties and enhance the stability of doped graphene electrodes, while the porous structure further increases the active surface. A bright future for graphene and graphene/Au NPs as a sensing material is thus expected.

Acknowledgement

The Centre National de la Recherche Scientifique (CNRS), the Lille1 University, the Hauts-de-France region and the CPER Photonics for Society are acknowledged for financial support.

References

1 Ratinac, K.R.; Yang, W.J.; Gooding, J.J.; *et al.* Graphene and related materials in electrochemical sensing. Electroanalysis 2011, 23 (4), 803–826.

2 Ambrosi, A.; Chua, C.K.; Bonanni, A.; Pumera, M. Electrochemistry of graphene and related materials. Chemical Reviews 2014, 114 (14), 7150–7188.

3 Akhavan, E.; Ghaderi, E. Graphene nanomesh promises extremely efficient in vivo photothermal therapy. Small 2013, 9 (21), 3593–3601.

4 Brownson, D.A.; Banks, C. E. Graphene electrochemistry: an overview of potential applications. Analyst 2010, 135 (11), 2768–2778.

5 Pumera, M. Electrochemistry of graphene: new horizons for sensing and energy storage. The Chemical Record 2009, 9 (4), 211–223.

6 Unwin, P.R.; Güell, A. G.; Zhang, G. Nanoscale electrochemistry of sp^2 carbon materials: From graphite and graphene to carbon nanotubes. Accounts of Chemical Research 2016, 49 (9), 2041–2048.

7 Szunerits, S.; Boukherroub, R. Sensing using localised surface plasmon resonance sensors. Chemical Communications 2012, 48, 8999–9010.

8 Mayer, K.M.; Hafner, J. H. Localized surface plasmon resonance sensors. Chemical Reviews 2011, 111 (6), 3828–3857.

9 Spadavecchia, J.; Barras, A.; Lyskawa, J.; *et al*. Approach for plasmonic based DNA sensing: amplification of the wavelength shift and simultaneous detection of the plasmon modes of gold nanostructures. Analytical Chemistry 2013, 85 (6), 3288–3296.

10 Szunerits, S.; Spadavecchia, J.; Boukherroub, R. Surface plasmon resonance: signal amplification using colloidal gold nanoparticles for enhanced sensitivity. Reviews in Analytical Chemistry 2014, 33 (3), 153–164.

11 Zagorodko, O.; Spadavecchia, J.; Yanguas Serrano, A.; *et al*. Highly sensitive detection of DNA hybridization on commercialized graphene coated surface plasmon resonance interfaces. Analytical Chemistry 2014, 86, 11211–11216.

12 Pissuwan, D.; Cortie, C.H.; Valenzuela, S.M.; Cortie, M.B. Functionalised gold nanoparticles for controlling pathogenic bacteria. Trends in Biotechnology 2010, 28 (4), 207–213.

13 Boisselier, E.; Astruc, D. Gold nanoparticles in nanomedicine: preparations, imaging, diagnostics, therapies and toxicity. Chemical Society Reviews 2009, 38, 1759–1782.

14 Pissuwan, D.; Valenzuela, S. M.; Cortie, M.B. Prospects for gold nanorod particles in diagnostic and therapeutic applications. Biotechnology and Genetic Engineering Reviews 2008, 25 (1), 93–112.

15 Gosh, P.; Han, G.; De, M.; *et al*. Gold nanoparticles in delivery applications. Advanced Drug Delivery Reviews 2008, 60 (11), 1307–1315.

16 Bechet, D.; Couleaud, P.; Frochot, C.; *et al*. Nanoparticles as vehicles for delivery of photodynamic therapy agents. Trends in Biotechnology 2008, 26 (11), 612–621.

17 Turchniuk, K.; Boukherroub, R.; Szunerits, S. Gold-graphene nanocomposites for sensing and biomedical applications. Journal of Materials Chemistry B 2015, 3, 4301–4324

18 Turcheniuk, K.; Hage, C.-H.; Spadavecchia, J.; *et al*. Plasmonic photothermal destruction of uropathogenic E. coli with reduced graphene oxide and core/shell nanocomposites of gold nanorods/reduced graphene oxide. Journal of Materials Chemistry B 2015, 3 (3), 375–386.

19 Kim, T. H.; Lee, K. B.; Choi, J. W. 3D graphene oxide-encapsulated gold nanoparticles to detect neural stem cell differentiation. Biomaterials 2013, 34 (34), 8660–8670.

20 Maurer, T.; Nicolas, R.; Lévêque , G.; *et al*. Enhancing LSPR sensitivity of Au gratings through graphene coupling to Au film. Plasmonics 2014, 9 (3), 507–512.

21 Schedin, F.; Lidorikis, E.; Lombardo, A.; *et al*. Surface-enhanced Raman spectroscopy of graphene. ACS Nano 2010, 4 (10), 5617–5626.

22 Zhou, H.; Qiu, C.; Liu, Z.; *et al*. Thickness-dependent morphologies of gold on n-layer graphenes. Journal of the American Chemical Society 2010, 132, 944–946.

23 Peng, H.-P.; Hu, Y.; Liu, P.; *et al.* Label-free electrochemical DNA biosensor for rapid detection of mutidrug resistance gene based on Au nanoparticles/toluidine blue–graphene oxide nanocomposites. Sensors and Actuators B: Chemical 2015, 207, 269–276.

24 Yu, Y.; Chen, Z.; He, S.; *et al.* Direct electron transfer of glucose oxidase and biosensing for glucose based on PDDA-capped gold nanoparticle modified graphene/multi-walled carbon nanotubes electrode. Biosensors and Bioelectronics 2014, 52 (15), 147–152.

25 Ma, W.; Li, X.; Han, D.; *et al.* Decoration of electro-reduced graphene oxide with uniform gold nanoparticles based on in situ diazonium chemistry and their application in methanol oxidation. Journal of Electroanalytical Chemistry 2013, 690, 111–116.

26 Liu, G.; Qi, M.; Zhang, Y.; *et al.* Nanocomposites of gold nanoparticles and graphene oxide towards an stable label-free electrochemical immunosensor for detection of cardiac marker troponin-I. Analytica Chimica Acta 2016, 909, 1–8.

27 Goncalves, G.; Marques, P.A.; Granadeiro, C.M.; *et al.* Surface modification of graphene nanosheets with gold nanoparticles: the role of oxygen moieties at graphene surface on gold nucleation and growth. Chemistry of Materials 2009, 21 (20), 4796–4802.

28 Shan, C.S.; Yang, H.F.; Han, D.X.; *et al.* Graphene/Au NPs/chitosan nanocomposites films for glucose biosensing. Biosensors and Bioelectronics 2010, 25 (5), 1070–1074.

29 Sahu, S. R.; Devi, M. M.; Mukherjee, P.; *et al.* Optical property characterization of novel graphene-X (X = Ag, Au and Cu) nanoparticle hybrids. Journal of Nanomaterials 2013, 6, 232409; DOI: 10.1155/2013/232409.

30 Zhang, P.; Huang, Y.; Lu, X.; *et al.* One-step synthesis of large-scale graphene film doped with gold nanoparticles at liquid-air interface for electrochemistry and raman detection applications. Langmuir 2014, 30 (29), 8980–8989.

31 Zhang, Z.; Chen, H.; Xing, C.; *et al.* Sodium citrate: A universal reducing agent for reduction/decoration of graphene oxide with au nanoparticles. Nano Research 2011, 4 (6), 599–611.

32 Movahed, S K.; Fakharian, M.; Dabiri, M.; Bazgir, A. Gold nanoparticle decorated reduced graphene oxide sheets with high catalytic activity for Ullmann homocoupling. RSC Advances 2014, 4 (10), 5243–5247.

33 Muszynski, R.; Seger, B.; Kamat, P.V. Decorating graphene sheets with gold nanoparticles. Journal of Physical Chemistry C 2008, 112 (14), 5263–5266.

34 Huang, X.; Li, H.; Li, S.; *et al.* Synthesis of gold square-like plates from ultrathin gold square sheets: the evolution of structure phase and shape. Angewandte Chemie International Edition 2011, 50 (51), 12245–12248.

35 Iliut, M.; Leordean, C.; Canpean, V.; *et al.* A new green, ascorbic acid-assisted method for versatile synthesis of Au-graphene hybrids as efficient surface-enhanced Raman scattering platforms. Journal of Materials Chemistry C 2013, 1, 4094–4104.

36 Zhou, L.; Gu, H.; Wang, C.; Zhang, J. Study on the synthesis and surface enhanced Raman spectroscopy of graphene-based nanocomposites decorated with noble metal nanoparticles. Colloids and Surfaces A: Physicochemical and Engineering Aspects 2013, 430, 103–109.

37 Sharma, P.; Darabdhara, G.; Reddy, T.M.; *et al.* Synthesis, characterization and catalytic application of Au NPs-reduced graphene oxide composites material: an eco-friendly approach. Catalysis Communications 2013, 40, 139–144.

38 Fu, W. L.; Zhen, S. J.; Huang, C. Z. One-pot green synthesis of graphene oxide/gold nanocomposites as SERS substrates for malachite green detection. Analyst 2013, 138 (10), 3075–3081.

39 Le, Z.; Liu, Z.; Qian, Y.; Wang, C. A facile and efficient approach to decoration of graphene nanosheets with gold nanoparticles. Applied Surface Science 2012, 258 (14), 5348–5353.

40 Yang, K.; Zhang, S.; Zhang, G.; *et al*. Graphene in mice: ultrahigh in vivo tumor uptake and efficient photothermal therapy. Nano Letters 2010, 10, 3318–3323.

41 Fu, C.; Kuang, Y.; Huang, Z.; *et al*. Electrochemical co-reduction synthesis of graphene/ Au nanocomposites in ionic liquid and their electrochemical activity. Chemical Physics Letters 2010, 499 (4), 250–253.

42 Vinodgopal, K.; Neppolian, B.; Lightcap, I.V.; *et al*. Sonolytic design of graphene-Au nanocomposites: simultaneous and sequential reduction of graphene oxide and Au (III). Journal of Physical Chemistry Letters 2010, 1 (13), 1987–1993.

43 Li, X.-R.; Li, X.-L.; Xu, M.-C.; *et al*. Gold nanodendries on graphene oxide nanosheets for oxygen reduction reaction. Journal of Materials Chemistry A 2014, 2 (6), 1697–1703.

44 Song, M.; Xu, J.; Wu, C. The effect of surface functionalization on the immobilization of gold nanoparticles on graphene sheets. Journal of Nanotechnology 2012, 2012, 5.

45 Wang, F.-B.; Wang, J.; Shao, L.; *et al*. Hybrids of gold nanoparticles highly dispersed on graphene for the oxygen reduction reaction. Electrochemistry Communications 2014, 38, 82–85.

46 Wang, Y.; Zhang, S.; Du, D.; *et al*. Self assembly of acetylcholinesterase on a gold nanoparticles–graphene nanosheet hybrid for organophosphate pesticide detection using polyelectrolyte as a linker. Journal of Materials Chemistry 2011, 21, 5319–5325.

47 Kong, B.S.; Geng, J.; Jung, H. T. Layer-by-layer assembly of graphene and gold nanoparticles by vacuum filtration and spontaneous reduction of gold ions. Chemical Communications 2009, 2174–2176.

48 Wang, N.; Lin, M.; Dai, H.; Ma, H. Functionalized gold nanoparticles/reduced graphene oxide nanocomposites for ultrasensitive electrochemical sensing of mercury ions based on thymine-mercury-thymine structure. Biosensors and Bioelectronics 2016, 79, 320–326.

49 Yang, J.; Strickler, J.R.; Gunasekaran, S. Indium tin oxide-coated glass modified with reduced graphene oxide sheets and gold nanoparticles as disposable working electrodes for dopamine sensing in meat samples. Nanoscale 2012, 4 (15), 4594–4602.

50 Yang, J.; Deng, S.; Lei, J.; *et al*. Electrochemical synthesis of reduced graphene sheet-AuPd alloy nanoparticle composites for enzymatic biosensing. Biosensors and Bioelectronics 2011, 29, 159–166.

51 Wang, Z.; Zhang, J.; Yin, Z.; *et al*. Fabrication of nanoelectrode ensemble by electrodeposition of gold nanoparticles on single-layer graphene oxide sheets. Nanoscale 2012, 4, 2728–2733.

52 Sun, X.; Ji, J.; Jiang, D.; *et al*. Development of novel electrocehmical sensor using pheochromocytoma cells and its assessement of arylamic cytotocixity. Biosensors and Bioelectronics 2013, 44, 122–126.

53 Wang, L.; Hua, E.; Liang, M.; *et al*. Graphene sheets, polyaniline and AuNPs based DNA sensor for electrochemical determination of BCR/ABL fusion gene with functional hairpin probe. Biosensors and Bioelectronics 2014, 51, 201–207.

54 Zaijun, L.; Xiulan, S.; Qianfang, X.; *et al*. Green and controllable strategy to fabricate well-dispersed graphene-gold nanocomposite film as sensing materials for the detection of hydroquinone and resorcinol with electrodeposition. Electrochimica Acta 2012, 85, 42–48.

55 Yue, X.; Song, W.; Zhu, W.; *et al.* In situ surface electrochemical co-reduction route towards controllable construction of AuNPs/ERGO electrochemical sensing platform for simultaneous determination of BHA and TBHQ. Electrochimica Acta 2015, 182, 847–855.

56 Zhou, M.; Wang, Y.; Zhai, Y.; *et al.* Controlled synthesis of large-area and patterned electrochemically reduced graphene oxide films. Chemistry: A European Journal 2009, 15 (25), 6116–6120.

57 Ramesha, G. K.; Sampath, S. Electrochemical reduction of oriented graphene oxide films: An in situ raman spectroelectrochemical study. The Journal of Physical Chemistry C 2009, 113 (19), 7985–7989.

58 Chng, E. L. K.; Pumera, M. Solid-state electrochemistry of graphene oxides: Absolute quantification of reducible groups using voltammetry. Chemistry: An Asian Journal 2011, 6 (11), 2899–2901.

59 Pumera, M.; Ambrosi, A.; Chng, E.L. Impurities in graphenes and carbon nanotubes and their influence on the redox properties. Chemical Science 2012, 3 (12), 3347–3355.

60 Xue, Y.; Zhao, H.; Wu, Z.; *et al.* The comparison of different gold nanoparticles/ graphene nanosheets hybrid nanocomposites in electrochemical performance and the construction of a sensitive uric acid electrochemical sensor with novel hybrid nanocomposites. Biosensors and Bioelectronics 2011, 29 (1), 102–108.

61 Alwarappan, S.; Erdem, A.; Liu, C.; Li, C.-Z. Probing the electrochemical properties of graphene nanosheets for biosensing applications. The Journal of Physical Chemistry C 2009, 113 (20), 8853–8857.

62 Xu, L.Q.; Yang, W.J.; Neoh, K.-G.; *et al.* Dopamine-induced reduction and functionalization of graphene oxide nanosheets. Macromolecules 2010, 43, 8336–8339.

63 Shang, N. G.; Papakonstantinou, P.; McMullan, M.; *et al.* Catalyst-free efficient growth, orientation and biosensing properties of multilayer graphene nanoflake films with sharp edge planes. Advanced Functional Materials 2008, 18 (21), 3506–3514.

64 Wang, Y.; Li, Y.; Tang, L.; *et al.* Application of graphene-modified electrode for selective detection of dopamine. Electrochemistry Communications 2009, 11 (4), 889–892.

65 Tang, L.; Wang, Y.; Li, Y.; *et al.* Preparation, structure, and electrochemical propeties of redcued graphene sheet films. Advanced Functional Materials 2009, 19, 2782–2789.

66 Sun, C.; Lee, H.; Yang, J.; Wu, C. The simultaneous electrochemical detection of ascorbic acid, dopamine, and uric acid using graphene/size-selected Pt nanocomposites. Biosensors and Bioelectronics 2011, 26, 3450–3455.

67 Mao, Y.; Bao, Y.; Gan, S.; *et al.* Electrochemical sensor for dopamine based on a novel graphene-molecular imprinted polymers composite recognition element. Biosensors and Bioelectronics 2011, 28, 291–297.

68 Sun, C.; Chang, C.; Lee, H.; *et al.* Microwave-assisted synthesis of a core-shell MWCNT/GONR heterostructure for the electrochemical detection of ascorbic acid, dopamine, and uric acid. ACS Nano 2011, 5, 7788–7795.

69 Liu, S.; Yan, J.; He, G.; *et al.* Layer-by-layer assembled multilayer films of reduced graphene oxide/gold nanoparticles for the electrochemical detection of dopamine. Journal of Electroanalytical Chemistry 2012, 672, 40–44.

70 Liu, M.; Chen, Q.; Lai, C.; *et al.* A double signal amplification platform for ultrasensitive and simultaneous detection of ascorbic acid, dopamine, uric acid and acetaminophen based on a nanocomposite of ferrocene thiolate stabilized $Fe_3O_4@Au$ nanoparticles with graphene sheet. Biosensors and Bioelectronics 2013, 48, 75–81.

71 Juanjuan, Z.; Ruiyi, L.; Zaijun, L.; *et al.* Synthesis of nitrogen-doped activated graphene aerogel/gold nanoparticles and its application for electrochemical detection of hydroquinone and o-dihydroxybenzene. Nanoscale 2014, 6, 5458–5466.

72 Zhang, Y.; Yang, A.; Zhang, X.; *et al.* Highly selective and sensitive biosensor for cysteine detection based on in situ synthesis of gold nanoparticles/graphene nanocomposites. Colloids and Surfaces A: Physicochemical and Engineering Aspects 2013, 436, 815–822.

73 Xue, C.; Wang, X.; Zhu, W.; *et al.* Electrochemical serotonin sensing interface based on double-layered membrane of reduced graphene oxide/polyaniline nanocomposites and molecularly imprinted polymers embedded with gold nanoparticles. Sensors and Actuators B: Chemical 2014, 196, 57–63.

74 Yang, F.; Wang, P.; Wang, R.; *et al.* Label free electrochemical aptasensor for ultrasensitive detection of ractopamine. . Biosensors and Bioelectronics 2016, 77, 347–352.

75 Vassilyev, Y.B.; Khazova, O.A.; Nikolaeva, N.N. Kinetics and mechanism of glucose electrooxidation on different electrode-catalysts: Part I. Adsorption and oxidation on platinum. Journal of Electroanalytical Chemistry and Interfacial Electrochemistry 1985, 196 (1), 105–125.

76 Makovos, E.B.; Liu, C.C. A cyclic-voltammetric study of glucose oxidation on a gold electrode. Bioelectrochemistry and Bioenergetics 1986, 15 (2), 157–165.

77 Adzic, R. R.; Hsiao, M. W.; Yeager, E. B. Electrochemical oxidation of glucose on single crystal gold surfaces. Journal of Electroanalytical Chemistry and Interfacial Electrochemistry 1989, 260 (2), 475–485.

78 Kong, F.-Y.; Li, X.-R.; Zhao, W.-W.; *et al.* Graphene oxide–thionine–Au nanostructure composites: preparation and applications in non-enzymatic glucose sensing. Electrochemistry Communications 2012, 14 (1), 59–62.

79 Ismail, N. S.; Le, Q. H.; Yoshikawa, H.; *et al.* Development of non-enzymatic electrochemical glucose sensor based on graphene oxide nanoribbon–gold nanoparticle hybrid. Electrochimica Acta 2014, 146, 98–105.

80 Kuila, T.; Bose, S.; Khanra, P.; *et al.* Recent advances in graphene-based biosensors. Biosensors and Bioelectronics 2011, 26, 4637–4648.

81 Parlak, O.; Tiwari, A.; Turner, A.P.; Tiwari, A. Template-directed hierarchical self-assembly of graphene based hybrid structure for electrochemical biosensing. Biosensors and Bioelectronics 2013, 49, 53–62.

82 Dharuman, V.; Hahn, J.H.; Jayakumar, K.; Teng, W. Electrochemically reduced graphene–gold nano particle composite on indium tin oxide for label free immuno sensing of estradiol. Electrochimica Acta 2013, 114, 590–597.

83 Borowiec, J.; Wang, R.; Zhu, L.; Zhang, J. Synthesis of nitrogen-doped graphene nanosheets decorated with gold nanoparticles as an improved sensor for electrochemical determination of chloramphenicol. Electrochimica Acta 2013, 99, 138–144.

84 Karthik, R.; Govindasamy, M.; Chen, S.-M.; *et al.* Green synthesized gold nanoparticles decorated graphene oxide for sensitive determination of chloramphenicol in milk, powdered milk, honey and eye drops. Journal of Colloid and Interface Science 2016, 475, 46–56.

85 Wang, P.; Liu, Z.-G.; Chen, X.; *et al.* UV irradiation synthesis of an Au-graphene nanocomposite with enhanced electrochemical sensing properties. Journal of Materials Chemistry A 2013, 1, 9189–9195.

86 Yu, Y.; Cao, Q.; Zhaou, M.; Cui, H. A novel homogeneous label-free aptasensor for 2,4,6-trinitrotoluene detection based on an assembly strategy of electrochemiluminescent graphene oxide with gold nanoparticles and aptamer. Biosensors and Bioelectronics 2013, 43, 137–142.

87 Deore, B.A.; Freund, M.S. Reactivity of poly(anilineboronic acid) with NAD+ and NADH. Chemistry of Materials 2005, 17 (11), 2918–2923.

88 Govindhan, M.; Amiri, M.; Chen, A. Au nanoparticle/graphene nanocomposite as a platform for the sensitive detection of NADH in human urine. Biosensors and Bioelectronics 2015, 66, 474–80.

89 Istrate, O.-M.; Rotariu, L.; Marinescu, V.E.; Bala, C. NADH sensing platform based on electrochemically generated reduced graphene oxide–gold nanoparticles composite stabilized with poly (allylamine hydrochloride). Sensors and Actuators B: Chemical 2016, 223, 697–704.

90 Fang, Y.; Guo, S.; Zhu, C.; *et al.* Self-assembly of cationic polyelectrolyte-functionalized graphene nanosheets and gold nanoparticles: A two-dimensional heterostructure for hydrogen peroxide sensing. Langmuir 2010, 26, 11277–11282.

91 Wang, Q.; Wang, Q.; Li, M.; *et al.* One-step synthesis of Au nanoparticle-graphene composites using tyrosine: electrocatalytic and catalytic properties. New Journal of Chemistry 2016, 40 (6), 5473–5482.

92 Xi, Q.; Chen, X.; Evans, D. G.; Yang, W. Gold nanoparticle-embedded porous graphene thin films fabricated via layer-by-layer self-assembly and subsequent thermal annealing for electrochemical sensing. Langmuir 2012, 28 (25), 9885–9892.

93 Chang, H.; Wang, X.; Shiu, K. K.; *et al.* Layer-by-layer assembly of graphene, Au and poly(toluidine blue O) films sensor for evaluation of oxidative stress of tumor cells elicited by hydrogen peroxide. Biosensors and Bioelectronics 2013, 41, 789–794.

94 Ju, J.; Chen, W. In Situ growth of surfactant-free gold nanoparticles on nitrogen-doped graphene quantum dots for electrochemical detection of hydrogen peroxide in biological environments. Analytical Chemistry 2014, 87, 1903–1910.

95 Maji, S.K.; Sreejith, S.; Mandal, A.K.; *et al.* Immobilizing gold nanoparticles in mesoporous silica covered reduced graphene oxide: a hybrid material for cancer cell detection through hydrogen peroxide sensing. ACS Applied Materials and Interfaces 2014, 6, 13648–13656.

96 Zhou, N.; Li, J.; Chen, H.; *et al.* A functional graphene oxide-ionic liquid composites-gold nanoparticle sensing platform for ultrasensitive electrochemical detection of Hg^{2+}. Analyst 2013, 138, 1091–1097.

97 Zhang, B.; Guo, L.-H. Highly sensitive and selective photoelectrochemical DNA sensor for the deteciton of Hg2+ in aqueous solutions. Biosensors and Bioelectronics 2012, 37, 112–115.

98 Ponnusamy, V. K.; Mani, V.; Chen, S.-M.; *et al.* Rapid microwave assisted synthesis of graphene nanosheets/polyethyleneimine/gold nanoparticle composite and its application to the selective electrochemical determination of dopamine. Talanta 2014, 120, 148–157.

99 Tian, X.; Cheng, C.; Yuan, H.; *et al.* Simultaneous determination of l-ascorbic acid, dopamine and uric acid with gold nanoparticles–β-cyclodextrin–graphene-modified electrode by square wave voltammetry. Talanta 2012, 93, 79–85.

100 Song, J.; Xu, L.; Xing, R.; *et al.* Synthesis of Au/graphene oxide composites for selective and sensitive electrochemical detection of ascorbic acid. Scientific Reports 2014, 4, 7515.

101 Yang, F.; Wang, P.; Wang, R.; *et al.* Label free electrochemical aptasensor for ultrasensitive detection of ractopamine. Biosensors and Bioelectronics 2016, 77, 347–352.

102 Ma, X.; Chen, M. Electrochemical sensor based on graphene doped gold nanoparticles modified electrode for detection of diethylstilboestrol. Sensors and Actuators B: Chemical 2015, 215, 445–450.

103 Pruneanu, S.; Pogacean, F.; Biris, A.R.; *et al.* Novel graphene-gold nanoparticle modified electrodes for the high sensitivity electrochemical spectroscopy detection and analysis of carbamazepine. Journal of Physical Chemistry C 2011, 115 (47), 23387–23394.

104 Linting, Z.; Ruiyi, L.; Zaijun, L.; *et al.* An immunosensor for ultrasensitive detection of aflatoxin B1 with an enhanced electrochemical performance based on graphene/ conducting polymer/gold nanoparticles/the ionic liquid composite film on modified gold electrode with electrodeposition. Sensors and Actuators B: Chemical 2012, 174, 359–365.

105 Zhou, L.; Wang, J.; Li, D.; Li, Y. An electrochemical aptasensor based on gold nanoparticles dotted graphene modified glassy carbon electrode for label-free detection of bisphenol A in milk samples. Food Chemistry 2014, 162, 34–40.

106 Niu, X.; Yang, W.; Wang, G.; *et al.* A novel electrochemical sensor of bisphenol A based on stacked graphene nanofibers/gold nanoparticles composite modified glassy carbon electrode. Electrochimica Acta 2013, 98, 167–175.

107 Wang, X.; You, Z.; Cheng, Y.; *et al.* Application of nanosized gold and graphene modified carbon ionic liquid electrode for the sensitive electrochemical determination of folic acid. Journal of Molecular Liquids 2015, 204, 112–117.

108 Li, J.; Feng, H.; Li, J.; *et al.* Fabrication of gold nanoparticles-decorated reduced graphene oxide as a high performance electrochemical sensing platform for the detection of toxicant Sudan I. Electrochimica Acta 2015, 167, 226–236.

109 Gupta, V. K.; Atar, N.; Yola, M. L.; *et al.* A novel glucose biosensor platform based on Ag@AuNPs modified graphene oxide nanocomposite and SERS application. Journal of Colloid and Interface Science 2013, 406, 231–237.

110 Thanh, T.D.; Balamurugan, J.; Lee, S.H.; *et al.* Effective seed-assisted synthesis of gold nanoparticles anchored nitrogen-doped graphene for electrochemical detection of glucose and dopamine. Biosensors and Bioelectronics 2016, 81, 259–267.

111 Xue, K.; Zhou, S.; Shi, H.; *et al.* A novel amperometric glucose biosensor based on ternary gold nanoparticles/polypyrrole/reduced graphene oxide nanocomposite. Sensors and Actuators B: Chemical 2014, 203, 412–416.

112 Dhara, K.; Ramachandran, T.; Nair, B.G.; Babu, T.G. , Au nanoparticles decorated reduced graphene oxide for the fabrication of disposable nonenzymatic hydrogen peroxide sensor. Journal of Electroanalytical Chemistry 2016, 764, 64–70.

113 Song, H.; Ni, Y.; Kokot, S. A novel electrochemical biosensor based on the hemin-graphene nano-sheets and gold nano-particles hybrid film for the analysis of hydrogen peroxide. Analytica Chimica Acta 2013, 788, 24–31.

114 Jia, N.; Huang, B.; Chen, L.; *et al.* A simple non-enzymatic hydrogen peroxide sensor using gold nanoparticles-graphene-chitosan modified electrode. Sensors and Actuators B: Chemical 2014, 195, 165–170.

115 Gu, C.-J.; Kong, F.-Y.; Chen, Z.-D.; *et al.* Reduced graphene oxide-Hemin-Au nanohybrids: Facile one-pot synthesis and enhanced electrocatalytic activity towards the reduction of hydrogen peroxide. Biosensors and Bioelectronics 2016, 78, 300–307.

116 Xie, L.; Xu, Y.; Cao, X. Hydrogen peroxide biosensor based on hemoglobin immobilized at graphene, flower-like zinc oxide, and gold nanoparticles nanocomposite modified glassy carbon electrode. Colloids and Surfaces B: Biointerfaces 2013, 107, 245–250.

117 Yang, X.; Ouyang, Y.; Wu, F.; *et al.* Size controllable preparation of gold nanoparticles loading on graphene sheets@ cerium oxide nanocomposites modified gold electrode for nonenzymatic hydrogen peroxide detection. Sensors and Actuators B: Chemical 2017, 238, 40–47.

118 Li, S.-J.; Shi, Y.-F.; Liu, L.; *et al.* Electrostatic self-assembly for preparation of sulfonated graphene/gold nanoparticle hybrids and their application for hydrogen peroxide sensing. Electrochimica Acta 2012, 85, 628–635.

119 Wang, S.; Wang, Y.; Zhou, L.; *et al.* Fabrication of an effective electrochemical platform based on graphene and AuNPs for high sensitive detection of trace Cu 2+. Electrochimica Acta 2014, 132, 7–14.

120 Jiang, J.; Fan, W.; Du, X. Nitrite electrochemical biosensing based on coupled graphene and gold nanoparticles. Biosensors and Bioelectronics 2014, 51, 343–348.

121 Xu, M.-Q.; Wu, J.-F.; Zhao, G.-C. Direct electrochemistry of hemoglobin at a graphene gold nanoparticle composite film for nitric oxide biosensing. Sensors 2013, 13 (6), 7492.

122 Ting, S.L.; Guo, C.X.; Leong, K.C.; *et al.* Gold nanoparticles decorated reduced graphene oxide for detecting the presence and cellular release of nitric oxide. Electrochimica Acta 2013, 111, 441–446.

123 Liu, J. Adsorption of DNA onto gold nanoparticles and graphene oxide: surface science and applications. Physical Chemistry Chemical Physics 2012, 14, 10485–10496.

124 Liang, J.; Chen, Z.; Guo, L.; Li, L. Electrochemical sensing of L-histidine based on structure-switching DNAzymes and gold nanoparticle–graphene nanosheet composites. Chemical Communications 2011, 47, 5476–5478.

125 Mao, S.; Lu, G.; Yu, K.; *et al.* Specific protein detection using thermally reduced graphene oxide sheet decorated with gold nanoparticle-antibody conjugates. Advanced Materials 2010, 22 (32), 3521–3526.

126 Zhong, Z.; Wu, W.; Wang, D.; *et al.* Nanogold-enwrapped graphene nanocomposites as trace labels for sensitivity enhancement of electrochemical immunosensors in clinical immunoassays: Carcinoembryonic antigen as a model. Biosensors and Bioelectronics 2010, 25 (10), 2379–2383.

127 Sun, G.; Lu, J.; Ge, S.; *et al.* Ultrasensitive electrochemical immunoassay for carcinoembryonic antigen based on three-dimensional macroporous gold nanoparticles/graphene composite platform and multienzyme functionalized nanoporous silver label. Analytica Chimica Acta 2013, 775, 85–92.

128 Yang, F.; Yang, Z.; Zhuo, Y.; *et al.* Ultrasensitive electrochemical immunosensor for carbohydrate antigen 19-9 using Au/porous graphene nanocomposites as platform and Au@Pd core/shell bimetallic functionalized graphene nanocomposites as signal enhancers. Biosensors and Bioelectronics 2015, 66, 356–362.

129 Gao, Y.-S.; Xu, J.-K.; Lu, L.-M.; *et al.* A label-free electrochemical immunosensor for carcinoembryonic antigen detection on a graphene platform doped with poly(3,4-ethylenedioxythiophene)/Au nanoparticles. RSC Advances 2015, 5, 86910–86918.

130 Liu, Z.; Wang, Y.; Guo, Y.; Dong, C. Label-free electrochemical aptasensor for carcino-embryonic antigen based on ternary nanocomposite of gold nanoparticles, hemin and graphene. Electroanalysis 2016, 28 (5), 1023–1028.

131 Pang, X.; Li, J.; Zhaoi, Y.; *et al.* Label-free electrochemiluminescent immunosensor for detection of carcinoembryonic antigen based on nanocomposites of GO/MWCNTs-COOH/Au@ CeO_2. ACS Applied Materials and Interfaces 2015, 7, 19260–19267.

132 Azzouzi, S.; Rotariu, L.; Benito, A.M.; *et al.* A novel amperometric biosensor based on gold nanoparticles anchored on reduced graphene oxide for sensitive detection of L-lactate tumor biomarker. Biosensors and Bioelectronics 2015, 69, 208–286.

133 Xue, Q.; Liu, Z.; Guo, Y.; Guo, S. Cyclodextrin functionalized graphene–gold nanoparticle hybrids with strong supramolecular capability for electrochemical thrombin aptasensor. Biosensors and Bioelectronics 2015, 68, 429–436.

134 Zhu, W.; Li, X.; Wang, Q.; *et al.* Ru(bpy)$_3^{2+}$/nanoporous silver-based electrochemiluminescence immunosensor for alpha fetoprotein enhanced by gold nanoparticles decorated black carbon intercalated reduced graphene oxide. Scientific Reports 2016, 6, 20348.

135 Chen, Z.; Zhang, C.; Li, X.; *et al.* Aptasensor for electrochemical sensing of angiogenin based on electrode modified by cationic polyelectrolyte-functionalized graphene/gold nanoparticles composites. Biosensors and Bioelectronics 2015, 65, 232–237.

136 Yu, S.; Cao, X.; Yu, M. Electrochemical immunoassay based on gold nanoparticles and reduced graphene oxide functionalized carbon ionic liquid electrode. Microchemical Journal 2012, 103, 125–130.

137 Zhang, J.; Sun, Y. P.; Xu, B.; *et al.* A novel surface plasmon resonance biosensor based on graphene oxide decorated with gold nanorod–antibody conjugates for determination of transferrin. Biosensors and Bioelectronics 2013, 45, 230–236.

138 Li, T.; Liu, Z.; Wang, L.; Guo, Y. Gold nanoparticles/Orange II functionalized graphene nanohybrid based electrochemical aptasensor for label-free determination of insulin. RSC Advances 2016, 6 (36), 30732–30738.

139 Sun, X. J., Z.; Wang, H.; Zhao, H. Highly sensitive detection of peptide hormone prolactin using gold nanoparticles-graphene nanocomposite modified electrode. International Journal of Electrochemical Science 2015, 10, 9714–9724.

140 Wang, L.; Xu, M.; Han, L.; *et al.* Graphene enhanced electron transfer at aptamer modified electrode and its application in biosensing. Analytical Chemistry 2012, 84 (17), 7301–7307.

141 Ruiyi, L.; Ling, L.; Hongxia, B.; Zaijun, L. Nitrogen-doped multiple graphene aerogel/gold nanostar as the electrochemical sensing platform for ultrasensitive detection of circulating free DNA in human serum. Biosensors and Bioelectronics 2016, 79, 457–466.

142 Lei, Y.; Yang, F.; Tang, L.; *et al*. Identification of chinese herbs using a sequencing-free nanostructured electrochemical DNA biosensor. Sensors 2015, 15 (12), 29882–29892.

143 Wei, Q.; Xiang, Z.; He, J.; *et al*. Dumbbell-like Au-Fe$_3$O$_4$ nanoparticles as label for the preparation of electrochemical immunosensors. Biosensors and Bioelectronics 2010, 26 (2), 627–631.

144 Wang, Y.; Ping, J.; Ye, Z.; *et al*. Impedimetric immunosensor based on gold nanoparticles modified graphene paper for label-free detection of Escherichia coli O157:H7. Biosensors and Bioelectronics 2013, 49, 492–498.

145 Liu, C.; Jiang, D.; Xiang, G.; *et al*. An electrochemical DNA biosensor for the detection of Mycobacterium tuberculosis, based on signal amplification of graphene and a gold nanoparticle-polyaniline nanocomposite. Analyst 2014, 139 (21), 5460–5465.

7

Recent Advances in Electrochemical Biosensors Based on Fullerene-C60 Nano-structured Platforms

Sanaz Pilehvar and Karolien De Wael

AXES Research Group, Department of Chemistry, University of Antwerp, Belgium

7.1 Introduction

Bio-nanotechnology is an emerging new field of nanotechnology and combines knowledge from engineering, physics, and molecular engineering with biology, chemistry, and biotechnology aimed at the development of novel devices such as biosensors, nanomedicines, and bio-photonics [1]. A biosensor is an analytical device that consists of a biological recognition element in direct spatial contact with a transduction element which ensures the rapid and accurate conversion of the biological events into a measurable signal [2]. However, the discovery of rich nanomaterials has opened up new opportunities in the field of biosensing research and offer significant advantages over conventional biodiagnostic systems in terms of sensitivity and selectivity [2, 3]. Among various nanostructured materials, carbon nanomaterials have been receiving great attention owing to their exceptional electrical, thermal, chemical, and mechanical properties and have found application in different areas as composite materials, energy storage and conversion, sensors, drug delivery, field emission devices, and nanoscale electronic components [4–6]. Moreover, the possibility to customize their synthesis with attached functional groups or to assemble them into three-dimensional arrays has allowed researchers to design high surface area catalysts and materials with high photochemical and electrochemical activity. Their exceptional electrochemical properties lead to their wide application for designing catalysts for hydrogenation, biosensors, and fuel cells [6]. The wide application of carbon nanomaterial for is construction of biosensors partly motivated by their ability to improve electron-transfer kinetics, high surface-to-volume ratios, and biocompatibility [6, 7]. In addition, the use of nanomaterials can help to address some of the key challenges in the development of biosensors, such as sensitive interaction of an analyte with biosensor surface, efficient transduction of the biorecognition event, and reduced response times. Various kinds of zero-, one-, two-, and three-dimensional carbon nanomaterials have been used. Examples of such materials include carbon nanotubes, nanowires, nanoparticles, nanoclusters, graphene, etc. [8]. Fullerene is a very promising member of carbon nanostructure family. The closed cage, nearly-spherical C60 and related analogues have attracted great interest in recent years. Multiple redox states, stability in many redox forms, easy functionalization,

Nanocarbons for Electroanalysis, First Edition.
Edited by Sabine Szunerits, Rabah Boukherroub, Alison Downard and Jun-Jie Zhu.

signal mediation, and light-induced switching are among their exceptional properties. In different applications, fullerenes have been used for the development of superconductors, (bio)sensors, catalysts, optical and electronic devices [9, 10]. Their superior electrochemical characteristics combined with unique physiochemical properties enable the wide application of fullerenes in the design of novel biosensor systems [11]. It is the aim of this review to present the most recent and relevant contributions in the development of biosensors based on fullerene-C60 and different biological components. A brief introduction and history of fullerene-C60 is first presented in section 7.1.2. Available methods for synthesis and functionalization of fullerene-C60 are mentioned in section 7.1.3. Finally, we briefly outline the current status and future direction for electrochemical biosensors based on fullerene-C60, especially, fullerene-C60 as an immobilizing platform for DNA. Recently, Afreen *et al.* introduced a review on functionalized fullerene-C60 as nanomediators for construction of glucose and urea biosensors [12]. However, the present review covers all aspects of biosensors based on fullerene-C60.

7.1.1 Basics and History of Fullerene (C60)

Fullerene is built up of fused pentagons and hexagons forming a curved structure. The smallest stable, and the most abundant, fullerene obtained by the usual preparation method is the Ih-symmetrical buckminsterfullerene C60. The next stable homologue is C70 followed by higher fullerenes C74, C76, C78, C80, C82, C84, and so on [13, 14]. Since the discovery of fullerenes, buckminsterfullerene (C60) has fascinated a large number of researchers due to its remarkable stability and electrochemical properties. The stability of the C60 molecules is due to the geodesic and electronic bonding present in its structure (Figure 7.1). In 1966, Deadalus (also known as D.E.H. Jones) considered the possibility of making a large hollow carbon cage (giant fullerene). Later on, in 1970, Osawa first proposed the spherical Ih-symmetric football structure of the C60 molecule. In 1984, it was observed that upon laser vaporization of graphite large carbon clusters of Cn with n = 30–190 can be produced. The breakthrough in the discovery of the fullerene happened in 1985 when Kroto and Smalley proved the presence of C60 and C70, which can be produced under specific clustering conditions. The second breakthrough in fullerene research was achieved by Kratschmer and Huffman. They invented the laboratory analogues of interstellar dust by vaporization of graphite rods in a helium

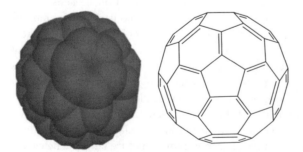

Figure 7.1 Schematic representation of C60. *Source*: Baghchi 2013[1]. Reproduced with permission of John Wiley and Sons.

atmosphere and observed that upon choosing the right helium pressure, the IR spectrum shows four sharp strong absorption lines which were attributed to C60 [11, 15].

Each carbon in a fullerene-C60 atom is bonded to three others and is sp^2 hybridized. The C60 molecule has two bond lengths, the (6,6) ring bonds can be considered as double bonds and are shorter than the (6,5) bonds. C60 is not "superaromatic" as it tends to avoid double bonds in the pentagonal rings, resulting in poor electron delocalization. Therefore, C60 structure behaves like an electron-deficient alkene, and reacts readily with electron-rich species. The estimated values of electron affinity (EA) (2.7 eV) and ionization potential (IP) (7.8 eV) of C60 indicate that it can easily contribute to the electron transfer reaction and reveal very rich electrochemistry which makes them attractive candidates for electroanalytical applications [7, 11, 16].

7.1.2 Synthesis of Fullerene

It was initially shown that the production of fullerene is achievable by means of an irradiating laser beam on a graphite rod placed in a helium atmosphere [17]. However, the overall yield rate of fullerene was insufficient for its potential applications in various industrial fields. Therefore, different production methods have been developed for sufficient production of fullerene [18–20]. The second proposed method is based on laser ablation of graphite in a helium atmosphere. In the laser ablation method, materials are removed from a solid surface by irradiating it with a laser beam. During the laser irradiation of graphite, materials are evaporated and their vapors are converted to plasma. Upon cooling the gas, the vaporized atom tends to combine and form fullerene [17]. The arc discharge process is an alternative method, where the vaporization of the input carbon source is achieved by the electric arc formed between two electrodes [17, 18]. They can also be produced by the non-equilibrium plasma method, where a non-equilibrium gas phase in the glow discharge is induced by the non-equilibrium plasma and fullerene is generated without the need for high temperatures [21].

7.1.3 Functionalization of Fullerene

Early studies on the C60 molecular structure showed that this carbon allotrope could undergo electron deficient polyolefin reactions. The (6,6) bonds have greater double bond character and are shorter than the (5,6) bonds and, thus, used to functionalize C60 by nucleophilic, radical additions, as well as cycloadditions [22]. Many reactions have been developed for the functionalization of C60, which consists of cyclopropanation (the Bingel reaction), (4 + 2) cycloaddition (the Diels–Alder reaction) and (3 + 2) cycloaddition (the Patro reaction), (2 + 2) cycloaddition. The Bingel reaction has been frequently used to prepare C60 derivatives in which a halo ester or ketone is first deprotonated by a base and subsequently added to one of the double bonds in C60 resulting in an anionic intermediate that reacts further into a cyclopropanated C60 derivative. In addition, cyclopropanation reactions have shown to be an efficient method for the preparation of fullerene derivatives with wide application in material science and biological applications (Scheme 7.1a) [23, 24]. Additionally, the double bonds exist in C60 can react with different dienes by Diels–Alder reaction (Scheme 1b). The main drawback of Diels–Alder reaction is low thermal stability of formed product [23, 25]. In another reaction (the Prato reaction), an azomethine ylide, generated in situ by the decarboxylation of iminium salts derived from the condensation of α-amino acids with

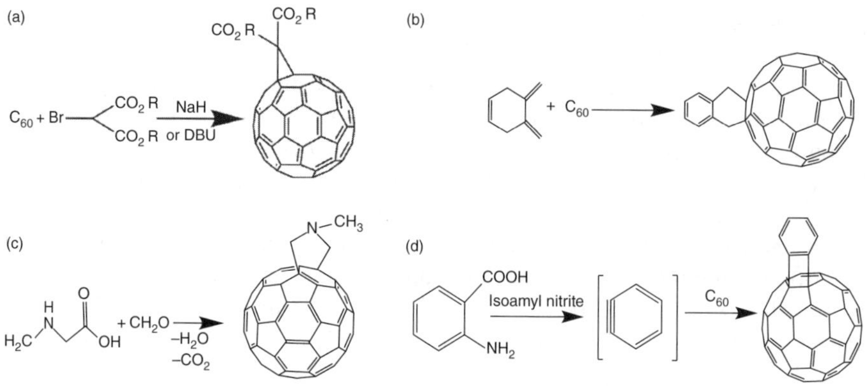

Scheme 7.1 (a) The Bingel reaction; (b) the Diels–Alder reaction; (c) the Prato reaction; and (d) the cycloaddition reaction.

aldehydes or ketones, react with fullerene to produce [3 + 2] cycloadduct (Scheme 7.1c) [23]. A wide variety of functionalization of fullerene molecule is possible by means of the Patro reaction. In another way, the addition of benzyne to C60 leads to the formation of [2 + 2] cycloadducts (cycloaddition) (Scheme 7.1d) [26]. However, among available methods for functionalization of fullerene, cycloaddition reactions have emerged as very useful and well-established methods for functionalization of one or several of the fullerene double bonds.

7.2 Modification of Electrodes with Fullerenes

The idea of introducing C60 chemically modified electrode (CME) was first reported by Compton and co-workers in 1992 [27]. They prepared C60-based CMEs by immobilizing C60 films by drop coating onto surfaces of the electrodes, which were then coated with Nafion as protecting films. It was observed that the current signal improved compared to those using C60 dissolved in solution. The electrochemical behavior of the C60-CMEs, in non-aqueous and aqueous solutions, has been widely investigated, suggesting the possibility of their electroanalytical applications [28]. Currently, fullerene-based CMEs are prepared in several different ways. The most common method is drop coating the electrode by using a fullerene solution of a volatile solvent [29–31]. Electrochemical deposition was also used for modification of the electrode surface by fullerene film [32]. Moreover, fullerene-based CMEs can be prepared by electro-polymerization where the formed fullerene units are connected by polymer side chains or via epoxide formation [33]. Alternative method for C60 films preparation is the self-assembled monolayer (SAM) films using either thiols or silane derivatives of C60 on the electrode surfaces [34].

Fullerene-C60 are widely used for construction of electrochemical biosensors. Generally, electrochemical biosensors are analytical devices which consist of a bioreceptor, an electrochemical active interface, a transducer element to convert biological reaction to an electrical signal, and a signal processor [35]. The principle of the electrochemical biosensors is based on the specific interaction between the analyte and

Figure 7.2 HOMO and LUMO gap in fullerene-C60.

biorecognition element which is also associated with a better correlation between the bioreceptor and the transducer surface [36–38]. Utilization of different kinds of nanomaterials leads to the important improvements in these aspects. For this reason, nanomaterials have been widely used in the construction of biosensors in order to improve the sensing performance of the biosensors [2]. The ability of signal mediation, easy functionalization, and light-induced switching lead to the fact that fullerene be considered as a new and attractive element in the fabrication of biosensors. Different biomolecules or organic ligands can be immobilized to the shell of fullerene by adsorption or covalent attachment [39]. Fullerenes are not harmful to biological component, they can locate the closest distance to the active site of biomolecules. In this way, they can make close arrangements with biomolecules and easily accept or donate electrons [9, 14]. Furthermore, they are small enough and provide ideal substrate for absorbing energy, taking up electrons and releasing them with ease to a transducer. Their high electron-accepting property is due to a low-lying, triply-degenerate, lowest unoccupied molecular orbital (LUMO) which is around 1.8 eV above its five-fold degenerate highest occupied molecular orbital (HOMO) (Figure 7.2) [40].

Carbon nanotubes (CNT) were also widely used to chemically functionalized electrodes due to their remarkable electrical, chemical, mechanical, and structural properties. It was shown that the chemically modified CNT would make the electrodes more sensitive and selective in detection applications. Furthermore, CNTs have advantages over other carbon nanomaterials, as they exhibit superior electrocatalytic properties [28].

7.2.1 Fullerene (C60)-DNA Hybrid

In general, combining two or more different materials via interaction forces leads to the appearance of novel hybrid materials with unique properties. In addition, the nanoscale properties of biomolecules hold great promise, and their characteristic in nanoscale is different from their bulk complements. Biomolecules offer great compatibility and suitability to form bio-nano hybrid structures and significant efforts have been made to form the nano-bio hybrid systems for various applications [41, 42]. There are various approaches available to create hybrid materials consist of biomacromolecules and nanomaterials. One approach is based on connecting the molecules through covalent bonding and the other approach is adsorption (or wrapping) of a material onto the surface of the other materials via supramolecular interactions, encapsulation or groove of the other molecule [43–45].

Among the various biomolecules, DNA attracted a great attention due to its superior properties such as structural regularity, biocompatibility, and unique double helix structure, which leads a range of outstanding properties that are hard to find in other biomolecules [46]. Furthermore, DNA is a potential material for combining with other chemicals, especially with nanomaterials by different interactions. Taking into account the advantages of DNA and carbon nanomaterials, the combination of DNA and carbon nanomaterials offers unique advantages for different application. Hence, the combination of DNA with new carbon allotropes is a skillful and challenging area that can lead to the development of novel nano-biomaterials with exceptional properties for a variety of potential applications such as gas sensors and catalysts, as well as electronic and optical devices, sensitive biosensors, and biochips [46, 47].

7.2.1.1 Interaction of DNA with Fullerene

The interaction between DNA and other species plays an important role in life science since it is in direct contact with the transcription of DNA, mutation of genes, origins of diseases, and molecular recognition studies [45]. Cassell *et al.* studied the interaction between DNA and fullerene. For this purpose, N,N-dimethylpyrrolidinium iodide is used as a complexing agent to form DNA/fullerene complexes through the phosphate groups of the DNA backbone, which was imaged by TEM. It was shown that the complexation of free fullerene with DNA is sterically permitted and surfactants can be used in order to prevent the DNA/fullerene hybrids from aggregation [48]. Later on, Pang *et al.* [49] studied the interaction of DNA with fullerene-C60 in depth. The method used was based on the double-stranded DNA (dsDNA) modified gold electrodes (dsDNA/Au) in combination with the electrochemical method for investigation of the interactions between C60 derivatives and DNA. They have chosen $[Co(phen)_3]^{3+/2+}$ as an appropriate electroactive indicator, which can interact electrostatically and intercalatively with dsDNA, to characterize the interactions. In the presence of dsDNA, the peak currents related to $[Co(phen)_3]^{3+/2+}$ decreased due to its interaction with dsDNA and then recovered significantly in the presence of $H_{10}C_{60}(NHCH_2CH_2OH)_{10}$. Electrochemical studies with dsDNA-modified gold electrodes, shows that the C60 derivative could interact strongly with dsDNA, with binding sites of the major groove and the phosphate backbone of dsDNA. The interaction between dsDNA and $H_{10}C_{60}(NHCH_2CH_2OH)_{10}$ was attributed to the interaction between the delocalized π electrons of $H_{10}C_{60}(NHCH_2CH_2OH)_{10}$ and DNA and the binding of $H_{10}C_{60}(NHCH_2CH_2OH)_{10}$ to the major grooves of the double helix. It is believed that $H_{10}C_{60}(NHCH_2CH_2OH)_{10}$, in the protonated form, interacts electrostatically with the negatively-charged phosphate backbone of the dsDNA. It can also access the major groove of the double helix and interact with the delocalized π system of bases of dsDNA. When fullerene is electrically neutral, the electrostatic interaction with the dsDNA vanishes and the π–π interactions is present. In addition, it was shown that the binding and dissociation of $H_{10}C_{60}(NHCH_2CH_2OH)_{10}$ to the dsDNA is a reversible process [49]. So far, it was believed that the water-soluble C60 molecules only bind to the major grooves and the free ends of the dsDNA. However, extensive simulations indicated that the association of hydrophobic C60 can also occur at the minor groove sites and no complexation occurs at the major grooves. The free ends of the double-strand DNA fragment are the hydrophobic regions which favor the diffusion of hydrophobic fullerenes toward their docking sites [50]. In addition, calculation of the binding energy showed that the hybrid C60-DNA complexes are energetically favorable compared to the unpaired molecules. The self-association of C60 molecules in the presence of DNA molecules revealed that the self-association between C60 molecules occurs in the early stages of simulation. However, after 5 ns of simulation, one of the C60 molecule binds to one end of the DNA. Visual observation of the obtained results from simulation showed that the overall shape of the dsDNA molecule is not affected by the association of C60. However, the association C60 has more impact on the DNA structure when more hydrophobic contacting surfaces are exposed at the end of double-strand DNA. The binding between the C60 and DNA molecules is attributed to the hydrophobic interaction between the C60 and hydrophobic sites on the DNA [50]. In another study, it is reported that the C60 molecule binds to the ssDNA molecule with a binding energy of about $-1.6\,eV$, showing that fullerenes can strongly bind to nucleotides [51]. However, the mobility of DNA and their interaction with water molecules which often present in real physical systems were not taking into these calculations.

7.2.1.2 Fullerene for DNA Biosensing

The preparation of DNA hybridization sensors involves the attachment of oligonucleotide probes on the surface of the electrode. DNA immobilization step has been considered as a fundamental step in the fabrication of DNA biosensing [52, 53]. Various electrode materials, such as gold, carbon paste, glassy carbon, carbon fibers, and screen printed electrodes, have been utilized to immobilize the DNA. Despite, carbon nanomaterials such as C60 are compounds that have attracted much interest as the materials for DNA sensors and biosensors because of their unique properties. Shiraishi *et al.* demonstrated a new procedure of immobilizing DNA onto a fullerene impregnated screen printed electrode (FISPE) for detection of 16S rDNA, extracted from *Escherichia coli* [54]. The integrated FISPE was a mixture of ink and fullerene solution, which is modified with probe DNA in the next step. The efficiency of the developed method was tested by detecting 46S rDNA of *Escherichia coli* by means of the modified electrode with perfectly matched probes. It is shown that the reduction peak of $Co(phen)_3^{+3}$ (electroactive hybridization indicator) is enhanced only on the perfectly matched probes modified electrode after hybridization due to the accumulation of indicator into the hybrid between perfectly matched probe and rDNA of target. In addition, it was observed that the electrochemical response of $Co(phen)_3^{+3}$ accumulated in the hybrid was improved when using FISPE, which based on the author's opinion shows that the probe DNA was immobilized in a high concentration onto the air plasma activated FISPE surface.

Other carbon nanomaterials were also used for the development of new (bio)sensing systems for applications in the food industry, environmental monitoring, and clinical diagnostics. For example, recently CNT-modified arrays have been used to detect DNA targets by combining the CNT nanoelectrode array with $Ru(bpy)_2^{+3}$ mediated guanine oxidation [55]. In another study, a MWCNT-COOH modified glassy carbon was used in combination with an amino functionalized oligonucleotide probe and pulse voltammetric transduction method [56]. Recently, an indicator-free AC impedance measurements of DNA hybridization based on DNA probe-doped polypyrrole film over a MWCNT layer reported by Cai *et al.* [57] A five-fold sensitivity enhancement was observed compared to analogous measurements without CNT. However, most of the cases suffer from the feasibility of scale-up conditions, due to the low yield and expensive experimental procedures. The sample inhomogeneity in CNT samples due to the different production procedures, further limits their application.

7.2.1.3 Fullerene as an Immobilization Platform

Nanosized materials can be used as potential building blocks to construct higher ordered supramolecular architectures for designing the highly-sensitive biosensing platform. For example, the working electrode modified with partially reduced fullerene-C60 show exceptional properties, such as high electroactive surface area, excellent electronic conductivity, and good biocompatibility [58, 59]. Zhang *et al.* have developed a technique to disperse fullerene-C60 nanotubes (FNTs) homogeneously into aqueous solution by forming a kind of complex with ssDNA [58]. The FNT/DNA was modified onto the surface of the GCE by air-drying/adsorption, enabling the electrochemical analysis of the modified electrode with voltammetric technique. The modified electrode was employed for the electrochemical detection of dopamine (DA) in the presence of ascorbic acid. The interaction of FNT with DNA was studied by UV-Vis measurements. The observed red shift attributed to the weak binding between the two, and it was shown that π–π stacking and hydrophobic interaction contribute in the formation of FNT/DNA hybrid. It is believed that the strong physisorption of DNA onto the FNTs via a wrapping mechanism prevent the FNT/DNA from precipitation upon adding water or organic solvent. Obtained SEM images of the surface FNT/DNA modified electrode proved the formation of uniform films. In another study, Gugoasa *et al.* investigated the influence of dsDNA which is physically immobilized on the multi-walled carbon nanotubes (MWCNT), synthetic monocrystalline diamond (DP) and fullerenes-C60 on the detection of three different neutransmitters such DA, epinenephrine and norepinephrine [60]. Optimized working condition for dsDNA biosensors was found to be a value of 4.0 for pH and the 0.1 mol/L KNO_3. It has been shown that the highest improvement of the signal for the DA was recorded when dsDNA was immobilized on DP. While the larger working concentration and the lowest LOD were obtained when dsDNA has been immobilized on MWCNT, both LOD and LOQ decreased when dsDNA was immobilized on fullerene-C60. This occurrence attributed to the fact that the immobilization matrix has a very important contribution to the biosensor performance.

Several studies show that not only the nature of the material, but also the geometry of the substances at the molecular level has the effect on the behavior of the biosensors [60]. However, the obtained fullerene-C60 by simply stirring or ultrasonication treatments was not suitable for biomedical applications because of their aggregation properties. To solve these limitations, the covalent binding of nano-C60 to aminoacids, hydroxyl groups, carboxyl groups etc., which can increase the nanoparticle's ability to interact with the biological environment can be performed [61–63]. On the other hand, the synthesis of the functionalized C60 with non-covalent interaction based on supramolecular chemistry would preserve the original structure and electrochemical properties of C60. Supramolecular chemistry is the chemistry of the intermolecular bond, aims at developing highly complex chemical system components in interacting by non-covalent intermolecular forces [64]. A new supramolecular method is developed by Han *et al.* for preparation of thiol and amino functionalized C60 nanoparticles with better water solubility and larger active surface area [64]. For this purpose, the amino functionalized 3,4,9,10-perylenetetracarboxylic dianhydride (PTC-NH2) was used as a π electron compound which can be bond to the surface of C60 via supramolecular interaction. Then, the prussian blue carried gold nanoparticles (Au@PBNPs) were interacted with C60 nanoparticles (Au@PB/C60) and the detection aptamers for platelet-derived growth factor B-chain (PDGF-BB) was labeled by Au@PB/C60 and the coupled with alkaline phosphatase for electrochemical aptasensing (Scheme 7.2a). The combination of fullerene-C60 and AuNPs have been used for immobilization of a large amount of

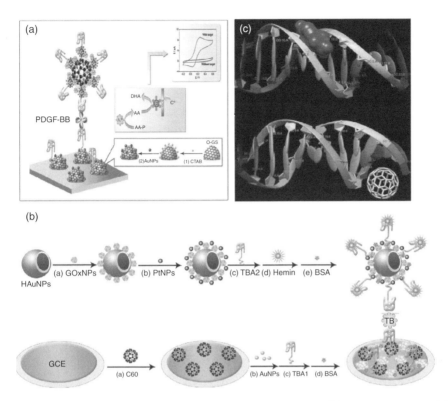

Scheme 7.2 (a) Schematic illustration of the stepwise aptasensor fabrication process and the dual signal amplification mechanism. *Source*: Han 2013 [64]. Reproduced with permission of Elsevier. (b) schematic diagram of fabrication and detection of the ECL aptasensor. *Source*: Zhuo 2014 [65]. Reproduced with permission of Elsevier. and (c) results of molecular modeling related to (a) groove binding of small molecules to the minor groove of dsDNA and (b) groove binding of fullerene-C60 to the major groove of dsDNA. *Source*: Gholivand 2014 [66]. Reproduced with permission of Elsevier.

capture aptamers on the surface of the electrode. The obtained SEM and TEM images showed that the Au@PBNPs were adsorbed uniformly and tightly on the C60 nanoparticles. The performance of developed aptasensor was investigated by detecting PBGF-BB standard solutions (Table 7.1).

Electrochemiluminescence (ECL) is a powerful analytical tool for the detection of clinical samples. A peroxydisulfate/oxygen ($S_2O_8^{2-}/O_2$) system is widely used for amplification of ECL signals where the dissolved O_2 can serve as a co-reactant [67]. The enzymatic reaction can catalyze in situ generation of the dissolved O_2 [63, 65]. Zhao *et al.* [65] developed a sandwich-type aptasensor based on mimicking bi-enzyme cascade catalysis to in situ generate the co-reactant of dissolved O_2 for signal amplification to detect thrombin (TB). In this study, gold nanoparticles (AuNPs) were utilized as carriers to immobilize glucose oxidase nanoparticles (GOxNPs) and platinum nanoparticles (PtNPs). GOxNPs could catalyze the glucose to generate H_2O_2, which could be further catalyzed by hemin/G-quadruplex and PtNPs, in order to in situ generate dissolved O_2 with high concentration. The detection aptamer of thrombin (TBA2) was

Table 7.1 Comparison of different fullerene-C_{60} modified biosensors.

Receptor	Analyte	Linear Range	Sensitivity	LOD	References
ssDNA	Dopamine	2–160 μM	—	0.6 μM	[58]
ssDNA	PDGF-BB	0.002–40 nM	—	0.6 pM	[64]
ssDNA	Thrombin	1 μM–10 nM	—	0.3 fM	[65]
ssDNA	16S rDNA	—	—	—	[54]
dsDNA	CD	0.1–25.0 nM	$0.0235\ \mu A \cdot nM^{-1}$	0.03 nM	[66]
dsDNA	Dopamine	10^{-5}–10^{-2} M	$100\ nA \cdot nM^{-1}$	1.2 μM	[60]
dsDNA	Epinephrine	10^{-6}–10^{-2} M	$100\ nA \cdot nM^{-1}$	0.1 μM	[60]
dsDNA	Norepinephrine	10^{-5}–10^{-2} M	$0.1\ nA \cdot nM^{-1}$	2.3 μM	[60]
Anti-IgG	IgG	—	$1.25 \times 10^{2}\ Hz/(mg/mL)$	—	[69]
Anti-Hb	Hb	—	$1.5 \times 10^{4}\ Hz$	$<10^{-4}\ mg/mL$	[69]
Anti-*E. coli*	*Escherichia coli* O157:H7	3.2×101 to 3.2×106 CFU/mL	—	15 CFU/mL	[72]
GOD-Chit	Glucose	0.05–1 mM	—	694 ± 8 μM	[80]
cobalt(II) hexacyanoferrate-GOD	Glucose	0–8 mM	$5.60 \times 10^{2}\ nA/mM$	1.6 μm	[82]
Glucose oxidase	Glucose	—	$5.9 \times 10^{2}\ Hz/\Delta log\ M$	3.9×10^{-5} M	[87]
Urease	Urea	1.2 mM–0.042 mM	$59.67 \pm 0.91\ mV/dcade$	—	[91]
AuNPs-TVL	Laccase	0.03–0.30 M	—	0.006 mM	[88]
Anti-*E. coli*	*Escherichia coli* O157:H7	3.2×101 to 3.2×106 CFU/mL	—	15 CFU/mL	[72]

immobilized on the PtNPs/GOxNPs/AuNPs and hemin was intercalated into the TBA2 to obtain the hemin/G-quadruplex/PtNPs/GOxNPs/AuNPs nanocomplexes, which was utilized as signal tags (Scheme 7.2b). The surface of glassy carbon is modified with C60 and electrochemical deposited AuNPs for further immobilization of thiol-termi-nated thrombin capture aptamer (TBA1). The TBA1, TB, and TBA2 make a sandwich-type structure. The zero-dimensional nano-C60 was shown to enhance the immobilization of nanoparticles but also amplified the ECL signal owing to its large specific surface area. The developed aptasensor is characterized by the ECL measure-ments. The bare GCE showed relatively low ECL intensity in the low concentration level of dissolved O_2. The ECL intensity of the bare GCE was enhanced in the presence of dissolved O_2. The ECL intensity was increased when using nano-C60 was coated onto the electrode, due to the enrichment effect of nano-C60 on peroxydisulfate lumi-nescence. Electrodepositing of AuNPs was further enhanced the ECL intensity since it accelerate the electron transfer in ECL reaction. However, the ECL intensity decreased successively when TBA1 were immobilized onto the electrode. The ECL signal dropped again after the incubation of modified electrode with the target analyte of TB. The ECL aptasensor also evaluated by CV in 0.1 M PBS. While the relatively low CV intensity was obtained at bare GCE, the CV intensity reduced when the electrode was coated with C60 due to its low electrical conductivity (Table 7.1). In another report, Gholivand *et al.* studied the mechanism of the prevention of Parkinson's disease by means of Carbiodopa (CD) drug at a double-stranded DNA (dsDNA) and fullerene-C60-modified glassy car-bon electrode (dsDNA/FLR/GCE) by cyclic voltammetry [66]. They have used multi-variate analysis to distinguish the complex system. Firstly, the effect of pH on the electrochemical system has been studied which showed that a value of 4.0 for pH resulted in higher sensitivity of the system. The cyclic voltammograms recorded at dif-ferent electrodes showed that the electrocatalytic behavior toward oxidation of CD at FLR/GCE is improved noticeably in comparison with the bare GCE. When dsDNA was added to the CD solution both oxidation and reduction peaks decreased markedly and shifted to less and more positive potentials, respectively, which indicate that CD inter-acts with dsDNA. The electronic UV-Vis absorption spectroscopy was used to charac-terize the interaction between dsDNA and small molecules. The absence of red shift in the obtained spectra was attributed to the fact that the binding mode is not the interca-lative binding and it could be groove binding [66]. By means of all these observations, it has been suggested that small molecules, such as CD, interact with the minor groove, while large molecules (fullerene-C60) tend to interact with the major groove binding site of DNA. This phenomenon was earlier reported and further proved by molecular modelling, which is performed in this study (Scheme 7.2c).

7.2.2 Fullerene(C60)-Antibody Hybrid

Conventional immunosensors suffer from drawbacks, such as intrinsic complexity and the requirement for signal amplification, large sample size, and high cost. However, by integrating nano-scale carbon materials, most of these limitations can be overcome [68]. Especially, fullerene-C60 with conjugate π electrons are considered as electro-philic molecules, and they can be attacked by electron-donating molecules, such as amines, antibodies, and enzymes. A sensitive immobilized C60-antibody-coated pie-zoelectric crystal sensor, based on C60-anti-human IgG and C60-anti-hemoglobin,

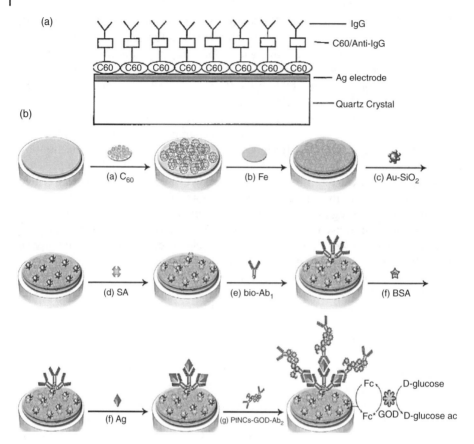

Scheme 7.3 (a) Diagrams of the C60-anti-human IgG-coated quartz crystal electrode for IgG; (b) the fabrication of the electrochemical immunosensor for *Escherichia coli* O157:H7. *Source*: Li 2013 [72] and Pan 2004 [69]. Reproduced with permission of Elsevier.

were developed to detect IgG and hemoglobin in aqueous solutions (Scheme 7.3a) [69]. For this purpose, a fullerene C60-coated piezoelectric quartz crystal has been used to investigate the interaction between C60 and the antibody. The changes in the resonant frequency of the crystal are recorded which is directly related to the deposited mass [70]. The frequency changes, respond sensitively to the adsorption of anti-IgG onto the C60 coated crystals. The interaction between C60 and anti-IgG is found to be chemisorption with good reactivity. The effect of the C60 coating load on the frequency response of the C60 coated PZ crystal for anti-IgG in water was investigated. The PZ quartz crystal with more C60 coating exhibited a larger frequency shift, but the frequency shift of the C60-coated PZ sensor tends to level off with larger amounts of C60 coating. It was ascribed to the fact that C60 can only adsorp IgG on its surface to some extent. The obtained results have been revealed that the concentration of antibody, temperature, and pH have an impact on the response of the biosensor. The immobilized C60-anti-hemoglobin (C60-Hb)-coated piezoelectric quartz crystal hemoglobin biosensor was also developed to detect hemoglobin in solutions. The partially irreversible response of the C60-coated piezoelectric crystal for anti-hemoglobin was tested, suggesting the chemisorption and the good reactivity of anti-hemoglobin on C60 coated crystal. The immobilized C60-Hb coated piezoelectric crystal sensor exhibited linear response frequency to the concentration of hemoglobin with sensitivity of about 1.56×10^4 Hz $(mg\,mL^{-1})$ and detection limit of $<10^{-4}\,mg\,mL^{-1}$ to hemoglobin in solutions (Table 7.1) [71].

Recently, Li *et al.* reported development of a sensitive and efficient electrochemical immunosensor for amperometric detection of *Escherichia coli* O157:H7 (*E. coli* O157:H7) [72]. The immunosensing platform was first composed of fullerene, ferrocene, and thiolated chitosan composite nano-layer (C60/Fc/CHI–SH) and then AuNPs coated SiO_2 nanocomposites were assembled on the thiolated layer. Next, the large amount of avidin was coated on the Au-SiO_2 surface, which was used to immobilize biotinylated capture antibodies of *E. coli* O157:H7 (bio-Ab1). For signal amplification, the glucose oxidase (GOD)-loaded platinum nanochains (PtNCs) were used as a tracing tag to label signal antibodies (Ab2) (Scheme 7.3b). It has been shown that Au-SiO_2 embedded C60/Fc/CHI–SH provide a biocompatible platform for increasing the surface area to capture a large amount of SA/bio-Ab1 and Ab2 and GOD multi-functionalized PtNCs nanocomposites for improved sensitivity.

7.2.3 Fullerene(C60)-Protein Hybrid

Direct electron transfer of biological redox proteins plays an important role in elucidating the intrinsic thermodynamic characteristics of biological systems and designing new kinds of biosensors or biomedical devices [73]. Fullerenes-C60 are ideal nanomaterials for absorbing energy, taking up electrons and releasing them to the transducer. They are small enough to locate at closest distance to the active site of the catalytic enzyme, which makes the electron transfer easier. Moreover, they are not harmful to biological material and proteins [74–76]. The interaction between the enzyme and the nanomaterial surface can be accomplished by a covalent or non-covalent bond. The improved stability, accessibility, and selectivity, as well as the reduced leaching, can be achieved through covalent bonding because the location of the biomolecule can be controlled [77]. Moreover, several types of immobilization methods have been developed for biomolecules. These methods include entrapment, encapsulation, covalent binding, cross-linking, and adsorption [78, 79].

7.2.3.1 Enzymes
Glucose oxidase: The determination of glucose is medically important for diagnosis of diabetes since the low absorption of glucose can lead to diabetes. In general, glucose is being detected by an electrochemical method with an immobilized glucose oxidase (GOD) enzyme. Glucose oxidase (GOD) is a glycoprotein which catalyzes the electron transfer from glucose to oxygen with the byproduct of gluconic acid and hydrogen peroxide [80–82]. The preparation of the immobilized GOD enzyme surface is a crucial step in the development of electrochemical glucose sensors. In most cases, GOD enzymes were immobilized by the entrapment of GOD in polymers or macromolecules, e.g., polyvinyl alcohol, agar, collagen, cellulose triacetate, gelatin, and Nafion [83–85]. On the other hand, the covalently coupled enzymes results in the formation of highly stable bonds between enzyme and matrix [86]. Electron-releasing molecules such as amines can attack fullerene-C60 with 60 π electrons. Therefore, the NH group

containing enzyme molecules is expected to bond chemically to the fullerene -C60 molecule, resulting in the formation of stable, immobilized C60-enzymes [82].

Chuang *et al.* reported fullerene-C60/GOD enzyme immobilized platform to catalyze the oxidation of glucose and produce gluconic acid, which was detected by a C60-coated PZ quartz crystal sensor for glucose [87]. The C60-GOD platform was characterized by FT-IR spectroscopy, which showed absorption peaks at $1148 \, cm^{-1}$ and $1600 \, cm^{-1}$ of GOD and $525–570 \, cm^{-1}$ for fullerene-C60. The activity of the synthetic C60/GOD was investigated by means of the oxygen electrode detector because GOD catalyze the oxidation of glucose, which results in the consumption of oxygen. The effect of the amount of the immobilized glucose oxidase on the oxidation rate was investigated. It was shown that the consumption of oxygen is linearly proportional to the number of pieces of immobilized enzyme. The obtained results showed that only C60 coated crystals with immobilized enzyme responded sensitively to glucose. The studies on the effect of the amount of C60 coating on the frequency response of the PZ glucose sensor with the immobilized GOD enzyme shows that the thicker C60 coating exhibits a better response but, with a larger amount of coating, it is leveled off. The pH and temperature effect on the activity of the immobilized enzyme C60-glucose showed that an optimum pH of 7 and 30°C for temperature is suitable for the glucose oxidase activity (Table 7.1).

In another report, direct electrochemistry of GOD was achieved with GOD-hydroxyl fullerenes (HFs) modified glassy carbon electrode which protected with a chitosan membrane [80]. The formed GOD-HFs nanoparticles in the chitosan membrane was characterized by TEM images, which showed the average size of 20 nm for GOD-HFs nanoparticles. It has been shown that while no redox peak was observed at bare was bare GCE, Chit/GOD/GCE and Chit/HFs/GCE, a pair of well-defined redox peaks was appeared at the Chit/GOD-HFs/GCE. The CVs remained unchanged after successive potential cycle, showing that the formed Chit/GOD-HFs film was stable on the GC electrode. In addition, the obtained K_m value was lower than that of conventional values, showing a strong interaction and higher affinity of glucose for the modified electrode.

Lin *et al.* developed a mixed-valence cluster of cobalt(II) hexacyanoferrate and fullerene C60-enzyme-based electrochemical glucose sensor [82]. The C60-GOD was synthesized and applied with mixed-valence cobalt (II) hexacyanoferrate for analysis of glucose. Glucose in solution can be oxidized by C60-GOD-modified glassy carbon electrode, which is followed by the oxidation of the reduced C60-GOD by oxygen in the solution and the formation of H_2O_2. On the other hand, the cobalt(II) hexacyanoferrate $(Co_3[Fe(CN)_6]_2)_{(Red)}$ can oxidize by means of produced H_2O_2. At the end, the oxidized $(Co_3[Fe(CN)_6]_2)_{(Ox)}$ was reduced with an applied electrode voltage at 0.0 mV and the reduced current can be traced for the detection of glucose (Scheme 7. 4a).

It has been shown that the electrodes with immobilized enzymes (C60-GOD) shows better responses than the electrode with free enzyme. In addition, it was demonstrated that the electrode with a thicker cobalt (II) hexacyanoferrate coating produces a larger current response for the H_2O_2. However, the current response apparently tends to level off with larger amounts of cobalt (II) hexacyanoferrate coating. The C60-GOD/cobalt(II) hexacyanoferrate-modified electrode in solutions at a higher stirring rate exhibited a larger current response to the same concentration of glucose. However, the current response apparently tends to level off at a higher stirring rate. Moreover, an optimum current response is obtained around pH 6.2 and 30°C.

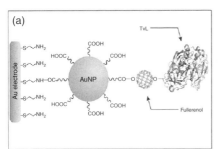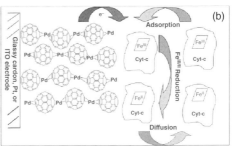

Scheme 7.4 (a) Au-SAM/AuNPs-Linker/Fullerenols/TvL composite material assembly, adapted from [88] and (b) proposed mechanism of cyt *c* immobilization and electrochemical reduction by C60-Pd polymer film modified electrode. *Source*: D'Souza 2005 [89]. Reproduced with permission of Elsevier.

Urease: Urea is one of the byproducts of protein metabolism. The precise detection of urea is crucial in various biomedical applications, glomerular filtration rate determination, and renal function tests. The urease enzyme could be employed for urea determination, whereby the urease catalyzes the hydrolysis of urea to form alkaline reaction products of NH^{+4} and CO_3^{-2}. The detection is based on pH changes resulted from by enzymatic reaction (7.1) [90]:

$$NH_2CONH_2 + H_2O \xrightarrow{\text{Urease}} NH_4^+ + CO_3^{2-} \tag{7.1}$$

Fullerenes have been also used in the fabrication of certain biosensors with different enzymes such as lipase and urease. Integrating of the fullerene molecule for construction of urea biosensing devices may enhance the sensitivity of the analytical method because it provides the high surface area-to-volume ratio for urease immobilization. A new way to construct a urea potentiometric biosensor has been developed by Saeedfar *et al.* [91]. The fullerene nanomaterial was functionalized with carboxyl groups by sonication, heat, and ultraviolet radiation. N,N′-dicyclohexylcarbodiimide (DCC) or N-(3-dimethylaminopropyl)-N′-ethylcarbodiimide hydrochloride (EDC) was utilized to immobilized urease enzyme onto carboxylic modified fullerenes (C60-COOH). It was observed that the lower sensitivity of the urea biosensor was obtained using water-insoluble DCC as a cross-linking agent instead of EDC. The immobilized urease catalyzed the hydrolysis of urea in the sample, which resulted in the production of OH^- ions. When the concentration of urea is low, the OH^- ion reacted with the buffer and the concentration of the buffer became important. Therefore, the buffer capacity could not maintain the pH and the sensitivity increased. When the concentration of buffer is high, the sensitivity of the biosensor decreased because of the OH^- ion reacted with buffer. The optimum pH range of the biosensor was obtained between pH 6.0 and 8.0. In another study, a fullerene-C60-coated piezoelectric quartz crystal urea sensor based on either solvated or immobilized urease was developed and applied to detect urea in aqueous solutions. The immobilized urease enzymes on fullerene-C60 shows lower sensitivity than that of the solvated urease detection (Table 7.1) [71].

Laccase: The high stability and bioactivity of the bio-electrochemical interfaces play a crucial role in the performance of laccase-based biosensors. The immobilization of enzymes on solid supports is one of the effective approaches to meet the requirement for a highly sensitive and stable biosensor. There is extensive interest to construct laccase biosensors in combination with nanomaterials, due to their unique properties [82]. The fullerene-C60 nanoparticles provide a suitable micro-environment for enzyme immobilization, maintaining their bioactivity, and accelerating the electron transfer between their redox active center and transducer surfaces [88]. An electrochemical biosensing platform based on the coupling of two different nanostructured materials (gold nanoparticles and fullerenols), has been developed and characterized by Lanzellotto *et al.* [88]. The proposed methodology was based on a multilayer material consisting in AuNPs, fullereneols, and Trametes versicolor laccase (TvL) assembled layer by layer onto a gold electrode surface (Scheme 7.4b). A linear dependence has been obtained between the voltammetric peak currents and the potential scan rate which attributed to the immobilization of the redox protein. The calculated electron transfer rate constant (k_s) values show the higher amount of immobilized TvL on nano-structures-modified electrodes compared to the gold electrode due to the increased roughness of the electrode surface. It was believed that the presence of nanostructured material increases the protein loading due to high surface-active and provide an ideal microenvironment for proteins. Microscopic characterization of the electrode surface before and after modification with TvL has been performed by scanning tunneling microscopy. Before enzyme immobilization, several nanoparticles of 15 nm are observed and, after modification with TvL, a huge increase of particles size is detected (35 nm). In addition, it was observed that Michaelis constant ($K_{M,app}$) decreases after introducing the AuNPs and fullerenol, suggesting an increased affinity of the enzymatic for the substrate. It is ascribed to the fact that the fullerenes provide a suitable microenvironment for the protein immobilization and induce the protein molecule mobility in order to correctly orient its redox centers in order to achieve a proper electron transfer.

Carbon nanotubes (single-walled carbon nanotubes (SWCN) and multi-walled carbon nanotubes (MWCN)), with high surface area, high adsorption capacity, and rapid desorbability are widely used for construction of enzyme electrodes. One of the most recent example is reported by Barberis *et al.*, where simultaneous amperometric detection of ascorbic acid (AA) and antioxidant capacity has been performed based on fullerenes-C60/C70 or nanotubes-modified graphite sensor-biosensor systems, and ascorbate oxidase. It was reported that the combination of fullerene and ascorbate oxidase enzyme resulted in the complete AA shielding and in the highest selecting capacity toward AA while nanotubes only increase sensitivity without ability to discriminate between different compounds [92]. Authors hypothesized that fullerenes absorb more enzyme during dips, so that they can oxidize more AA before it reaches the transducer surface.

7.2.3.2 Redox Active Proteins

Direct electrochemistry of hemoglobin (Hb) immobilized on fullerene-nitrogen doped carbon nanotubes (C60–NCNT)/Chitosan (CHIT) composite matrix is reported by Sheng *et al.* [93]. The developed C60-NCNT/CHIT modified electrode was utilized for the determination of H_2O_2. The obtained TEM image of NCNT shows that after immobilization, some C60 amorphous nanoparticles with the size of ~4 nm were found visible inside NCNT. The recorded FTIR spectra showed that the relative shifts of the peaks which are ascribable to the π electron interaction between C60 and NCNT. The amide I and II bands related to C60-NCNT/Hb have similar shapes to that of free Hb indicating that Hb is successfully immobilized on C60–NCNT. The electrochemical measurements show no redox peaks in the cyclic voltammograms of the bare GCE but the background current increase at C60-NCNT/GC electrode. At Hb/NCNT/CHIT/ GC electrode, there is only one cathodic peak, which can be observed from the CV. After immobilization of Hb on the C60–NCNT/CHIT/GC electrode, a pair of well-defined redox peaks related to the Hb and (Fe^{III}/Fe^{II}) are observed.

Cytochrome c: Cytochrome c (cyt c) is a heme containing metalloprotein located in the inter membrane space of mitochondria. It has a low molecular weight ($M_w = 12\,400$ D) with a single polypeptide chain of 104 amino acid residues covalently attached to the heme moiety. It plays a key role in the biological respiratory chain, whose function is to transfer electrons between cytochrome c reductase (complex III) and cytochrome c oxidase (complex IV) [94].

One example of the application of the fullerene film modified electrodes for immobilizing a cyt c has been reported by d'Souza *et al.* [89]. Two types of fullerene film modified electrodes were utilized for immobilization of cyt c. One involves an electrochemically-conditioned fullerene drop-coated film electrode and the other an electro-polymerized fullerene, cross-linked with palladium acetate complex film electrode. The immobilization of cyt c on the fullerene film modified electrode was examined by piezoelectric microgravimetry at a quartz crystal microbalance. It was shown that upon addition of cyt c the frequency decreased to reach plateaus. In addition, the blue shift and the broadness of the bands observed in UV–Vis spectra was attributed to the cyt c molecules which are tightly packed on the electrode surface. The proposed mechanism of the cyt c immobilization is illustrated in Scheme 7.4c. It is claimed that cyt c is immobilized by one or more of following ways: (1) electrostatically binding of the electron deficient C60 molecules with the cyt c molecules; (2) negative charge on the C60 film electrostatically bind with positively-charged parts of cyt c protein; and (3) surface structure of C60 or C60-palladium may affect the immobilization of cyt c.

The CV behavior of cyt c immobilized on the C60 drop-coated film GCE and the C60-Pd polymer film modified electrode shows that upon addition of cyt c to the solution, a cathodic peak appeared at $E_{p,c} = -400$ mV versus Ag/AgCl. When the potential scan reversed, an anodic peak at $E_{p,a} = 50$ mV versus Ag/AgCl was also observed indicating the reversible and slow electron transfer process. When the equilibrium occurs at C60-Pd polymer film-modified GCE in the cyt c buffer solution, the CV peaks were still present showing the stable immobilization of cyt c onto the C60-Pd polymer film-modified electrode. The effect of the C60-Pd polymer film thickness on CV properties of the immobilized cyt c was examined which suggests that the amount of immobilized cyt c increased with the increase of the C60-Pd film thickness. Csisza'r *et al.* utilized C60 fullerene film modified electrodes for the electrochemical reactions of cyt c [95]. They have investigated the electrochemical behavior of fullerene films in the neutral state, which are porous intrinsic semiconductors. They can be reduced to form semiconductor or conducting salts. They assumed that partially reduced fullerene films have a structure with a pole, or negatively-charged outside and an apolar inside. The porosity of the films was estimated in two ways: firstly, by means of measuring the oxidation of gold and the reduction of the oxide in phosphate buffer. The oxidation and reduction waves can be suppressed by the presence of fullerene films. The other method is based on chronocoulometry method which allows the calculation of the electrode surface area. The partial reduction produces irreversibly small amounts of $C60^-$, and/or $C60^{2-}$ intermediates and the film becomes a cation exchanger. When partially-reduced fullerene films used, the electrochemical response of cyt c became much better. In the case of thin film, the half-wave potential of quasi-reversible reaction was 285 mV, which is close to the standard redox potential of native cyt c (260 mV). It was also shown that with thicker films, the catalytic activity of cyt c is lower. In addition, the presence of partially reduced fullerene films stabilized the electrochemical reaction of cyt c. The response on reduced and then oxidized films was also investigated. The oxidized film was still apparently coherent and did not show any signal of cyt c. A better response was observed if the films were porous and partially charged. Generally, the neutral fullerene films lack the charge, and fully reduced or oxidized films lack the porous character. If the fullerene film reduction was carried out in the presence of Na^+, resulted in completely inactive electrode because it converts the films mainly to semiconducting Na_6C60, which cannot participate in the reaction of cyt c. If the reduction of C60 film was carried out in the presence of K^+, the electrodes showed short-lived transient responses again, as on bare electrodes or with neutral fullerene films.

7.3 Conclusions and Future Prospects

Recently, nanostructured materials have been significantly used to create state-of-the-art electrochemical biosensors with enhanced performance (Figure 7.3). They provide the analytical devices with the ability of miniaturization and reduced response time, and cost effectiveness for applications in clinical diagnosis. Among different nanomaterials, carbon nanomaterials hold potential promise as a material for designing a new generation of biosensors due to their unique characteristics. Recently, fullerene-C60 contributed greatly to the field of biosensing and bio-nanotechnology. The unique electrochemical and physicochemical properties, together with biocompatibility characteristics of fullerene, allow its wide use for designing the highly sensitive chemical/biosensors.

In this review, we presented the most recent applications of fullerene-C60 based electrochemical biosensors which employed various kinds of biomolecules. Especially,

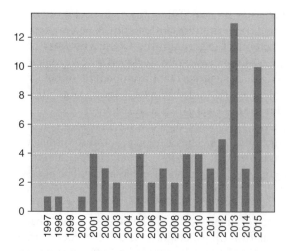

Figure 7.3 Nanostructured materials: Published items in each year. *Source*: www.webofknowledge.com.

electrochemical biosensors based on the interaction between fullerene-C60 and DNA has been reviewed in depth. It has been shown that fullerene-C60 has been widely utilized for improving the sensitivity of electrochemical biosensors. Not only they provide a suitable immobilization platform for DNA and antibodies, but they also have the ability to induce in redox-active proteins a proper orientation which leads to better electron transfer properties. Therefore, fullerene-C60 can be easily extended to immobilize and obtain direct electrochemistry of enzymes and proteins. However, the range of applications is still limited and further investigation is required. Easy functionalizations and high surface area of fullerene can be utilized for designing more sensitive biosensing devices with high stability. The recent developments of electrochemical biosensors based on fullerene-C60 may bring many researchers to use other analogues of fullerene-C60 in the construction of electrochemical biosensors. Furthermore, multiple functionalizations of these kinds of nanomaterials may lead to the improved performance of biosensors. On the other hand, taking into account the biocompatibility of fullerene-C60, different kind of biomolecules such as microoganisms, organelle, and cells can be easily integrated in the biosensors fabrication. Moreover, fullerene-based biosensors could be integrated within bio-chips with on-board electronics. This will lead to fabrication of devices which are small, low-cost, and with simple operation procedure. Therefore, electrochemical biosensors based on fullerene-C60 with their cost-effectiveness and suitability for microfabrication can be expected to become increasingly popular in the near future.

References

1 Baghchi, M.; Moriyama, H.; Shahidi, F. Bio-Nanotechnology: A Revolution in Food, Biomedical and Health Sciences. Wiley-Blackwell, Hoboken, NJ; 2013.

2 Jianrong, C.; Yuqing, M.; Nongyue, H.; Xiaohua, W.; Sijiao, L. Nanotechnology and biosensors. Biotechnology Advances 2004, 22, 505–518.

3 Vashist, S.; Venkatesh, A.G.; Mitsakakis, K.; *et al.* Nanotechnology-based biosensors and diagnostics: Technology push versus industrial/healthcare requirements. BioNanoScience 2012, 2, 115–126.

4 Thostenson, E.T.; Ren, Z.; Chou, T.-W. Advances in the science and technology of carbon nanotubes and their composites: A review. Composite Science and Technology 2001, 61, 1899–1912.

5 Gooding, J.J. Nanostructuring electrodes with carbon nanotubes: A review on electrochemistry and applications for sensing. Electrochimica Acta 2005, 50, 3049–3060.

6 Vairavapandian, D.; Vichchulada, P.; Lay, M.D. Preparation and modification of carbon nanotubes: Review of recent advances and applications in catalysis and sensing. Analytica Chimica Acta 2008, 626, 119–129.

7 Jariwala, D.; Sangwan, V.K.; Lauhon, L.J.; *et al.* Carbon nanomaterials for electronics, optoelectronics, photovoltaics, and sensing. Chemical Society Review 2013, 42, 2824–2860.

8 Gogotsi, Y.; Presser, V. Carbon Nanomaterial. Taylor and Francis Group, New York; 2014.

9 Bosi, S.; da Ros, T.; Spalluto, G.; Prato, M. Fullerene derivatives: An attractive tool for biological applications. European Journal of Medicinal Chemistry. 2003, 38, 913–923.

10 Dresselhaus, M.S.; Dresselhaus, G.; Eklund, P.C. Science of Fullerenes and Carbon Nanotubes: Their Properties and Applications. Elsevier Science, London; 1996.

11 Langa, F.; Nierengarten, J.F. Fullerenes: Principles and Applications. The Royal Society of Chemistry, Cambridge, UK; 2007.

12 Afreen, S.; Muthoosamy, K.; Manickam, S.; Hashim, U. Functionalized fullerene (C_{60}) as a potential nanomediator in the fabrication of highly sensitive biosensors. Biosensors and Bioelectronics 2015, 63, 354–364.

13 Dresselhaus, M.S.; Dresslhaus, G. Fullerenes and fullerene derived solids as electronic materials. Annual Review of Material Research 1995, 25, 487–523.

14 Jensen, A.W.; Wilson, S.R.; Schuster, D.I. Biological applications of fullerenes. Bioorganic and medicinal chemistry 1996, 4, 767–779.

15 Smalley, R.E. Discovering the fullerenes. Review of Modern Physics 1997, 69, 723–730.

16 Shinar, J.; Vardeny, Z.V.; Kafafi, Z.H. Optical and Electronic Properties of Fullerene and Fullerene-Based Materials. Marcel Dekker, New York; 2000.

17. Lieber, C.M.; Chen, C.-C. Preparation of fullerenes and fullerene-based materials, in Solid bState Physics, (eds Henry, E., Frans, S.) volume 48, pp. 109–148. Academic Press, Waltham, MA; 1994.

18 Scott, L.T. Methods for the chemical synthesis of fullerenes. Angewandte Chemie International Edition 2004, 43, 4994–5007.

19 Dai, L.; Mau, A.W. Controlled synthesis and modification of carbon nanotubes and C_{60}: Carbon nanostructures for advanced polymeric composite materials. Advanced Materials 2001, 13, 899–913.

20 Kharlamov, A.I.; Bondarenko, M.E.; Kirillova, N.V. New method for synthesis of fullerenes and fullerene hydrides from benzene. Russian Journal of Applied Chemistry 2012, 85, 233–238.

21 Inomata, K.; Aoki, N.; Koinuma, H. Production of fullerenes by low temperature plasma chemical vaper deposition under atmospheric pressure. Japanese Journal of Applied Physics 1994, 33, 197–199.

22 Hirsch, A. Functionalization of fullerenes and carbon nanotubes. Physica Status Solidi B 2006, 243, 3209–3212.

23 Caballero, R.; de la Cruz, P.; Langa, F. basic principles of the chemical reactivity of fullerenes, chapter 3 in Fullerenes: Principles and Applications (2). The Royal Society of Chemistry, London; 2012; pp. 66–124.

24 Bingel, C. Cyclopropanierung von fullerenen. Chemische Berichte 1993, 126, 1957–1959.

25 Rotello, V.M.; Howard, J.B.; Yadav, T.; *et al.* Isolation of fullerene products from flames: Structure and synthesis of the C_{60}-cyclopentadiene adduct. Tetrahedron Letters 1993, 34, 1561–1562.

26 Arena, F.; Bullo, F.; Conti, F.; *et al.* Synthesis and epr studies of radicals and biradical anions of C_{60} nitroxide derivatives. Journal of the American Chemical Society 1997, 119, 789–795.

27 Compton, R.G.; Spackman, R.A.; Wellington, R.G.; *et al.* A C_{60} modified electrode: Electrochemical formation of tetra-butylammonium salts of C_{60} anions. Journal of Electroanalytical Chemistry 1992, 327, 337–341.

28 Sherigara, B.S.; Kutner, W.; D'Souza, F. Electrocatalytic properties and sensor applications of fullerenes and carbon nanotubes. Electroanalysis 2003, 15, 753–772.

29 Lokesh, S.V.; Sherigara, B.S.; Mahesh, J.H.M.; Mascarenhas, R.J. Electrochemical reactivity of C_{60} modified carbon paste electrode by physical vapor deposition method. International Journal of Electrochemical Science 2008, 3, 578–587.

30 Compton, R.G.; Spackman, R.A.; Riley, D.J.; *et al*. Voltammetry at C_{60}-modified electrodes. Journal of Electroanalytical Chemistry 1993, 344, 235–247.

31 Jehoulet, C.; Obeng, Y.S.; Kim, Y.T.; *et al*. Electrochemistry and langmuir trough studies of fullerene C_{60} and C_{70} films. Journal of the American Chemical Society 1992, 114, 4237–4247.

32 Chang, C.-L.; Hu, C.-W.; Tseng, C.-Y.; *et al*. Ambipolar freestanding triphenylamine/ fullerene thin-film by electrochemical deposition and its read-writable properties by electrochemical treatments. Electrochimica Acta 2014, 116, 69–77.

33 Bedioui, F.; Devynck, J.; Bied-Charreton, C. Immobilization of metalloporphyrins in electropolymerized films: Design and applications. Accounts of Chemical Research 1995, 28, 30–36.

34 Imahori, H.; Azuma, T.; Ajavakom, A.; *et al*. An investigation of photocurrent generation by gold electrodes modified with self-assembled monolayers of C_{60}. Journal of Physical Chemisty B 1999, 103, 7233–7237.

35 Lowe, C.R. Biosensors. Trends in Biotechnology 1984, 2, 59–65.

36 Buck, R.P.; Hatfield, W.E.; Umana, M.; Bowden, E.F. Biosensors Technology: Fundamentals and Applications; Marcel Dekker, New York; 1990.

37 Turner, A.P.F. Biosensors: Sense and sensibility. Chemical Society Reviews 2013, 42, 3184–3196.

38 Guo, S.; Dong, S. Biomolecule-nanoparticle hybrids for electrochemical biosensors. Trends in Analytical Chemistry 2009, 28, 96–109.

39 Hirxch, A.; Brettreich, M. Fullerene: Chemistry and Reactions. Wiley, Weinheim, Germany, 2005.

40 Guldi, D.M.; Prato, M. Excited-state properties of C_{60} fullerene derivatives. Accounts of Chemical Research 2000, 33, 695–703.

41 Wang, J. Nanomaterial-based electrochemical biosensors. Analyst 2005, 130, 421–426.

42 Li, J.; Wu, N. Biosensors Based on Nanomaterial and Nanodevices. CRC Press, Boca Raton, FL; 2014.

43 Katz, E.; Willner, I. Integrated nanoparticle-biomolecule hybrid systems: Synthesis, properties, and applications. Angewandte Chemie International Edition 2004, 43, 6042–6108.

44 Kerman, K.; Saito, M.; Tamiya, E.; *et al*. Nanomaterial-based electrochemical biosensors for medical applications. Trends in Analytical Chemistry 2008, 27, 585–592.

45 Yang, X.; Ebrahimi, A.; Li, A.; Cui, Q. Fullerene-biomolecule conjugate and their biomedical applications. International Journal of Nanomedicine 2014, 9, 77–92.

46 Hu, C.; Zhang, Y.; Bao, G.; *et al*. DNA functionalized single-walled carbon nanotubes for electrochemical detection. Journal of Physical Chemistry B 2005, 109, 20072–20076.

47 Nalwa, H.S. Handbook of Nanostructured Biomaterials and their Applications in Nanobiotechnology volume 2. American Science Publishers, California USA; 2005.

48 Cassell, A.M.; Scrivens, W.A.; Tour, J.M. Assembly of DNA/fullerene hybrid materials. Angewandte Chemie International Edition1998, 37, 1528–1531.

49 Pang, D.-W.; Zhao, Y.-D.; Fang, P.-F.; *et al*. Interactions between DNA and a water-soluble C_{60} derivative studied by surface-based electrochemical methods. Journal of Electroanalytical Chemistry 2004, 567, 339–349.

50 Zhao, X.; Striolo, A.; Cummings, P.T. C_{60} binds to and deforms nucleotides. Biophysical Journal. 2005, 89, 3856–3862.

51 Alshehri, M.; Cox, B.; Hill, J. C_{60} fullerene binding to DNA. European Physical Journal B 2014, 87, 1–11.

52 Gooding, J.J. Electrochemical DNA hybridization biosensors. Electroanalysis 2002, 14, 1149–1156.

53 Kerman, K.; Kaboyashi, M.; Tamiya, E. Recent trend in electrochemical DNA biosensor technology. Measurement Science and Technology 2004, 15, 1–11.

54 Shiraishi, H.; Itoh, T.; Hayashi, H.; *et al.* Electrochemical detection of E. coli 16s rDNA sequence using air-plasma-activated fullerene-impregnated screen printed electrodes. Bioelectrochemistry 2007, 70, 481–487.

55 Lin, Y.; Lu, F.; Tu, Y.; Ren, Z. Glucose biosensors based on carbon nanotube nanoelectrode ensembles. Nano Letters 2004, 4, 191–195.

56 Cai, H.; Cao, X.; Jiang, Y.; He, P.; Fang, Y. Carbon nanotube-enhanced electrochemical DNA biosensor for DNA hybridization detection. Analytical and Bioanalytical Chemistry 2003, 375, 287–293.

57 Cai, H.; Xu, Y.; He, P.-G.; Fang, Y.-Z. Indicator free DNA hybridization detection by impedance measurement based on the DNA-doped conducting polymer film formed on the carbon nanotube modified electrode. Electroanalysis 2003, 15, 1864–1870.

58 Zhang, X.; Qu, Y.; Piao, G.; Zhao, J.; Jiao, K. Reduced working electrode based on fullerene C_{60} nanotubes@DNA: Characterization and application. Materials Science and Engineering B 2010, 175, 159–163.

59 Xing, H.; Wong, N.Y.; Xiang, Y.; Lu, Y. DNA aptamer functionalized nanomaterials for intracellular analysis, cancer cell imaging and drug delivery. Current Opinion in Chemical Biology 2012, 16, 429–435.

60 Gugoasa, L.A.; Stefan-van Staden, R.I.; Alexandru Ciucu, A.; van Staden Jacobus, F. Influenceof physical immobilization of DSDNA on carbon based matrices of electrochemical sensors. Current Pharmaceutical Analysis 2014, 10, 20–29.

61 Mashayekhi, H.; Ghosh, S.; Du, P.; Xing, B. Effect of natural organic matter on aggregation behavior of C_{60} fullerene in water. Journal of Colloid and Interface Science 2012, 374, 111–117.

62 Li, W.; Yuan, R.; Chai, Y. Determination of glucose using pseudobienzyme channeling based on sugar-lectin biospecific interactions in a novel organic-inorganic composite matrix. Journal of Physical Chemistry C 2010, 114, 21397–21404.

63. aZhuo, Y.; Yuan, P.-X.; Yuan, R.; Chai, Y.-Q.; Hong, C.-L. Nanostructured conductive material containing ferrocenyl for reagentless amperometric immunosensors. Biomaterials 2008, 29, b1501–1508.

64 Han, J.; Zhuo, Y.; Chai, Y.; *et al.* Multi-labeled functionalized C_{60} nanohybrid as tracing tag for ultrasensitive electrochemical aptasensing. Biosensors and Bioelectronics 2013, 46, 74–79.

65 Zhuo, Y.; Ma, M.-N.; Chai, Y.-Q.; Zhao, M.; Yuan, R. Amplified electrochemiluminescent aptasensor using mimicking bi-enzyme nanocomplexes as signal enhancement. Analytical Chimica Acta 2014, 809, 47–53.

66 Gholivand, M.-B.; Jalalvand, A.R.; Goicoechea, H.C. Multivariate analysis for resolving interactions of carbidopa with dsDNA at a fullerene- C_{60}/GCE. International Journal of Biology Macromolecules 2014, 69, 369–381.

67. Wang, H.; Yuan, R.; Chai, Y.; Niu, H.; Cao, Y.; Liu, H. Bi-enzyme synergetic catalysis toin situ generate coreactant of peroxydisulfate solution for ultrasensitive electrochemiluminescence immunoassay. Biosensors and Bioelectronics 2012, 37, 6–10.

68 Rusling, J.F. Nanomaterials-based electrochemical immunosensors for proteins. Chemical Record 2012, 12, 164–176.

69 Pan, N.-Y.; Shih, J.-S. Piezoelectric crystal immunosensors based on immobilized fullerene C_{60}-antibodies. Sensors and Actuators B 2004, 98, 180–187.

70 Barnes, C.; d'Silva, C.; Jones, J.P.; Lewis, T.J. The theory of operation of piezoelectric quartz crystal sensors for biochemical application. Sensors and Actuators A 1992, 31, 159–163.

71 Chou, F.F.; Chang, H.W.; Li, T.L.; Shih, J.S. Piezoelectric crystal/surface acoustic wave biosensors based on fullerene C_{60} and enzymes/antibodies/proteins. Journal of the Iranian Chemical Society 2008, 5, 1–15.

72 Li, Y.; Fang, L.; Cheng, P.; *et al*. An electrochemical immunosensor for sensitive detection of Escherichia coli o157:H7 using C_{60} based biocompatible platform and enzyme functionalized Pt nanochains tracing tag. Biosensors and Bioelectronics 2013, 49, 485–491.

73 Hu, N. Direct electrochemistry of redox proteins or enzymes at various film electrodes and their possible applications in monitoring some pollutants. Pure and Applied Chemistry 2001, 73, 1979–1991.

74 Kurz, A.; Halliwell, C.M.; Davis, J.J.; *et al*. A fullerene-modified protein. Chemical Communications 1998, 433–434.

75 Braun, M.; Atalick, S.; Guldi, D.M.; *et al*. Electrostatic complexation and photoinduced electron transfer between Zn-cytochrome c and polyanionic fullerene dendrimers. Chemistry: A European Journal. 2003, 9, 3867–3875.

76 Witte, P.; Beuerle, F.; Hartnagel, U.; *et al*. Water solubility, antioxidant activity and cytochrome C binding of four families of exohedral adducts of C_{60} and C_{70}. Organic and Biomolecular Chemistry 2007, 5, 3599–3613.

77 Yin, Y.; Lü, Y.; Wu, P.; Cai, C. Direct electrochemistry of redox proteins and enzymes promoted by carbon nanotubes. Sensors 2005, 5, 220–234.

78 Lukehart, L.M.; Scott, R.A. Nanomaterials: Inorganic and Bioinorganic Perpectives; John Wiley and Sons Ltd.: London, UK, 2008.

79 Guisan, J.M. Methods in Biotechnology: Immobilization of Enzymes and Cells volume 22. Humana Press Inc., Totawa, NJ; 2006.

80 Gao, Y.-F.; Yang, T.; Yang, X.-L.; *et al*. Direct electrochemistry of glucose oxidase and glucose biosensing on a hydroxyl fullerenes modified glassy carbon electrode. Biosensors and Bioelectronics 2014, 60, 30–34.

81 Hecht, H.J.; Kalisz, H.M.; Hendle, J.; *et al*. Crystal structure of glucose oxidase from Aspergillus niger refined at 2.3 Åreslution. Journal of Molecular Biology 1993, 229, 153–172.

82 Lin, L.-H.; Shih, J.-S. Immobilized fullerene C_{60}-enzyme-based electrochemical glucose sensor. Journal of the Chinese Chemical Society 2011, 58, 228–235.

83 Künzelmann, U.; Böttcher, H. Biosensor properties of glucose oxidase immobilized within SiO_2 gels. Sensors and Actuators B 1997, 39, 222–228.

84 Higuchi, A.; Hara, M.; Yun, K.-S.; Tak, T.-M. Recognition of substrates by membrane potential of immobilized glucose oxidase membranes. Journal of Applied Polymer Science 1994, 51, 1735–1739.

85 Liu, B.; Hu, R.; Deng, J. Fabrication of an amperometric biosensor based on the immobilization of glucose oxidase in a modified molecular sieve matrix. Analyst 1997, 122, 821–826.

86 Tischer, W.; Wedekind, F. Immobilized enzymes: Methods and applications, in Biocatalysis:From Discovery to Application (Fessner, W.-D., Archelas, A., Demirjian, D.C., Furstoss, R., *et al.* eds). Springer, Berlin; 1999, volume 200, pp. 95–126.

87 Chuang, C.-W.; Shih, J.-S. Preparation and application of immobilized C_{60}-glucose oxidase enzyme in fullerene C_{60}-coated piezoelectric quartz crystal glucose sensor. Sensors and Actuators B 2001, 81, 1–8.

88 Lanzellotto, C.; Favero, G.; Antonelli, M.L.; *et al.* Nanostructured enzymatic biosensor based on fullerene and gold nanoparticles: Preparation, characterization and analytical applications. Biosensors and Bioelectronics 2014, 55, 430–437.

89 D'Souza, F.; Rogers, L.M.; O'Dell, E.S.; *et al.* Immobilization and electrochemical redox behavior of cytochrome c on fullerene film-modified electrodes. Bioelectrochemistry 2005, 66, 35–40.

90 Soares, J.C.; Brisolari, A.; Rodrigues, V.D.C.; *et al.* Amperometric urea biosensors based on the entrapment of urease in polypyrrole films. Reactive Functional Polymers 2012, 72, 148–152.

91 Saeedfar, K.; Heng, L.; Ling, T.; Rezayi, M. Potentiometric urea biosensor based on an immobilised fullerene-urease bio-conjugate. Sensors 2013, 13, 16851–16866.

92 Barberis, A.; Spissu, Y.; Fadda, A.; *et al.* Simultaneous amperometric detection of ascorbic acid and antioxidant capacity in orange, blueberry and kiwi juice, by a telemetric system coupled with a fullerene- or nanotubes-modified ascorbate subtractive biosensor. Biosensors and Bioelectronics 2015, 67, 214–223.

93 Sheng, Q.; Liu, R.; Zheng, J. Fullerene–nitrogen doped carbon nanotubes for the direct electrochemistry of hemoglobin and its application in biosensing. Bioelectrochemistry 2013, 94, 39–46.

94 Moore, G.R.; Pettigrew, G.W. Cytochromes c, Evolutionary, Structural, and Physiochemical Aspects. Springer-Verlag, New York; 1990.

95 Csiszár, M.; Szűcs, Á.; Tölgyesi, M.; *et al.* Electrochemical reactions of cytochrome c on electrodes modified by fullerene films. Journal of Electroanalytical Chemistry 2001, 497, 69–74.

8

Micro- and Nano-structured Diamond in Electrochemistry: Fabrication and Application

Fang Gao and Christoph E. Nebel

Fraunhofer Institute, Freiburg, Germany

8.1 Introduction

Synthetic diamond has been studied extensively during the last 30 years due to its outstanding chemical/physical properties such as wide bandgap, high thermal conductivity, and extreme hardness. Compared to bulk diamond, diamond materials with micro- and nanostructured surfaces provide a facile way to fine tune diamond properties such as field emission, hydrophilicity, and specific surface. In this chapter, the fabrication and application of micro- and nanostructured diamond will be discussed.

The fabrication method of diamond nanostructures can be divided into two categories: top-down etching and bottom-growth. The early work on 3D micro-structured diamond dates back to mid-1990s, using chemical vapor infiltration (CVI) techniques. In this technology, carbon or carbide fibers were typically used as the growth template. Almost in parallel, reactive ion etching (RIE) was applied to achieve diamond surface nanostructuring. After that the diamond surface nanostructures, typically vertically aligned diamond nanowires (or nanorods) has been mainly fabricated using top-down plasma etching techniques. In recent year, the templated diamond growth has gained increasing attention due to the wide choice of template, mask-free production, and unlimited surface enlargement. In this chapter, the development and main techniques used in these two approaches will be elaborated. Nevertheless, other less common methods, such as catalytic etching by metal particles, steam activation and selective materials removal will also be discussed.

As indicated by the title, this chapter will mainly deal with the application of micro- and nanostructured diamond in electrochemistry. In these applications, the advantage of nanostructured diamond can be divided into three aspects: 1) providing enlarged surface area for charge storage and catalyst deposition; 2) tip-enhanced electrochemical reactions used in sensing applications; 3) diamond membranes with micro- or nanopores can be applied in the electrochemistry separation and purification applications. Examples and explanations on these applications will be given in this chapter.

Nanocarbons for Electroanalysis, First Edition.
Edited by Sabine Szunerits, Rabah Boukherroub, Alison Downard and Jun-Jie Zhu.

8.2 Fabrication Method of Diamond Nanostructures

8.2.1 Reactive Ion Etching

Reactive ion etching is a dry etching technique using reactive plasma to remove material from a surface. Schematic illustrations of such techniques are shown in Figure 8.1. The plasma is ignited either capacitively (Figure 8.1a) or inductively (Figure 8.1b). In both cases, a radio frequency (RF) voltage is added between the anode (the top and the wall of the reaction chamber) and cathode (the sample platter) to accelerate the electrons up and down; in the meantime, the massive ions are relatively unaffected. When electrons hit the anode, they are fed out to the ground. However, because the sample platter is (direct current) DC insulated, a negative potential will build up when electrons touch the cathode. In the capacitively coupled plasma (CCP) etching, the energy of plasma is coupled with the negative bias of the sample, i.e. when a high-density plasma is built, a high bias is unavoidable. As a result, the freedom to adjust the etching parameters is very limited. In the latter case, however, the plasma energy is decoupled from the sample bias. Therefore, high-density plasma is allowed without the danger to overcharge the sample.

The RIE techniques on diamond were developed in late 1990s. In 1997 Shiomi reported the diamond surface etch via CF_4/O_2 plasma [1]. Using sputtered Al with a thickness of 0.4 μm as the etching shadow mask, diamond surface microstructures were formed. In this research, it was found out that the ratio of CF_4 and O_2 in the etching atmosphere plays an important role in the etching selectivity between diamond and Al. however, an abnormal phenomenon was also reported: when pure O_2 plasma was used to etch polycrystalline diamond (PCD), a high density of diamond nanowires will appear. The explanation at that time was local surface phase transition induced by ion bombardment introduced self-masking effect to the etching process. These mask-free or self-masking phenomena were repeatedly used in the diamond nanowire formation processes [2, 3]. However, more recent research has shown that these effects are likely due to the oxide residues in the etching chamber acting as the etching mask, because

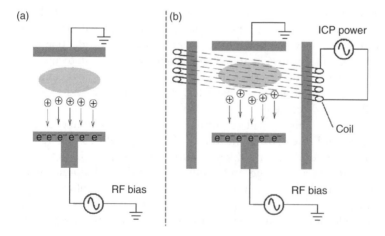

Figure 8.1 Schematic illustration showing different RIE mechanisms: (a) capacitively coupled plasma etching; (b) inductively coupled plasma (ICP) etching.

high-resolution transmission electron microscopy (HRTEM) has revealed the existence of a thin layer of amorphous oxide deposited all over these nanowires [4].

Metal nanoparticles deposited on diamond surface are often used as etching masks. In 2008, Zou *et al.* reported the formation of gold nanoparticle on diamond surface via a thermal dewetting method [5]. The diamond sample coated by 5 nm Au layer was heated up to 850°C in Ar/H$_2$ plasma. The Au nanoparticles were formed in situ via the self-organization of the molten thin layer, and they were used as the shadow mask in the later RIE etching. In this method, diamond nanopillars of a high density of ~10^9 cm^{-2} were formed with diameters of ~30 nm and heights of ~400 nm. A higher density and aspect ratio was later achieved by using more resistive etching masks and more aniso-tropic etching techniques. In 2010, Smirnov *et al.*, reported the high aspect ratio (~30) and high density diamond nanowires fabricated using Ni nanoparticles as etching mask and ICP etching techniques [6]. The method is shown in Figure 8.2 (a): a thin Ni film of 1 nm was evaporated on the PCD surface, and the layer was later thermally melted to form a dense layer of nickel particles (~10^{10} cm^{-2}, Figure 8.2b). The diamond wires were formed by using these particles as shadow mask in an ICP etching process (Figure 8.2c). Finally the mask residues were removed by wet-chemical cleaning. Dimensions of nano-wires were: height (1200 ± 200) nm, width (35 ± 5) nm, density ~1010 cm^{-2}.

Electrochemical deposition provides another method to controllably deposition metal nanoparticle as mask for etching processes. Gao *et al.* has shown that the density and size of the Pt nanoparticles electrochemically deposited on diamond electrode sur-face can be fine-tuned via surface pretreatment and deposition time [7]. The density is tunable in the range of $3.3 \pm 0.5 - 12.7 \pm 1.5 \times 10^9$ cm^{-2}, and the size can be adjusted between 18 ± 5 and 46 ± 9 nm. These particles can be used as etching masks for diamond column fabrication [8].

Alternatively, mask can also be formed using self-organized colloid particles. Periodically organized 2D structures of spherical particles can be obtained using capil-lary force and water evaporation [9, 10]. Okuyama *et al.* reported the fabrication of

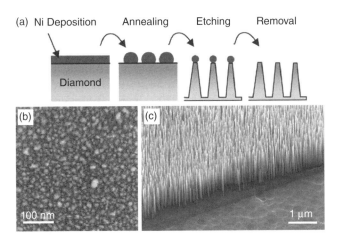

Figure 8.2 (a) Scheme showing the fabrication method of diamond nanowires; (b) an SEM image showing nickel nanoparticles from the dewetting of a 1 nm thick nickel film on diamond; (c) an SEM image showing the diamond nanowire after etching and nickel removal.

well-ordered diamond micropillars using self-organized SiO_2 spheres [11]. In this research, SiO_2 particles of 1 μm diameter was coated on the flat nucleation side of the PCD sample, forming a self-organized monolayer of a 2D hexagonal close-packing of SiO_2 spheres. By consequential O_2 plasma etching, this structure is transferred onto the underlying diamond substrate. Similarly, Yang *et al.* used self-organized diamond nano-particles as etching masks [12]. In this case, surface dipole interactions between the diamond seeds and the diamond substrate result in the attachment of seeds onto the substrate [13]. The diamond seeds with a typical size of 8–10 nm were dispersed in water and treated with ultrasonication. The diamond to be etched was then immersed in the suspension. The density of the particle attached to the diamond surface can be adjusted by the immersing time and the concentration of the suspension. After nano-particle attachment, the diamond surface was etched in oxygen-rich plasma. In this process, diamond nanowires of 10 nm length and an average separation of 11 nm were obtained with an etching time of 10 s.

Besides nanowires, nanoporous materials can also be fabricated using top-down etching method. RIE with shadow mask is a pattern transferring technique in principal. Therefore, more sophisticated nanoporous structures can be obtained if a nanoporous mask is used. For example, porous anodic aluminum oxide (AAO) has been studied for more than 50 years [14]. The pore size and distribution can be well-controlled by solution composition, applied potential and temperature [15]. It has been widely used as shadow masks for metal deposition [16] and pattern transfer on Si [17]. In 2000, Masuda *et al.* developed a fabrication method of diamond honeycomb electrode using porous AAO as the etching mask [18]. In the later work, the same group also realized and characterized porous diamond electrodes with a wide range of pore sizes and depths using this method [19, 20]. By etching through the diamond layer, diamond membrane was also fabricated [21].

8.2.2 Templated Growth

On the contrary to top-down etching method diamond surface can also be fabricated via templated-growth. In the 1990s in order to enhance the nucleation density, people used porous silicon as the growth substrate [22–25]. After that other porous materials, such as porous titanium [26], Zeolite [27] and porous carbide [28] were also used in for these reason. Normally, planar films were obtained. However, porous diamond films are sometimes obtained [29–31]. In these methods, the porous substrates are used, no matter whether intentionally or not, as the growth template. The diamond film grown on top of the porous substrate inherited the porous structures. A more typical templated-growth method came in the mid-1990s. Chemical vapor infiltration (CVI) was used in this method [32, 33]. In CVI, microwave plasma or a hot-filament will generate reactive species above the porous substrate. Chemical vapor deposition (CVD) in porous structures is enhanced by the manipulated gas flow which send the carbon and hydrogen radicals through the porous sample. Although there are successful reports on diamond coating up to millimeter-depth into the porous template [34], there is no evidence showing that this technique is able to coat nanoporous materials. A more 'modern' growth method came in 1999. Demkowicz *et al.* report the plasma CVD growth of diamond on a SiC-whisker compact [35]. In this research, a diamond seeding procedure was performed by adjusting the pH to 12.7. Diamond seeds with diameters of ~300 nm were attached to SiC whiskers for further overgrowth. With a coating thickness of ~300 nm, the diamond coating was formed up to a depth of ~50 μm. In 2001, Baranauskas *et al.* reported the templated diamond on natural pyrolized fibers [36]. In their research they compared the result of overgrowth with and without nanodiamond seeding process, and showed the necessity of the seeds for a complete coating.

Later developments of the 3D diamond coating follow the similar approach; the focus is mainly on tests on a large variety of materials for the growth template. Si-based materials due to their high melting point, chemical inertness, stable interface with carbon, as well as the suitable thermal expansions are widely applied as the growth template. In 2009, Luo *et al.* reported the growth of diamond on silicon nanowires using hot-filament CVD method [37]. The wires are vertically aligned and the coating covers the entire wires (~5 μm). In the same year, Kondo et al, reported the diamond fiber growth on quartz fiber filter [38]. Compared to the results of Luo *et al.*, a more complex template with interweaved SiO_2 fibers was used. The growth was carried out at a very low temperature of 500°C. The filter paper used has a thickness of 450 μm. However, the diamond growth only penetrated the top 15–20 μm of the template. Also, the coating shows a non-homogenous nature: the coating thickness at the surface can be one order of magnitude higher than deep in the template.

Nanomaterials based on sp^2 carbon are also candidates for the templated growth. Already in 2005, Terranova *et al.* reported nanodiamond coating on carbon nanotubes (CNT) [39]. The meaning of this work is limited by the fact that they were using a specially designed CVD system with nanocarbon as the carbon source. After that there has been no further report on diamond/CNT core-shell structures for more than five years. The difficulty in coating CNT with diamond lies in two aspects. The first is the difficulty for seeding. Hydrocarbons are known to be mostly hydrophobic. Therefore the water-based diamond colloid cannot properly penetrate. The second problem is more fundamental. The typical diamond growth condition requires the hydrogen plasma etching of the co-deposited graphitic carbon. Therefore, the CNT which acts as the growth template can be etched during the growth as well [40]. In 2012, Zou *et al.* reported the diamond/CNT teepee-like composite using optimized seeding and growth techniques [41]. In this study, the nanodiamond seeding was carried out using electrospray method. The methanol suspension of 5 nm nanodiamond was electrosprayed under 35 kV bias on to the grounded CNT substrate. After drying, the CNT array showed a teepee-like morphology. The diamond was deposited using 1% CH_4 in H_2 and ~1200 K.

Besides wire/fiber templates spherical compact from silica or opal lattice is also often used as the growth template. In 2012, Kurdyukov *et al.* reported the fabrication of porous diamond membranes using templated growth on synthetic opal film consists of SiO_2 beads (diameter: 520 ± 30 nm) [42]. The template lattice was formed via self-organization driven by the capillary force. To further enhance the stability of the lattice, high temperature (1000–1050°C) annealing was applied. In this research, up to 15 layers of silicon dioxide spheres were coated. Similar work has been reported by Kato *et al.* in the same year on boron-doped diamond foam for electrochemistry applications [43]. The template was fabricated by simply drop-casting SiO_2 spheres on boron-doped diamond substrate. The boron-doped diamond foam showed a strong Fano resonance in the Raman spectrum, which indicated a metallic conductivity [44, 45]. Also, a redox peak separation of ~59 mV for Hexaammineruthenium chloride is recorded, which shows that this material is suitable for electrochemistry.

Currently, the nonuniformity of the diamond coating in a 3D template is one of the most prominent problems for this growth technology. In order to obtain larger surface area in one growth, the diamond coating needs to penetrate as deep as possible. However, there is a contradiction in nature. If the radical chemistry in diamond growth is considered, one would find the following formula [46]:

$$C_DH + H^{\cdot} \leftrightarrow C_D^{\cdot} + H_2 \tag{8.1}$$

$$C_D^{\cdot} + CH_3^{\cdot} \leftrightarrow C_DCH_3 \tag{8.2}$$

Formula (8.1) is the activation process of a surface C-H site, and the formula (8.2) is the addition of a CH_3 group on the diamond surface. For the normal growth conditions, atomic hydrogen and methyl radicals need to be abundant. On a planar sample surface, this is normally the case. In 3D templates, however, this requirement is hardly satisfied. The reason is that when the radicals collide into the template, they lose the kinetic energy which is necessary to trigger the above mentioned reactions. For this reason, the diamond growth will decrease deeper into the template (Figure 8.3).

Concerning this problem, there is up to now no theoretical optimization about the growth condition regarding plasma power, gas pressure, methane concentration as well as the sample temperature. However, most recent works concentrated on lower temperature (<600°C) lower pressure (<30 mbar) growth conditions. In these conditions, the diamond growth is slow, and the mean-free-path for radicals is longer. Moreover, if the diamond growth is slow, the upper pores of the template will be closed slower, so that the growth on the lower parts will be less hindered. Also, by using low temperature more choices of template materials are available. In 2015, Ruffinatto *et al.* reported the coating of glass fiber filter paper up to a thickness of ~250 μm [48]. This deep thickness is possibly due to the optimized growth condition and the large (micrometer-sized)

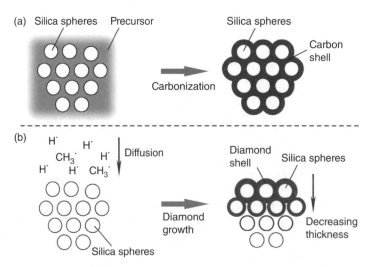

Figure 8.3 Schematic illustrations showing the difference between templated-growth of porous sp^2 carbon (a) and the templated-growth of diamond (b). *Source:* Gao 2015 [47]. Reproduced with permission of the American Chemical Society.

pores in the template. In the same year, Hébert *et al.* reported the coating of nanoporous conductive polymers using nanocrystalline with an ultralow-temperature growth technique (<450° C) [49]. In this study the authors emphasized the importance of high seeding density achieved by infiltration method. They claim that the seeds layer can prevent the etching effect of the hydrogen plasma on the polymer. The electrochemistry measurements showed that the background redox activities of the underlying porous polypyrrole layer were completely quenched after the diamond coating, showing that the coating is of pinhole-free quality.

Beside the optimization of the diamond growth parameters, another way to overcome the penetration problem in the templated growth is the layer-by-layer growth method [47]. Gao *et al.* reported an improve fabrication method of the diamond foam fabrication reported in reference [43]. Rather than depositing the template and diamond in one deposition, they deposit the composite in a layer-by-layer manner (Figure 8.4). In this method, the diffusion limitation shown in Figure 8.3 is relieved, and the thickness of the diamond has no theoretical upper limit.

Finally, there is a kind of less typical templated growth technique which can be called masked-growth. It resembles the templated growth in the way that a porous template is needed; however, the diamond will not grow on the template but on the unmasked area. Such methods are very similar to the metal deposition with shadow masks which has been often applied to obtain highly ordered metal dot patterns using ordered particles or porous AAO [16, 50]. In 2001, Matsuda *et al.* fabricated vertically aligned diamond nanocylinders using this method [51]. In this research, porous AAO was used as the mask for diamond cylinder growth. The bottom of the through-hole AAO film was seeded with nanodiamond (size: 50 nm). Afterwards, the AAO film was flipped over

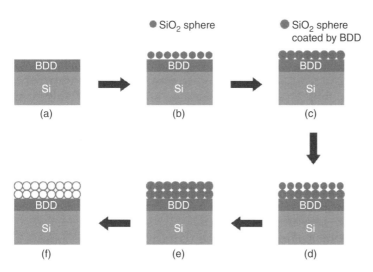

Figure 8.4 Schematic illustration showing the layer-by-layer growth technique for diamond foam electrodes showing: (a) boron-doped diamond (BDD) substrate growth on Si; (b) spin-coating of SiO$_2$ spheres on the substrate; (c) CVD diamond coating on SiO$_2$ templates; (d) spin-coating of the second SiO$_2$ layer, (e) the second CVD diamond coating on SiO$_2$ templates; (f) removal of SiO$_2$ templates. *Source*: Gao 2016 [47]. Reproduced with permission of the American Chemical Society. (*See color plate section for the color representation of this figure.*)

and the CVD deposition was carried out from the front side. The result showed that well-aligned polycrystalline diamond nanocylinders with a density of $4.6 \times 10^8 \, cm^{-2}$ were generated. The morphologies including the hexagonal close-packing and the average diameter were both inherited from the AAO mask, showing the effect of masked growth. By properly shaping the AAO masks, diamond cylinders with triangular and square cross sections can also be synthesized [52].

8.2.3 Surface Anisotropic Etching by Metal Catalyst

Diamond can react with H_2 at high temperatures with the help of metal catalyst particles. The etching mechanism was reported as early as 1993 by Ralchenko *et al.* [53]. They reported on the diamond patterning using Iron group elements (Fe, Co and Ni). The etching mechanism was explained in three steps: 1) carbon dissolution in metal, 2) diffusional transport to the metal-gas interface and 3) carbon desorption in the form of methane. They also showed that Fe has the strongest etching effect. This phenomenon was later discovered in nanoscale by Konishi *et al.* in 2006 [54]. They found that the Co nanoparticles generated by in-situ reduction from $Co(NO_3)_2$ in H_2 atmosphere could catalytically etch diamond surface in the same atmosphere and leave nanosized etch-pits on diamond surface. They also found that the geometry of the etch-pit is facet-dependent: on (111) facets the etch-pits were triangular or hexagonal (Figure 8.5a); on (100) facets, the etch-pits were rectangular (Figure 8.5b); on (110) facets, channels along the {111} direction will be formed (Figure 8.5c). This nanopit formation phenomenon was also reported later on other metal particles, including Ni [55], Fe [56, 57] and Pt [58]. Au nanoparticles are reported to be unable to etch diamond surface under heating and H_2 atmosphere [59].

In the fabrication of nanoporous structures using catalyzed etching, the density of the etch-pits is dependent on the density of metal nanoparticles. Therefore, particle density is an important parameter to control. Mehedi *et al.* have shown that if the particles are formed by thermal dewetting of a thin Ni metal film, the density of particles is linearly decreasing with the increasing thickness for films thinner than 5 nm [60]. However, the difference is within one order of magnitude ($10^{10} \, cm^{-2}$). For thickness more than 5 nm, the density drops heavily. Also, they found out that the etching stops at a depth of 400–500 nm into the diamond surface, due to the lack of hydrogen inside the deep pores [59].

A detailed study on the etching mechanism is published recently in reference [60]. In this study, diamond substrates were etched by Ni nanoparticles during a H_2 annealing process. The surface chemical composition was monitored via X-ray photoelectron spectroscopy (XPS) and electron energy loss spectroscopy (EELS) throughout the annealing process. No formation of nickel carbide was detected both before and after Ni removal. Moreover, the gas composition in the reaction chamber is also analyzed. There was a clear increase in the amount of methane in the gas mixture, and this increase is in accordance with the carbon loss calculated from the average depth and diameter of the nanopores. Therefore, the three-step etching mechanism is confirmed.

8.2.4 High Temperature Surface Etching

Even without metal catalyst diamond can react with H_2, water vapor and CO_2 at high temperatures. This phenomenon is discovered when researchers tries to understand

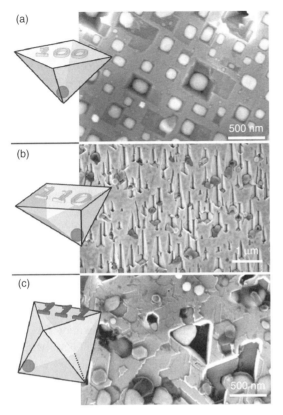

Figure 8.5 SEM images of observed pitting and channeling of (a): (100), (b): (110) and (c): (111) directed and etched single-crystal diamonds. The inset is a modeled octahedron which reflects symmetries of {111} oriented planes with indicated planes. Red circles indicate Ni. *Source*: Smirnov 2010 [55]. Reproduced with permission of AIP Publishing. (*See color plate section for the color representation of this figure.*)

the surface etch-pits (trigons) on nature diamonds [61]. There have been positive and negative etch-pits. Their formation and their relation with a variety of crystal defect have been reported [62]. However, because these etch-pits are shallow and micrometers in size, their effect on the local properties and the total surface area is small. Therefore, we will not go into the details of these effects in this chapter.

A practical catalyst-free etching or surface roughening method of diamond came in 2011. Ohashi *et al.* reported the diamond surface nanotexturing via the steam-activation method which is normally applied to activated carbon [63]. Temperatures between 600 and 900°C were applied to the sample. Water vapor was used as etchant. The etching effect started at 700°C: some triangular shaped etch-pits appeared on (111) facets. At higher temperatures rigorous corrosion of the BDD surface was observed. High density of columnar structures was obtained on the diamond surface. From the analysis of the outlet gas, CO and H_2 were confirmed to be the product. Therefore the etching mechanism is presumably:

$$C + H_2O \leftrightarrow CO + H_2 \tag{8.3}$$

An interesting consequence of the etching is the enhancement of the peak ratio between the diamond peak and the graphitic peak, showing a decreasing amount of sp^2 contentafter the process. Together with the observed fact that the (111) facet was more prone to etching, a hypothesis about the etching mechanism was deduced: the activation process first state with the sp^2-rich grain boundaries, and then the (111) facets which are close to grain boundaries are also selectively etched following the reaction (8.3). Some recent results published by the same group shows that boron-concentration has also an effect on the surface enlargement of diamond [64]. It is believed that in highly boron-doped samples, the proportion of (111) is larger than in lower boron-doped samples. Therefore, the etching has a stronger effect of the surface.

Other gases such as CO_2 and O_2 have also shown similar effects. Zhang *et al.* have reported that the activation process with CO_2 atmosphere at 800 and 900°C has a preferential etching on the (100) facets [65], via:

$$C + CO_2 \leftrightarrow 2CO \tag{8.4}$$

Because O_2 is a much stronger oxidant than H_2O and CO_2, direct annealing in O_2 containing atmosphere will have an uncontrollable etching effect on the sample in a relative short period. Therefore, special procedures are needed if O_2 is used for diamond etching. Kondo *et al.* shows that if this process is split into two phases, i.e. a high temperature (1000°C) graphitization process in Ar and a mild (425°C) oxidation in air, controlled and selective etching will happen on both (111) and (100) facets on the diamond surface [66].

8.2.5 Selective Material Removal

It has been shown for a long time that diamond is almost nondestructible wet- or electrochemically [67]. Diamond films can be used in acidic fluoride [68], alkaline [69]. Electrochemically, diamond has been used in the mixture of 1.0 M HNO_3 and 2.0 M NaCl at a current density of 0.5 A cm^{-2} for up to 12 h with no evidence of damage [70]. In H_2SO_4 solutions diamond electrodes have shown stability in a current range of 1–10 A cm^{-2} [71]. This information indicates the possibility to grow diamond together with some other materials and selectively remove the co-deposited materials after deposition. In this way, a porous diamond backbone will be fabricated.

The most commonly deposited non-diamond material in diamond growth is graphite. In the early research on diamond growth, it was known that graphitic carbon is co-deposited with diamond in CVD process [46, 72]. Normally, this graphitic growth needs to be minimized to enhance diamond quality. However, if a diamond sample is intentionally deposited with a high sp^2-content, and the non-diamond carbon is removed afterwards via selective etching, a way towards porous diamond can be found. This is shown by Kriele *et al.* in 2011 [73]. Using up to 20% methane in the gas mixture for diamond growth, highly graphitic diamond is deposited. Raman spectroscopy shows a very high G and D band showing the poor quality. By partially removing the sp^2 content in air at ~550°C nanopores can be generated on the diamond film. In this work, 150 nm thick diamond films were used. As a result, a nanoporous membrane was generated. Similar work was reported by Feng *et al.* [74]. However, they reported the selective removal from a thicker diamond film. The results showed that after etching away the non-diamond carbon, a fibrous diamond skeleton was formed. TEM shows that these fibers consisted of fine diamond grains. While the outer shape of the micrometer-size grains was kept, the diamond surface becomes highly porous.

The graphite removal can also happen in situ if etching gases are introduced in the growth atmosphere. In this method, diamond growth took place at the edge of no-growth region of C–H–O ternary diagram [75]. Diamond porous nanowire structures were generated directly from the growth. The authors used very different gas mixtures from a 'typical' diamond growth: very high methane concentration together with a high percentage of CO_2 (CH_4: CO_2: H_2=0.2: 0.8: 1). In this combination, the authors believed that the sp^2 carbon deposited by the high methane concentration was etched in situ by the excessive CO_2. This process results in very porous surface structures.

Recently, a new approach was reported by scientist working in the SiC field. Rather than depositing diamond and remove the by-product after the growth, they grow SiC and keep the by-product, which is diamond. Zhuang *et al.* reported the fabrication of free-standing diamond network by the selective removal of β-SiC in a SiC-diamond composite via wet-chemical etching (Figure 8.6) [76]. They varied the gas phase ratio between CH_4 and tetramethylsilane (TMS) to obtain different porosities in the network. When TMS/CH_4 ratio varies between 0.8 and 2.2% the porosity shifted from ~15 to ~70% almost linearly. This method shows a powerful tool in the fabrication of diamond porous membranes. From another view point, it is a kind of templated growth where the template is grown in situ. Therefore, the problem shown in Figure 8.3 is solved.

8.2.6 sp^2-Carbon Assisted Growth of Diamond Nanostructures

In some occasions, diamond nanostructures can be co-deposited with sp^2 carbon in some 'unusual' deposition conditions. For all these conditions, the sp^2 carbon grows

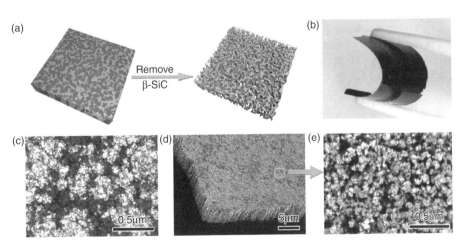

Figure 8.6 (a) Illustration of fabricating a diamond network from the composite film: the gray phase (β-SiC) is removed, leaving a yellow porous phase (diamond). (b) An optical photo of a flexible freestanding diamond network film. (c) SEM surface images of a nanocrystalline diamond/β-SiC composite film deposited with TMS/CH4 ratio of 1.5%. (d) Diamond network fabricated by etching the β-SiC phase from composite film shown in image c. (e) High-magnification SEM images of the surface of the film shown in image (d). *Source*: Zhuang 2015 [76]. Reproduced with permission of the American Chemical Society. (*See color plate section for the color representation of this figure.*)

simultaneously with diamond, and the non-diamond carbon acts as the 'template' for the diamond growth. In 2004, Sun *et al.* reported the growth of diamond nanorods during the H_2 plasma post-treatment of CNT [77]. This discovery was based on their earlier discovery of a high density nucleation on CNT during short time (<10 h) H_2 plasma treatment [78]. They found that if the treatment time was elongated to >20 h, diamond nanorods with diameters of 4–8 nm and length up to 200 nm were synthesized. Based on the observation that these nanorods are covered by a thin layer of amorphous carbon sheath, they deduced a possible growth mechanism (Figure 8.7): the amorphous sp^2 carbon clusters were generated first in the plasma. By continuous insertion of hydrogen, diamond nucleates were generated inside the clusters, which is followed by further crystal growth. During the diamond growth, the competing deposition of amorphous carbon will continue and wrap the outer surface of the as-grown diamond structures. In this mechanism, the amorphous carbon sheath plays a decisive role in the 1D growth of the diamond wire by confining the lateral growth.

A similar but more controllable growth method came later in 2010. Hsu *et al.* reported the diamond growth in atmospheric pressure CVD condition which is commonly used for CNT growth [79]. Similar to the previous work, they discovered the growth of diamond wires inside a CNT sheath. However, they wires are 60–90 nm in diameter and up to tens of micrometers in length. In this research, the importance of atomic hydrogen is also stressed. It is reported that the 12 h cooling process ($1.2°C\,min^{-1}$) was indispensable for the diamond wire synthesis.

In a recent report by Zhang *et al.* the diamond has been 'assembled' in side double wall CNT using diamantane dicarboxylic acid (DDA) as the building block [80]. The assembly process was realized by pulling individual DDA molecules inside the CNTs by a force resembling the capillary force, which is monitored and proved by high resolution transmission electron microscopy (HRTEM). The carboxylic groups in DDA were later removed by H_2 annealing at 600°C for 12 h. In this way, the diamantane part of DDA is connected forming a diamond nanowire with a diameter of 0.78 nm.

Besides the CNT-related growth, a N_2 containing deposition gas mixture will also generated wire-shaped morphology of the resulting diamond film. In 2007, Arenal *et al.* carried out a thorough research on ultrananocrystalline diamond (UNCD) growth

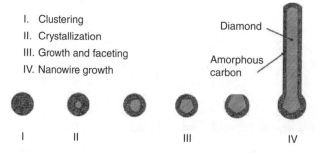

Figure 8.7 The proposed model for the formation of nanodiamonds, and the growth of diamond nanorods under hydrogen plasma irradiation of MWCNTs at high temperatures. Amorphous carbon clusters are formed in step I. The crystallization of diamond begins in the core of the carbon clusters (step II), followed by the diamond growth and faceting stage (step III). After the diamond nanocrystallites are faceted, diamond nano-rods begin to grow at the nanorod tips (step IV). Source : Sun 2004 [77]. Reproduced with permission of John Wiley and Sons.

under N_2 containing plasma [81]. They observed the formation of diamond nanowires with high density and uniform distribution at temperatures >800°C and with N_2 proportions >10% in the reaction gas mixture. HRTEM results showed that for the samples grown at 10% N_2, 800°C, the nanowires formed are 80–100 nm long with a core-shell structure. The diamond core is composed of 5 nm wide, 6–10 nm long nanocrystalline segments. This core is enveloped by sp^2 carbon, which is similar to the CNT-related cases. Almost the same results were reported in the same year by Vlasov *et al.* during the N-doped UNCD growth [82]. Also, similar to ref [77], these wires were elongated in the (110) direction, showing similar growth mechanism. In fact, the (110) preferential growth in C_2-dimer-dominated UNCD growth is confirmed theoretically [83]. The influence of N_2 concentration is likely to be the result of CN dimers which can preferentially attach to certain diamond facets [84].

8.2.7 High Pressure High Temperature (HPHT) Methods

Typically, the above mentioned techniques for porous diamond fabrication are based on CVD growth, or based on bulk diamond fabricated by CVD. Meanwhile, HPHT method has also been used to fabricate porous diamond. There are two approaches reported: (1) Starting from porous sp^2 carbon materials and using HPHT treatment to turn sp^2 into sp^3 carbon; in 2011, Zhang *et al.* fabricated monolithic transparent porous diamond crystals from mesoporous carbon CMK-8 via HPHT treatment (21 GPa, 1600°C) [85]. The aim of this research is to lower the sp^2-sp^3 conversion temperature by using mesoporous carbon. The diamond product inherited the porous structure, although specific surface of the porous diamond obtained was $33\,m^2g^{-1}$, much smaller than the starting porous carbon ($1250\,m^2g^{-1}$). (2) Using HPHT treatment to sinter diamond powder into porous bulk ceramic. Due to the high melting point of diamond, the sintering of diamond particle takes place also at high temperatures. In 2009, Zang *et al.* reported the fabrication of bulk BDD electrode using BDD particles [86]. Sintering was carried out at 1450°C, 6 GPa, with 15 wt% Fe–Co–B alloy powders as sintering catalyst. The BDD ceramic has 1–10 μm pores associated with grain boundaries; the porosity is measured to be 14%.

8.3 Application of Diamond Nanostructures in Electrochemistry

8.3.1 Biosensors Based on Nanostructured Diamond

The Biosensors is an important application for diamond-based nanomaterials. The appealing properties of diamond for biosensing include bio-compatibility [87–89], facile surface termination [90, 91] and functionalization [92–95], and easily cleaned and restored surface [96]. This application starts in 2008 with the work from Yang *et al.* using boron-doped diamond nanowires as a DNA sensing platform (Figure 8.8) [97]. Slightly earlier, they discovered that the electrochemical surface grafting of diamond nanowires with nitrophenyl linkers took place preferentially at the top of the wires [12]. Therefore, they continued the research by attaching single strand DNA (SS-DNA) on the top of the aminophenyl linker. The redox responses from ferro/ferricyanide are highly sensitive to the surface condition of the diamond nanowires. The peak current

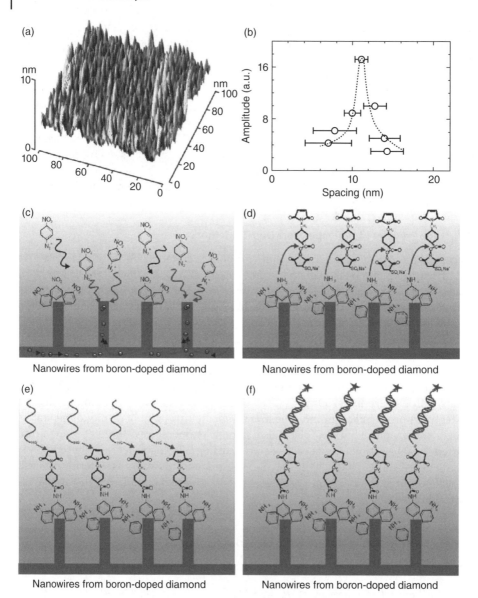

Figure 8.8 (a) Typical AFM image of diamond nanostructured surface; b) Fourier transformed surface properties of diamond nanostructured surface. From Fourier analysis, the average wire separation is about 11 nm with a narrow variation; (c)–(f) Schematic pictures of the biofunctionalization of vertically aligned diamond nanowires: (c) electrochemical grafting of nitrophenyl, (d) amination and crosslinker attachment, (e) probe DNA attachment, (e) DNA hybridization.

shrinks gradually as the phenyl linker (Figure 8.8c), cross-linker (Figure 8.8d), SS-DNA (Figure 8.8e), and the complementary DNA (which forms double strand DNA with the immobilized SS-DNA Figure 8.8f) were attached to the surface. With this technique, the detection limit for DNA is lowered to ~2 pM with good reproducibility and anti-interference properties against mismatching SS-DNA [98, 99].

Besides DNA sensing, diamond nanostrutures are also used in the sensing of a wide varieties of bio-organic chemicals. In 2009, Wei *et al.* reported the diamond grass electrode had an improved sensitivity towards the detection of uric acid and dopamine [2]. The electroxidation current was recorded from the electrode. Compared to a flat BDD electrode, the current as well as the reaction kinetics are both enhanced after surface modification. It is believed that the nanowire formation enhance the number of reactive site at the electrode surface (e.g. the tip of the wires), and thus a larger signal current.

In some cased, surface nanostructures do not only multiply the signal according to the surface-enlargement but also enable the electrode to detect substances which are otherwise not detectable. For glucose sensing, the flat BDD electrode is report to be nonreactive, the electroxidation current appear only after nanostructuring [37, 100]. Luo *et al.* reported the amperometric glucose sensor using diamond-coated silicon nanowires [37]; a sensitivity of $8.1\,\mu A\,mM^{-1}\,cm^{-2}$ with a limit of detection of $0.2\pm0.01\,\mu M$ was achieved. In a later research, diamond nanowires electrode was also used to detect tryptophan using differential pulse voltammetry [101]. The detection limit was $5\times10^{-7}\,M$ was obtained on BDD nanowires, as compared to $1\times10^{-5}\,M$ recorded on planar BDD electrodes [102].

It is worth noticing that voltammetric methods have seldom used in these detection. The reason is probably because of the heavily increased background current after the surface nanostructuring (we will come to this point again in the next section). In voltammetry techniques, the signal is limited by diffusion. Therefore, a further increase of the surface only increases the capacitive background proportional to the surface-enlargement but not the signal. Therefore, even if voltammetric method has been used, the background current need to be suppressed with pulse techniques [101]. This has been pointed out in the very early days of porous diamond electrode research [20] and repeatedly confirmed in later research [6, 43, 49, 103].

8.3.2 Energy Storage Based on Nanostructured Diamond

Due to the direct link between the electrical double layer capacitance and the surface enlargement, it is reasonable to use surface enlarged electrodes for energy storage devices. The idea is even more rationalized by the fact that the energy storage in double layer capacitors is proportional to the square of potential window: [104]

$$E = \frac{1}{2}CV^2 \tag{8.5}$$

$$P = \frac{V^2}{4R_s}, \tag{8.6}$$

where E, P, C, V and R_s are the energy, maximum power, capacitance, potential window and series resistance, respectively. As is pointed out, the diamond has so far the widest reported potential window reported in the aqueous electrolytes. Therefore, the attempt to use diamond as a potential supercapacitor material had already been made when the first known boron-doped diamond porous electrode was fabricated: Honda *et al.* reported the investigation of the diamond nanohoneycomb electrode for double layer capacitor applications in 2000 [105]. Using cyclic voltammetry and impedance methods in a three-electrode setup, they estimated that the double layer capacitance is ~200 times higher than a flat BDD electrode. The gravimetric capacitance was calculated to be $16\,F\,g^{-1}$, which is trivial, compared to sp^2 carbon based materials at that time [106]. Therefore, a generally negative conclusion was made on the idea of diamond-based supercapacitors. After that, the research on this topic is almost blank for about ten years until 2009. Kondo *et al.* made another attempt by using free standing boron-doped hollow fiber film for as supercapacitor electrode [38]. This time, a gravimetric capacitance of $13\,F\,g^{-1}$ was measured. This value is still small compared to sp^2 carbon materials [104]. Although the fabrication of the material is considerably easier than the previous results, there was no improvement in terms of gravimetric capacitance. Another prominent problem seen from this research is the large R_s. No information about the boron concentration in the diamond was given in the paper. However, the impedance spectroscopy shows an R_s in the order of $10^4\,\Omega\,cm^{-2}$, which shows the poor conductivity of the material.

During the last four years (2012–2015), the topic of diamond-based supercapacitor gained popularity due to a large variety of porous, surface-enlarged diamond electrode has been fabricate, including diamond foam [43, 107], diamond hollow fibers [47, 48], diamond-coated CNTs [103, 108] and diamond-coated conductive polymers [49]. These electrodes share some common properties like easy fabrication, applicable to large scale fabrication (up to 6-inch), high boron concentration (confirmed by Raman spectrum or electrochemical measurements). A summary on the fabrication of properties of these materials are shown in Table 8.1.

In 2015, the result published by Gao *et al.* shows the first diamond-based pouch-cell supercapacitor device based on free standing diamond paper [47]. The structure of the cell is shown in Figure 8.9a. Glass microfiber filters (GF/A, Whatman) was used as the separator and $3\,M\,NaClO_4$ was used as the aqueous electrolyte. The image of the device is shown in Figure 8.9b. Thanks to the two electrode device, many properties such as the potential window, series resistance, power and energy can be measured and calculated more reliably. The result shows a new understanding of the potential window of diamond. Instead of saying generally that diamond has a large potential window in aqueous solutions, the researchers find that it makes more sense to relate the potential window to the columbic efficiency of the device which is given by:

$$\text{Efficiency} = \frac{Q_{\text{discharging}}}{Q_{\text{charging}}} \times 100\%. \tag{8.7}$$

where $Q_{charging}$ and $Q_{discharging}$ are the amount of charge during charging and discharging processes, respectively. The result of the window opening test on the diamond device is shown in Figure 8.9c. It is clearly seen that the water splitting starts around 1.3 V. However, the current is still small even at a high voltage of 2.5 V due to the slow kinetics of water-splitting at diamond surface. As a result, the columbic efficiency of the device still exceeds 90% at 2.0 V (Figure 8.9d). On the other hand, although slow, the water splitting do consumes charges which are stored in the device. Therefore, when used in a potential window larger than 1.3 V, long-term energy storage is not possible. However, impedance spectroscopy shows that the relaxation time, which shows the highest working frequency of the given device [110], reaches 31.7 ms. This small relaxation time is

Table 8.1 Comparison between different diamond materials in literature.

Nanostructures	Fabrication Method	Thickness[1] (μm)	Surface-Enlargement	Areal Capacitance (mF cm^{-2})[2]	Ref.
Honeycomb	RIE etching with AAO mask	0.5	200	1.97	[105]
Hollow diamond fibers	Templated growth on quartz fiber filter	~20	~20	—	[38]
Diamond foam	Templated growth on quartz spheres	7	~40	—	[43]
Diamond-coated CNTs	Templated growth on CNT	3	116	0.58	[108]
Diamond nanowires	ICP etching with Ni nanoparticle mask	1	~10	—	[6]
Diamond foam	Layer-by-Layer Templated growth on quartz spheres	2.6	—	0.598	[107]
Diamond-coated CNTs	Templated growth on CNT	~40	~450	—	[103]
Diamond-coated silicon nanowires	Templated growth on Si Nanowires	~5	13	0.105	[109]
Diamond-coated porous polypyrrole	Ultralow-temperature Templated growth on Si Nanowires	~10	~300	3	[49]
Hollow diamond fibers	Templated growth on quartz fiber filter	~50	—	0.688 (based on two-electrode measurements)	[47]

[1] For aligned wire-structures, heights of the structures are shown
[2] Listed values are obtained in 3-electrode measurements unless otherwise noted

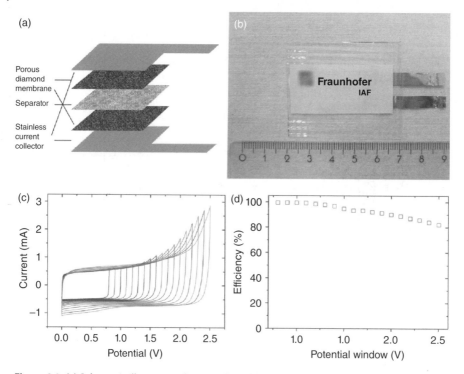

Figure 8.9 (a) Schematic illustration showing the laminated structure of a diamond-based pouch cell; (b) a photo of a diamond pouch-cell supercapacitor; the logo of the institute is intentionally blurred for copyright reasons. (c) window opening test for the diamond pouch cell in 3 M NaClO$_4$ at 1 V s^{-1} with potential windows from 0.8–2.5 V; (d) plot of the device efficiency against the potential window. *Source*: Yang 2014 [132]. Reproduced with permission of the Royal Society of Chemistry.

believed to be because of the high ion mobility in the aqueous electrolyte and the macroporous nature of the diamond paper. Moreover, the stability of the diamond-based device is proved i20 000 galvanostatic charge/discharging cycles. The capacitance drop only ~8%. Therefore, the device is well suitable for high-voltage and high-frequency application in aqueous solutions.

8.3.3 Catalyst Based on Nanostructured Diamond

Nanostructured catalyst is constantly of interest for material scientist due to the nano-sized effect and large specific surface [111]. However, nanosized particles are thermal dynamically unstable. Therefore, nanoparticles are often supported by other materials with also large specific surface so that the total surface energy is reduced [112, 113]. Carbon materials, often activated carbon are used for this purpose in commercially available product [114–117]. However, sp^2-carbon-based materials suffer from the electrochemical corrosion which is given by [118]:

$$C + 2H_2O \rightarrow CO_2 + 4H^+ + 4e^-, E^0 = 0.207 \text{ V vs SHE} \tag{8.8}$$

and

$$C + H_2O \rightarrow CO + 2H^+ + 2e^-, E^0 = 0.518 \text{ V vs SHE.} \tag{8.9}$$

On the other hand, diamond is resistive to electrochemical corrosion. Therefore, using diamond as a substitute for traditional porous carbon is a plausible idea.

Research on this topic has been carried out during the last decade on planar diamond electrodes [7, 119–123]. However, compared to sp^2-carbon-based materials, planar diamond electrode lacks the sufficient surface area which would make it a high-performance performance for catalyst. As a result, surface-enlarged diamond electrode should be used as a high-surface area support for catalyst. As early as 2001, Honda *et al.* electrodeposited Pt nanoparticles on nanoporous diamond honeycomb electrodes [124]. In their research, electrodes with surface roughness factors of 10.9 and 15.9 were used, and Pt surfaces equal to 3–4 times a planar Pt electrode were obtained. Due to the low Pt coverage on the porous electrode, the large surface area provided by the porous diamond is not fully used. Therefore, these results are comparable to later researches in which Pt nanoparticles were deposited onto planar diamond surfaces [7, 120].

In fact, the combination of porous diamond and catalytic nanoparticles has been reported only in very limited cases. The above-mentioned difficulty in uniformly depositing metal or other catalyst on the porous diamond electrode might have caused this situation. According to current results on electrodeposition on diamond electrodes, there are two main problems in achieving a uniform nanoparticle coating. One is the low nucleation density of metal deposition on diamond. Due to the inertness of a diamond surface, the nucleation can only happen on grain boundaries or other local defects [125]. In the case of a planar diamond electrode, pre-deposition treatment such as nanodiamond scratching [122], mechanical polishing [126] and wet-chemical seeding [7], have been applied to enhance the nucleation density. However, these methods are either difficult to implement on porous diamond or have not been tested so far. The second reason is the diffusion limitation in electrochemical deposition processes. During the deposition, reactive ions are depleted inside the porous matrix and the diffusing ions only arrive at the uppermost part of the porous structure. As a result, only the top of the electrode is coated, which results in a low coverage of deposits [12].

In 2015, Gao *et al.* partially solved this problem for vertically aligned structures [127]. They applied physical instead of chemical deposition for the coating to achieve high homogeneity on high-aspect ratio structures. In their research, DC sputtering has been used to deposit thin (1–3 nm) Pt layers on vertically aligned diamond nanowires (Figure 8.10). TEM results showed that the thin layers self-assemble into nanoparticles with diameters less than 10 nm due to the minimization of the surface energy (Figure 8.10c–d). Characterization of the catalytic activities shows the diamond-Pt composite showed a high specific area of $33 \text{ m}^2 \text{ g}^{-1}$ in terms of Pt which is in the same magnitude with traditional Pt/C catalyst [117]. The electrode also achieved 23 times enlargement of Pt activity compared to a planar Pt electrode. The value is the highest reported on Pt-diamond system so far. However, for other more complex diamond structures such as diamond foam and diamond hollow fibers, to achieve a dense and uniform catalyst coating still remains as a problem to solve.

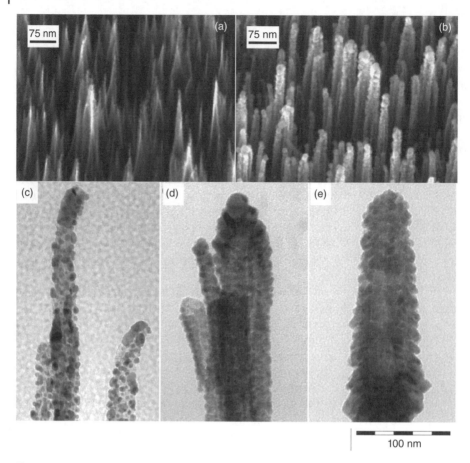

Figure 8.10 (a) SEM images of diamond nanowires and (b) diamond nanowires coated by 20 nm (nominal) Pt; (c) TEM images of diamond nanowires coated by Pt layers of nominal thicknesses of 20 nm, (d) 40 nm, and (e) 60 nm.

8.3.4 Diamond Porous Membranes for Chemical/Electrochemical Separation Processes

In recent years, due to the high demanding in robust membranes for separation and purification, researches on polymer including conductive polymers have attracted wide attention [128–130]. Compared to polymers, diamond has numerous advantages including the mechanical strength, chemical stability, wide potential window, as well as high conductivity (if highly boron-doped). The substantial development on the diamond nanostructuring during the last 15 years enables the fabrication of various porous diamond membranes with variable porosities. Therefore, the authors believe that functional diamond membranes will be an important topic in the diamond community for the coming years.

The fabrication of electrically conductive diamond membranes dates back to around the year 2000 when the through-hole diamond honeycomb electrode was fabricated [21]. However, the application was not clear until 2011 when Honda *et al.* reported the study on an electrically-switchable diamond-like carbon (DLC) membrane [131]. The membrane was fabricated via templated growth on a porous AAO substrate. The pore size could be tuned between 14 and 105 nm. By applying a potential on the membrane, the ion flux through the membrane can be selectively accelerated or hindered depending on the charge of the ions: ions with the same charge as the membrane will be repulsed and vice versa. Although this research was carried out on DLC, the application can be easily transferred to porous diamond membranes. Due to the better conductivity, chemical stability and wider potential window of diamond, the results may be further improved.

Besides potential, the surface termination of the membrane is another parameter to adjust in a diamond membrane. It has been shown that a diamond membrane can be tuned between superhydrophobic to hydrophilic by changing the surface from H- to O-termination. In 2014, Yang *et al.* reported the fabrication of diamond membrane by coating CVD diamond on a microstructured copper mesh [132]. Due to the hydrogen rich deposition gas mixture, as-grown membranes are H-terminated and thus superhydrophobic with a contact angle of >150° for water and other aqueous solutions. By putting a droplet of water–oil mixture onto the membrane, the water will be retained by the membrane while oil will go through it. In this way, the water is separated from oil (Figure 8.11). The surface-wettability can easily turn hydrophilic by air-annealing at 500° C as the surface become O-terminated.

More functionality can be realized by further surface-modification via organic linker molecules. The surface grafting of diamond can be realized by photochemistry [87], electrochemistry [133] and wet-chemistry methods [134]. With appropriate linker molecules, it is possible to terminated diamond surface by e.g. $-NH_2$, $-COOH$, and $-HSO_3$, and different surface functionality can be realized. Ruffinatto *et al.* showed an exemplary work on the functionalization of diamond fiber paper with aliphatic C_4 linkers [48]. The functionalization was realized by butylamine in an aqueous solution

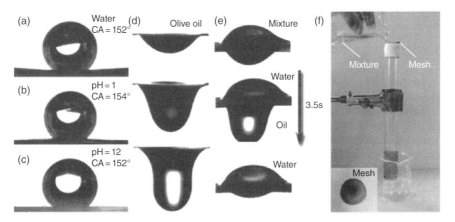

Figure 8.11 Photograph of water droplets with pH = 7 (a), acidic, pH = 1 (b) and basic, pH = 12 (c) dropped on diamond meshes showing a superhydrophobic wettability. The dynamic behaviors of a sole oil (d) and mixed water–oil droplet (e) on diamond mesh showing a quickly permeation of oil through the mesh and the separation of water–oil, respectively. (f) Photograph of a water–oil mixture separation device designed by the diamond mesh, indicating the fully water–oil separation. *Source:* Yang 2014 [132]. Reproduced with permission of the Royal Society of Chemistry.

with a pH of 10. The mechanism is a nucleophilic substitution where the ammonium moiety acts as the leaving group. The C_4 functionalized diamond membrane was used in protein extraction applications. The filtering experiments showed that the functionalized diamond membrane retains 20 times more protein than the one which is not functionalized.

8.4 Summary and Outlook

In this chapter, the fabrication and application of a large variety of porous diamond and diamond nanostructures are introduced. Diamond fabrication techniques, including vertical structures via RIE etching and more complex 3D structures using templated-growth, has been developed in great depth during the last two decades. However, there is still room for improvements. Diamond etching method starts with thick bulk diamond. Therefore, the phase purity (in terms of non-diamond carbon) is easier to control. However, the morphology is limited to vertical structures and to achieve high aspect ratio (>30) is not yet reported. For templated-growth methods, the coatings are normally NCD with a large proportion of grain boundaries. The quality of diamond is difficult to guarantee. One solution is to achieve high density seeding so that the coalescence of nuclei happens earlier during the growth. Hopefully in this way, thinner films with lower sp^2-rich grain boundary proportions can be achieved. However, the sp^2 carbon growth in the deeper part cannot be easily inhibited. Direct growth of pure, large aspect ratio diamond wires seems to be a good solution. However, it is not yet clear if it is possible to dope these wires. Also, current reports on direct diamond nanowire growth are mainly on the report of the phenomenon; the technology is not readily on the application level.

On the other hand, the application of micro- and nanostructured diamond covers almost every aspect of electrochemistry including sensor, energy storage/conversion and separation/purification. However, challenges and chances remain on this topic. The authors would like to particularly emphasize on the diamond membrane fabrication and modification. Nowadays, the fabrication of diamond membrane has already been reported by several groups in the community with methods which are easy to handle and reproduce. By appropriate surface termination, it is expected that diamond membranes can be qualified for the tasks of other polymer membranes such as selective ion permeability and specific adsorption of chemicals. Due to the numerous unique chemical and physical properties, diamond will be a promising material for a new generation of membrane systems. Researches in this direction will lead to a broad range of new applications. On the contrary, the research on sensors based on diamond nanomaterials is less reported in recent years. As is pointed out in this chapter, the diffusion limitation in electrochemistry is a principal problem for surface enlarged electrode for sensing application. In addition, the diamond surface is inert, which means specific adsorption/accumulation of analytes is rare. Therefore, the surface enlargement cannot provide positive influence in the sensing. However, it is still possible that porous diamond can be used as porous substrate for gas sensing where diffusion limitation is not an issue.

Acronyms

AAO	Anodized aluminun oxide
BDD	Boron-doped diamond
CNT	Carbon nanotube
CVD	Chemical vapor deposition
CVI	Chemical vapor infiltration
DC	Direct current
DDA	Diamantane dicarboxylic acid
DLC	Diamond-like carbon
DNA	Deoxyribonucleic acid
EELS	Electron energy loss spectroscopy
HPHT	High-pressure high-temperature
HRTEM	High-refsolution transmission electron microscopy
ICP	Inductively coupled plasma
NCD	Nanocrystalline diamond
PCD	Polycrystalline diamond
RF	Radio frequency
RIE	Reactive ion etching
SEM	Scanning electron microscopy
SS-DNA	Single-strand DNA
TEM	Transmission electron microscopy
TMS	Tetramethylsilane
UNCD	Ultrananocrystalline diamond
XPS	X-ray photoelectron spectroscopy

References

1 H. Shiomi. Reactive ion etching of diamond in O_2 and CF_4 plasma, and fabrication of porous diamond for field emitter cathodes. Japanese Journal of Applied Physics, 1997, Part 1, 36, 7745–7748.

2 M. Wei, C. Terashima, M. Lv, A. Fujishima, Z.Z. Gu. Boron-doped diamond nanograss array for electrochemical sensors. Chemical Communications, 2009, 3624–3626.

3 C. Terashima, K. Arihara, S. Okazaki, *et al.* Fabrication of vertically aligned diamond whiskers from highly boron-doped diamond by oxygen plasma etching, ACS Applied Material Interfaces, 2011, 3, 177–182.

4 Y. Coffinier, S. Szunerits, H. Drobecq, *et al.* Diamond nanowires for highly sensitive matrix-free mass spectrometry analysis of small molecules. Nanoscale, 2012, 4, 231–238.

5 Y.S. Zou, Y. Yang, W.J. Zhang, *et al.* Fabrication of diamond nanopillars and their arrays. Applied Physics Letters, 2008, 92, 053105.

6 W. Smirnov, A. Kriele, N. Yang, C.E. Nebel. Aligned diamond nano-wires: Fabrication and characterisation for advanced applications in bio- and electrochemistry. Diamond Related Materials, 2010, 19, 186–189.

7 F. Gao, N.J. Yang, W. Smirnov, *et al*. Size-controllable and homogeneous platinum nanoparticles on diamond using wet chemically assisted electrodeposition. Electrochimica Acta, 2013, 90, 445–451.

8 I. Shpilevaya, W. Smirnov, S. Hirsz, *et al*. Nanostructured diamond decorated with Pt particles: preparation and electrochemistry. RSC Advances, 2014, 4, 531–537.

9 N.D. Denkov, O.D. Velev, P.A. Kralchevsky, *et al*. Two-dimensional crystallization. Nature, 1993, 361, 26–26.

10 R. Micheletto, H. Fukuda, M. Ohtsu. A simple method for the production of a 2-dimensional, ordered array of small latex-particles. Langmuir, 1995, 11, 3333–3336.

11 S. Okuyama, S.I. Matsushita, A. Fujishima. Periodic submicrocylinder diamond surfaces using two-dimensional fine particle arrays. Langmuir, 2002, 18, 8282–8287.

12 N.J. Yang, H.S. Uetsuka, E. Sawa, C.E. Nebel. Vertically aligned nanowires from boron-doped diamond, Nano Letters, 2008, 8, 3572–3576.

13 O.A. Williams, J. Hees, C. Dieker, *et al*. Size-dependent reactivity of diamond nanoparticles. ACS Nano, 2010, 4, 4824–4830.

14 F. Keller, M.S. Hunter, D.L. Robinson. Structural features of oxide coatings on aluminum. Journal of the Electrochemical Society, 1953, 100, 411–419.

15 A.P. Li, F. Müller, A. Birner, *et al*. Hexagonal pore arrays with a 50–420 nm interpore distance formed by self-organization in anodic alumina. Journal of Applied Physics, 1998, 84, 6023.

16 H. Masuda, M. Satoh, Fabrication of gold nanodot array using anodic porous alumina as an evaporation mask, Japanese Journal of Applied Physics, 1996, Part 2, 35, L126–L129.

17 D. Crouse, Y.H. Lo, A.E. Miller, M. Crouse. Self-ordered pore structure of anodized aluminum on silicon and pattern transfer. Applied Physics Letters, 2000, 76, 49–51.

18 H. Masuda, M. Watanabe, K. Yasui, *et al*. Fabrication of a nanostructured diamond honeycomb film. Advanced Materials, 2000, 12, 444–447.

19 K. Honda, T.N. Rao, D.A. Tryk, *et al*. Impedance characteristics of the nanoporous honeycomb diamond electrodes for electrical double-layer capacitor applications. Journal of the Electrochemical Society, 2001, 148, A668–A679.

20 K. Honda, M. Yoshimura, R. Uchikado, *et al*. Electrochemical characteristics for redox systems at nano-honeycomb diamond. Electrochimica Acta, 2002, 47 4373–4385.

21 H. Masuda, K. Yasui, M. Watanabe, *et al*. Fabrication of through-hole diamond membranes by plasma etching using anodic porous alumina mask. Electrochemica Solid Status, 2001, 4, G101–G103.

22 G.O. Ke, Z.J. Xing, X.T. Yin, *et al*. Diamond thin-films deposited on porous silicon substrates. Vacuum, 1992, 43, 1043–1045.

23 S.B. Iyer, S. Srinivas. Diamond deposition on As-anodized porous silicon: some nucleation aspects. Thin Solid Films, 1997, 305, 259–265.

24 A.N. Obraztsov, I.Y. Pavlovsky, V.Y. Timoshenko. Diamond seed incorporation by electrochemical treatment of silicon substrate. Diamond Related Materials, 1997, 6, 1629–1632.

25 A.J. Fernandes, P.J. Ventura, R.F. Silva, M.C. Carmo. Porous silicon capping by CVD diamond. Vacuum, 1999, 52, 215–218.

26 N.A. Braga, C.A.A. Cairo, E.C. Almeida, *et al*. From micro to nanocrystalline transition in the diamond formation on porous pure titanium. Diamond Related Materials, 2008, 17, 1891–1896.

27 E. Titus, M.K. Singh, K.N.N. Unni, *et al*. Diamond nucleation and growth on zeolites, Diamond Related Materials 2003, 12, 1647–1652.

28 Q. Wei, M.N.R. Ashfold, Z.M. Yu, L. Ma. Fabrication of adherent porous diamond films on sintered WC-13 wt.%Co substrates by bias enhanced hot filament chemical vapour deposition. Physica Status Solidi A, 2011, 208, 2033–2037.

29 V. Baranauskas, A.C. Peterlevitz, D.C. Chang, S.F. Durrant. Method of porous diamond deposition on porous silicon. Applied Surface Science, 2001, 185, 108–113.

30 N.A. Braga, C.A.A. Cairo, M.R. Baldan, *et al*. Optimal parameters to produce high quality diamond films on 3D Porous Titanium substrates. Diamond Related Materials, 2011, 20, 31–35.

31 M. Ma, M. Chang, X. Li. Synthesis of boron-doped diamond/porous Ti composite materials — Effect of carbon concentration. Journal of Wuhan University of Technology-Materials Science Education, 2012, 27, 328–332.

32 A. Glaser, S.M. Rosiwal, B. Freels, R.F. Singer. Chemical vapor infiltration (CVI): Part I: a new technique to achieve diamond composites. Diamond Related Materials, 2004, 13, 834–838.

33 A. Glaser, S.M. Rosiwal, R.F. Singer, Chemical vapor infiltration (CVI): Part II: Infiltration of porous substrates with diamond by using a new designed hot-filament plant. Diamond Related Materials, 2006, 15, 49–54.

34 J.M. Ting, A.G. Lagounov, M.L. Lake. Chemical vapour infiltration of diamond into a porous carbon. Materials Science Letters, 1996, 15, 350–352.

35 P.A. Demkowicz, N.S. Bell, D.R. Gilbert, *et al*. Diamond-coated silicon carbide whiskers. Journal of the American Ceramics Society, 1999, 82, 1079–1081.

36 V. Baranauskas, A.C. Peterlevitz, H.J. Ceragioli, S.F. Durrant, Growth of diamond and carbon structures on natural pyrolyzed fibers, Thin Solid Films, 398 (2001) 260–264.

37 D. Luo, L. Wu, J. Zhi. Fabrication of boron-doped diamond nanorod forest electrodes and their application in nonenzymatic amperometric glucose biosensing. ACS Nano, 2009, 3, 2121–2128.

38 T. Kondo, S. Lee, K. Honda, T. Kawai. Conductive diamond hollow fiber membranes. Electrochemical Communications, 2009, 11, 1688–1691.

39 M.L. Terranova, S. Orlanducci, A. Fiori, *et al*. Controlled evolution of carbon nanotubes coated by nanodiamond: the realization of a new class of hybrid nanomaterials. Chemistry of Materials, 2005, 17, 3214–3220.

40 N. Shankar, N.G. Glumac, M.F. Yu, S.P. Vanka. Growth of nanodiamond/carbon-nanotube composites with hot filament chemical vapor deposition. Diamond Related Materials, 2008, 17, 79–83.

41 Y. Zou, P.W. May, S.M.C. Vieira, N.A. Fox. Field emission from diamond-coated multiwalled carbon nanotube 'teepee' structures. Journal of Applied Physics, 2012, 112, 044903.

42 D.A. Kurdyukov, N.A. Feoktistov, A.V. Nashchekin, *et al*. Ordered porous diamond films fabricated by colloidal crystal templating. Nanotechnology, 2012, 23, 015601.

43 H. Kato, J. Hees, R. Hoffmann, *et al*. Diamond foam electrodes for electrochemical applications. Electrochemical Communications, 2013, 33, 88–91.

44 S. Prawer, R.J. Nemanich. Raman spectroscopy of diamond and doped diamond. Philosophical Transactions of the Royal Society of London, Series A, 2004, 362, 2537–2565.

45 W. Gajewski, P. Achatz, O.A. Williams, *et al.* Electronic and optical properties of boron-doped nanocrystalline diamond films. Physics Review B, 2009, 79, 045206.

46 J.E. Butler, R.L. Woodin. Thin-film diamond growth mechanisms. Philosophical Transactions of the Royal Society of London, Series A, 1993, 342, 209–224.

47 F. Gao, C.E. Nebel. Diamond-based supercapacitors: Realization and properties. ACS Applied Material Interfaces, 2016, 8(42), 28244–28254.

48 S. Ruffinatto, H.A. Girard, F. Becher, *et al.* Diamond porous membranes: A material toward analytical chemistry. Diamond Related Materials, 2015, 55, 123–130.

49 C. Hébert, E. Scorsone, M. Mermoux, P. Bergonzo. Porous diamond with high electrochemical performance. Carbon, 2015, 90, 102–109.

50 H.W. Deckman, J.H. Dunsmuir. Natural lithography. Applied Physics Letters, 1982, 41, 377–379.

51 H. Masuda, T. Yanagishita, K. Yasui, *et al.* Synthesis of well-aligned diamond nanocylinders. Advanced Materials, 2001, 13, 247–249.

52 T. Yanagishita, K. Nishio, M. Nakao, *et al.* Synthesis of diamond cylinders with triangular and square cross sections using anodic porous alumina templates. Chemistry Letters, 2002, 31(10), 976–977.

53 V.G. Ralchenko, T.V. Kononenko, S.M. Pimenov, *et al.* Catalytic interaction of Fe, Ni and Pt with diamond films - Patterning applications. Diamond Related Materials, 1993, 2, 904–909.

54 S. Konishi, T. Ohashi, W. Sugimoto, Y. Takasu. Effect of the crystal plane on the catalytic etching behavior of diamond crystallites by cobalt nanoparticles. Chemical Letters, 2006, 35, 1216–1217.

55 W. Smirnov, J.J. Hees, D. Brink, *et al.* Anisotropic etching of diamond by molten Ni particles. Applied Physics Letters, 2010, 97, 073117.

56 T. Ohashi, W. Sugimoto, Y. Takasu. Catalytic etching of synthetic diamond crystallites by iron. Applied Surface Science, 2012, 258, 8128–8133.

57 J. Wang, L. Wan, J. Chen, J. Yan. Anisotropy of synthetic diamond in catalytic etching using iron powder. Applied Surface Science, 2015, 346, 388–393.

58 T. Ohashi, W. Sugimoto, Y. Takasu. Catalytic etching of {100}-oriented diamond coating with Fe, Co, Ni, and Pt nanoparticles under hydrogen. Diamond Related Materials, 2011, 20, 1165–1170.

59 H.A. Mehedi, C. Hebert, S. Ruffinatto, *et al.* Formation of oriented nanostructures in diamond using metallic nanoparticles. Nanotechnology, 23 (2012), 23, 455302.

60 H.A. Mehadi, J.C. Arnault, D. Eon, *et al.* Etching mechanism of diamond by Ni nanoparticles for fabrication of nanopores. Carbon, 2013, 59, 448–456.

61 F.C. Frank, K.E. Puttick, E.M. Wilks. Etch pits and trigons on diamond: I. Philosophical Magazine, 1958, 3, 1262–1272.

62 F.K. de Theije, E. van Veenendaal, W.J.P. van Enckevort, E. Vlieg. Oxidative etching of cleaved synthetic diamond {111} surfaces. Surface Science, 2001, 492, 91–105.

63 T. Ohashi, J. Zhang, Y. Takasu, W. Sugimoto. Steam activation of boron doped diamond electrodes. Electrochimica Acta, 2011, 56, 5599–5604.

64 J. Zhang, T. Nakai, M. Uno, *et al.* Effect of the boron content on the steam activation of boron-doped diamond electrodes. Carbon, 2013, 65, 206–213.

65 J.F. Zhang, T. Nakai, M. Uno, *et al.* Preferential {100} etching of boron-doped diamond electrodes and diamond particles by CO_2 activation. Carbon, 2014, 70, 207–214.

66 T. Kondo, Y. Kodama, S. Ikezoe, *et al.* Porous boron-doped diamond electrodes fabricated via two-step thermal treatment. Carbon, 2014, 77 783–789.

67 G.M. Swain, A.B. Anderson, J.C. Angus. Applications of diamond thin films in electrochemistry. MRS Bulletin, 1998, 23, 56–60.

68 G.M. Swain. The susceptibility to surface corrosion in acidic fluoride media: A comparison of diamond, HOPG, and glossy carbon electrodes. Journal of the Electrochemical Society, 1994, 141, 3382–3393.

69 R. deClements, G.M. Swain. The formation and electrochemical activity of microporous diamond thin film electrodes in concentrated KOH. Journal of the Electrochemical Society, 1997, 144, 856–866.

70 Q.Y. Chen, M.C. Granger, T.E. Lister, G.M. Swain. Morphological and microstructural stability of boron-doped diamond thin film electrodes in an acidic chloride medium at high anodic current densities. Journal of the Electrochemical Society, 1997, 144, 3806–3812.

71 N. Katsuki, S. Wakita, Y. Nishiki, *et al.* Electrolysis by using diamond thin film electrodes. Japanese Journal of Applied Physics 2, 1997, 36, L260–L263.

72 F. Celii. Diamond chemical vapor deposition. Annual Review of Physical Chemistry, 1991, 42, 643–684.

73 A. Kriele, O.A. Williams, M. Wolfer, *et al.* Formation of nano-pores in nano-crystalline diamond films. Chemical Physics Letters, 2011, 507, 253–259.

74 S. Feng, X. Li, P. He, *et al.* Porous structure diamond films with super-hydrophilic performance. Diamond Related Materials, 2015, 56, 36–41.

75 Š. Potocký, O. Babchenko, K. Hruška, A. Kromka. Linear antenna microwave plasma CVD diamond deposition at the edge of no-growth region of C–H–O ternary diagram. Physica Status Solidi (B), 2012, 249, 2612–2615.

76 H. Zhuang, N.J. Yang, H.Y. Fu, *et al.* Diamond network: Template-free fabrication and properties. ACS Appl. Material Interfaces, 2015, 7, 5384–5390.

77 L.T. Sun, J.L. Gong, D.Z. Zhu, *et al.* Diamond nanorods from carbon nanotubes. Advanced Materials, 2004, 16, 1849.

78 L.T. Sun, J.L. Gong, Z.Y. Zhu, *et al.* Nanocrystalline diamond from carbon nanotubes. Applied Physics Letters, 2004, 84, 2901–2903.

79 C.H. Hsu, S.G. Cloutier, S. Palefsky, J. Xu. Synthesis of diamond nanowires using atmospheric-pressure chemical vapor deposition. Nano Letters, 2010, 10, 3272–3276.

80 J.Y. Zhang, Z. Zhu, Y.Q. Feng, *et al.* Evidence of diamond nanowires formed inside carbon nanotubes from diamantane dicarboxylic acid. Angewandte Chemie International Edition, 2013, 52, 3717–3721.

81 R. Arenal, P. Bruno, D.J. Miller, *et al.* Diamond nanowires and the insulator-metal transition in ultrananocrystalline diamond films. Physics Review B, 2007, 75, 195431.

82 I.L. Vlasov, O.I. Lebedev, V.G. Ralchenko, *et al.* Hybrid diamond-graphite nanowires produced by microwave plasma chemical vapor deposition. Advanced Materials, 2007, 19, 4058.

83 P.C. Redfern, D.A. Horner, L.A. Curtiss, D.M. Gruen. Theoretical studies of growth of diamond (110) from dicarbon. Journal of Physical Chemistry, 1996, 100, 11654–11663.

84 K.J. Sankaran, J. Kurian, H.C. Chen, *et al.* Origin of a needle-like granular structure for ultrananocrystalline diamond films grown in a N-2/CH$_4$ plasma. Journal of Physics D Applied Physics, 2012, 45, 365303.

85 L. Zhang, P. Mohanty, N. Coombs, *et al.* Catalyst-free synthesis of transparent, mesoporous diamond monoliths from periodic mesoporous carbon CMK-8. Proceedings of the National Academy of Sciences USA, 2010, 107, 13593–13596.

86 J.B. Zang, Y.H. Wang, H. Huang, W.Q. Liu. Electrochemical characteristics of boron doped polycrystalline diamond electrode sintered by high pressure and high temperature. Journal of Applied Electrochemistry, 2009, 39, 1545–1551.

87 W. Yang, O. Auciello, J.E. Butler, *et al.* DNA-modified nanocrystalline diamond thin-films as stable, biologically active substrates. Natural Materials, 2002, 1, 253–257.

88 A. Hartl, E. Schmich, J.A. Garrido, *et al.* Protein-modified nanocrystalline diamond thin films for biosensor applications. Natural Materials, 2004, 3, 736–742.

89 R. Hoffmann, A. Kriele, H. Obloh, *et al.* The creation of a biomimetic interface between boron-doped diamond and immobilized proteins. Biomaterials, 2011, 32, 7325–7332.

90 A. Denisenko, C. Pietzka, A. Romanyuk, *et al.* The electronic surface barrier of boron-doped diamond by anodic oxidation. Journal of Applied Physics, 2008, 103, 014904.

91 R. Hoffmann, A. Kriele, H. Obloh, *et al.* Electrochemical hydrogen termination of boron-doped diamond. Applied Physics Letters, 2010, 97, 052103.

92 S. Nakabayashi, N. Ohta, A. Fujishima. Dye sensitization of synthetic p-type diamond electrode. Physical Chemistry Chemical Physics, 1999, 1, 3993–3997.

93 S.A. Yao, R.E. Ruther, L.H. Zhang, *et al.* Covalent attachment of catalyst molecules to conductive diamond: CO_2 reduction using 'smart' electrodes. Journal of the American Chemical Society, 2012, 134, 15632–15635.

94 I. Zegkinoglou, P.L. Cook, P.S. Johnson, *et al.* Electronic structure of diamond surfaces functionalized by Ru(tpy)(2). Journal of Physical Chemistry C, 2012, 116, 13877–13883.

95 W.S. Yeap, X. Liu, D. Bevk, *et al.* Functionalization of boron-doped nanocrystalline diamond with N3 dye molecules. ACS Applied Material Interfaces, 2014, 6, 10322–10329.

96 R. Hoffmann, H. Obloh, N. Tokuda, *et al.* Fractional surface termination of diamond by electrochemical oxidation. Langmuir, 2012, 28, 47–50.

97 N. Yang, H. Uetsuka, E. Osawa, C.E. Nebel. Vertically aligned diamond nanowires for DNA sensing. Angew. Chem. Int. Edit., 2008, 47, 5183–5185.

98 N.J. Yang, H. Uetsuka, C.E. Nebel. Biofunctionalization of vertically aligned diamond nanowires. Advanced Functional Materials, 2009, 19, 887–893.

99 N.J. Yang, H. Uetsuka, O.A. Williams, *et al.* Vertically aligned diamond nanowires: Fabrication, characterization, and application for DNA sensing. Physica Status Solidi A, 2009, 206, 2048–2056.

100 Q. Wang, P. Subramanian, M. Li, *et al.* Non-enzymatic glucose sensing on long and short diamond nanowire electrodes. Electrochemical Communications, 2013, 34, 286–290.

101 S. Szunerits, Y. Coffinier, E. Galopin, *et al.* Preparation of boron-doped diamond nanowires and their application for sensitive electrochemical detection of tryptophan. Electrochemical Communications, 2010, 12, 438–441.

102 G.H. Zhao, Y. Qi, Y. Tian. Simultaneous and direct determination of tryptophan and tyrosine at boron-doped diamond electrode. Electroanalalysis, 2006, 18, 830–834.

103 H. Zanin, P.W. May, D.J. Fermin, *et al.* Porous boron-doped diamond/carbon nanotube electrodes. ACS Applied Material Interfaces, 2014, 6, 990–995.

104 E. Frackowiak. Carbon materials for supercapacitor application. Physical Chemistry Chemical Physics, 2007, 9, 1774–1785.

105 K. Honda, T.N. Rao, D.A. Tryk, *et al.* Electrochemical characterization of the nanoporous honeycomb diamond electrode as an electrical double-layer capacitor. Journal of the Electrochemical Society, 2000, 147, 659–664.

106 R. Kötz, M. Carlen. Principles and applications of electrochemical capacitors. Electrochimica Acta, 2000, 45, 2483–2498.

107 F. Gao, M.T. Wolfer, C.E. Nebel. Highly porous diamond foam as a thin-film micro-supercapacitor material. Carbon, 2014, 80, 833–840.

108 C. Hebert, J.P. Mazellier, E. Scorsone, *et al.* Boosting the electrochemical properties of diamond electrodes using carbon nanotube scaffolds. Carbon, 2014, 71, 27–33.

109 F. Gao, G. Lewes-Malandrakis, M.T. Wolfer, *et al.* Diamond-coated silicon wires for supercapacitor applications in ionic liquids. Diamond Related Materials, 2015, 51, 1–6.

110 P.L. Taberna, P. Simon, J.F. Fauvarque. Electrochemical characteristics and impedance spectroscopy studies of carbon-carbon supercapacitors. Journal of the Electrochemical Society, 2003, 150, A292.

111 Y.G. Guo, J.S. Hu, L.J. Wan. Nanostructured materials for electrochemical energy conversion and storage devices. Advanced Materials, 2008, 20, 2878–2887.

112 A. Corma, H. Garcia. Supported gold nanoparticles as catalysts for organic reactions. Chemical Society Review, 2008, 37, 2096–2126.

113 M. Stratakis, H. Garcia. Catalysis by supported gold nanoparticles: Beyond aerobic oxidative processes. Chemical Review, 2012, 112, 4469–4506.

114 M. Ciureanu, H. Wang. Electrochemical impedance study of electrode-membrane assemblies in PEM fuel cells. I. Electro-oxidation of H_2 and H_2/CO mixtures on Pt-based gas-diffusion electrodes. Journal of the Electrochemical Society, 1999, 146, 4031–4040.

115 J. Perez, E.R. Gonzalez, E.A. Ticianelli. Oxygen electrocatalysis on thin porous coating rotating platinum electrodes. Electrochimica Acta, 1998, 44, 1329–1339.

116 G. Tamizhmani, J.P. Dodelet, D. Guay. Crystallite size effects of carbon-supported platinum on oxygen reduction in liquid acids. Journal of the Electrochemical Society, 1996, 143, 18–23.

117 A. Pozio, M. De Francesco, A. Cemmi, *et al.* Comparison of high surface Pt/C catalysts by cyclic voltammetry. Journal of Power Sources, 2002, 105, 13–19.

118 J. Kim, J. Lee, Y. Tak. Relationship between carbon corrosion and positive electrode potential in a proton-exchange membrane fuel cell during start/stop operation. Journal of Power Sources, 2009, 192, 674–678.

119 F. Gao, N.J. Yang, H. Obloh, C.E. Nebel. Shape-controlled platinum nanocrystals on boron-doped diamond. Electrochemical Communications, 2013, 30, 55–58.

120 F. Gao, N.J. Yang, C.E. Nebel. Highly stable platinum nanoparticles on diamond. Electrochimica Acta, 2013, 112, 493–499.

121 X. Lu, J. Hu, J.S. Foord, Q. Wang. Electrochemical deposition of Pt–Ru on diamond electrodes for the electrooxidation of methanol. Journal of Electroanalytical Chemistry, 2011, 654, 38–43.

122 J. Hu, X. Lu, J.S. Foord. Nanodiamond pretreatment for the modification of diamond electrodes by platinum nanoparticles. Electrochemical Communications, 2010, 12, 676–679.

123 J. Hu, X. Lu, J.S. Foord, Q. Wang. Electrochemical deposition of Pt nanoparticles on diamond substrates. Physica Status Solidi A, 2009, 206, 2057–2062.

124 K. Honda, M. Yoshimura, T.N. Rao, *et al.* Electrochemical properties of Pt-modified nano-honeycomb diamond electrodes. Journal of Electroanalytical Chemistry, 2001, 514, 35–50.

125 T. Brulle, A. Denisenko, H. Sternschulte, U. Stimming. Catalytic activity of platinum nanoparticles on highly boron-doped and 100-oriented epitaxial diamond towards HER and HOR. Physical Chemistry Chemical Physics : PCCP, 2011, 13, 12883–12891.

126 L. Hutton, M.E. Newton, P.R. Unwin, J.V. Macpherson. Amperometric oxygen sensor based on a platinum nanoparticle-modified polycrystalline boron doped diamond disk electrode. Anal. Chem., 2009, 81, 1023–1032.

127 F. Gao, R. Thomann, C.E. Nebel. Aligned Pt-diamond core-shell nanowires for electrochemical catalysis. Electrochemical Communications, 2015, 50, 32–35.

128 C.F. de Lannoy, D. Jassby, D.D. Davis, M.R. Wiesner. A highly electrically conductive polymer–multiwalled carbon nanotube nanocomposite membrane. Journal of Membrane Science, 2012, 415–416, 718–724.

129 R. He, M. Echeverri, D. Ward, Y. Zhu, T. Kyu. Highly conductive solvent-free polymer electrolyte membrane for lithium-ion batteries: Effect of prepolymer molecular weight. Journal of Membrane Science, 2016, 498, 208–217.

130 L. Liu, F. Zhao, J. Liu, F. Yang. Preparation of highly conductive cathodic membrane with graphene (oxide)/PPy and the membrane antifouling property in filtrating yeast suspensions in EMBR. Journal of Membrane Science, 2013, 437, 99–107.

131 K. Honda, M. Yoshimatsu, K. Kuriyama, *et al.* Electrically-switchable, permselective membranes prepared from nano-structured N-doped DLC. Diamond Related Materials, 2011, 20, 1110–1120.

132 Y.Z. Yang, H.D. Li, S.H. Cheng, *et al.* Robust diamond meshes with unique wettability properties. Chemical Communications, 2014, 50, 2900–2903.

133 H. Uetsuka, D. Shin, N. Tokuda, *et al.* Electrochemical grafting of boron-doped single-crystalline chemical vapor deposition diamond with nitrophenyl molecules. Langmuir, 2007, 23, 3466–3472.

134 A. Bongrain, C. Agnès, L. Rousseau, *et al.* High sensitivity of diamond resonant microcantilevers for direct detection in liquids as probed by molecular electrostatic surface interactions. Langmuir, 2011, 27, 12226–12234.

9

Electroanalysis with C₃N₄ and SiC Nanostructures

Mandana Amiri

University of Mohaghegh Ardabili, Ardabil, Iran

9.1   Introduction to g-C₃N₄

Carbon nitrides, a group of polymeric materials that are constructed of carbon and nitrogen atoms, have become important complementary structures next to carbon. Berzelius and Liebig were the first who studied carbon nitride (C_3N_4) in the 1830s, when they made 'melon,' a polymeric derivative of carbon and nitrogen (linear polymers of connected tri-s triazines via secondary nitrogen) [1]. There is still a high debate on the real entity of a C_3N_4 idealized composition graphitic material. Due to a lack of experimental data, many theoretical studies were performed [2–4] until graphitic C_3N_4 (g-C_3N_4) was proven experimentally to be the most stable allotrope under ambient conditions [5]. Graphite is taken as the primary structure and triazine (C_3N_3) (Scheme 9.1a) had been nominated as the initial building block of g-C_3N_4. However, tri-s-triazine (heptazine) rings, which resemble the hypothetical polymer melon are another possible building block, and have recently shown to be energetically favorable over the triazine-based structure (Scheme 9.1b) [6].

Although the microstructures of g-C_3N_4 and graphite are similar, they differ largely in their physicochemical properties. The basic physical difference one can find in their colors, with g-C_3N_4 being yellow while graphite being black. The electronic and optical properties are totally different for both materials (Table 9.1). While graphite is an excellent conductor, g-C_3N_4 is known to be a wide-band semiconductor [7]. g-C_3N_4 is stable under different physicochemical condition. Some of its physical-bio-chemical stability includes: peculiar thermal stability, super hardness, low density, water resistivity, good biocompatibility and fluorescence/persistent luminescence nature and special electronic structure incorporating a medium band gap (2.7 eV) which induces exceptional photocatalytic, photoelectrochemical and optical properties [8, 9].

A number of studies have reported simple UV–Vis absorption and photoluminescence methods for investigating the special electronic properties of g-C_3N_4 [11, 12]. Figure 9.1a shows a typical absorption spectrum of g-C_3N_4, showing a band gap at about

Nanocarbons for Electroanalysis, First Edition.
Edited by Sabine Szunerits, Rabah Boukherroub, Alison Downard and Jun-Jie Zhu.
© 2017 John Wiley & Sons Ltd. Published 2017 by John Wiley & Sons Ltd.

420 nm, therefore making the material slightly yellow. The band gap energy (E_g) of the prepared samples can be estimated using Tauc's equation:

$$\alpha h\nu = B\,(h\nu - E_g)^{n/2}$$

where

α = absorption coefficient
ν = light frequency
B = proportionality constant

The value of n depends on the characteristics of the transition in the semiconductor. The E_g values were estimated by extrapolation of the linear part of the curves obtained by plotting $(\alpha h\nu)^2$ versus $h\nu$ [13]. The photoluminescence (PL) spectrum of g-C$_3$N$_4$ (Figure 9.1b) reveals a broad peak at $\lambda = 455$ nm with an excitation wavelength of $\lambda = 375$ nm.

Due to the presence of hydrogen and lone pair in the outer-electron shell of nitrogen, g-C$_3$N$_4$ has rich surface properties that can be applied to catalysis, such as electron-rich properties, H-bonding motifs, basic surface functionalities, etc. In addition, its stability in acidic, basic and neutral solvents, protic and aprotic solvents (water, alcohols, dimethylformamide, tetrahydrofuran, diethyl ether and toluene) together with high thermal stability (up to 600°C in air) allows the material to function both in liquid or

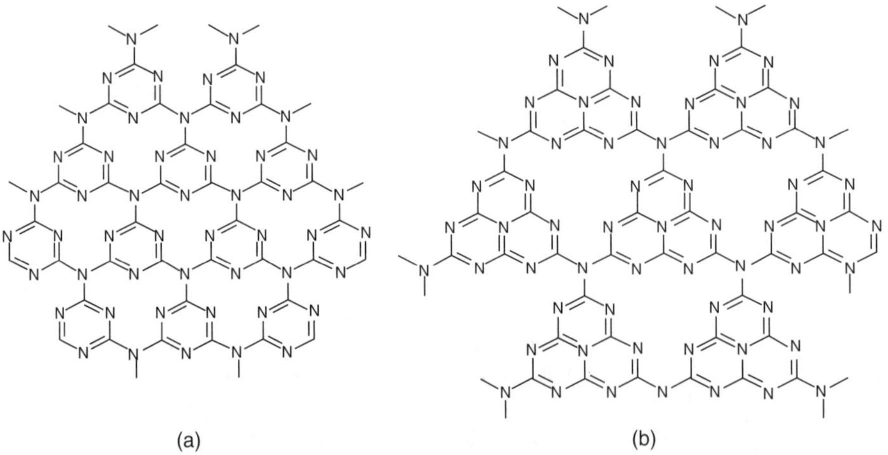

(a)   (b)

Scheme 9.1 The structure of (a) s-triazine: C$_3$N$_3$ (b) tri-s-trazine (heptazin): C$_3$N$_4$.

Table 9.1 Comparison of properties of graphite and g-C$_3$N$_4$.

	Color	Conductivity	Elastic modulus (Gpa)	Florescence	Band gap (eV)	Thermal stability (°C)
g-C$_3$N$_4$	Yellow	No	320 [10]	Yes	~2.7	600 [10]
Graphite	Black	Yes	8–15	No	~zero	700

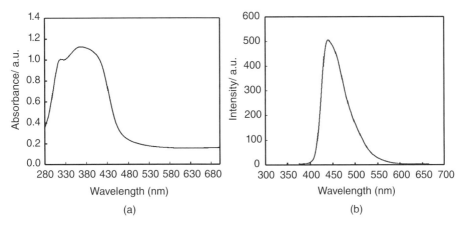

Figure 9.1 (a) UV-visible spectrum. (b) photoluminescence spectrum of g-C₃N₄.

gaseous mixtures and at high temperatures, g-C_3N_4 has thus attracted huge interest for catalysis [9], photocatalysis [7, 14], energy conversion and storage [15], lithium-ion battery [16], electrogenerated chemiluminescence [17] electrochemical sensor [18], hydrogenation reactions [19], NO decomposition [20] and fluorescent sensor [21].

9.2 Synthesis of g-C₃N₄

Generally, structures such as triazine and heptazine derivatives (e.g. cyanamide, dicyandiamide guanidine hydrochloride, urea, thiourea, and melamine,) which are oxygen-free compounds and rich in reactive nitrogen and contain pre-bonded C–N core are the most common precursors applied for the chemical synthesis of g-C_3N_4 [9].

In case of melamine as precursor material [22], the reaction is initiated by a poly-addition reaction followed by a poly-condensation step where the precursors are primarily condensing towards melamine. In the second step ammonia is eliminated through condensation. Melamine derivatives are essentially formed up to 350°C. At around 520°C, condensation occurs, networks of the final polymeric C_3N_4 (See Scheme 9.2), with the rest of material becoming unstable slightly above 600°C are formed. Overheating up to 700°C results in clean- up: the residue-free disappearance of the material via generation of nitrogen and cyano fragments.

Electrochemistry has also shown to be of interest for the fast synthesis of ultrathin g-C_3N_4. In this method, melamine is electrolyzed in NaOH for 40 min (Figure 9.2). This method has many advantages such as operation at room temperature, no need to use of toxic and aggressive reagents, as well as being fast. Although the thickness of ultrathin g-C_3N_4 is about 2 nm, it has excellent dispersion stability in water. As the applied voltages exceed the electrochemical potential window of water, anodic oxidation of water happens and hydroxyl and oxygen radicals are produced. These radicals oxidize melamine molecule to melamine radical anions in basic media. Radical–radical attachment leads to the formation of dimmers, like is the case of aniline polymerization. The dimers are further oxidized and then matured to oligomers via coupling of melamine radicals

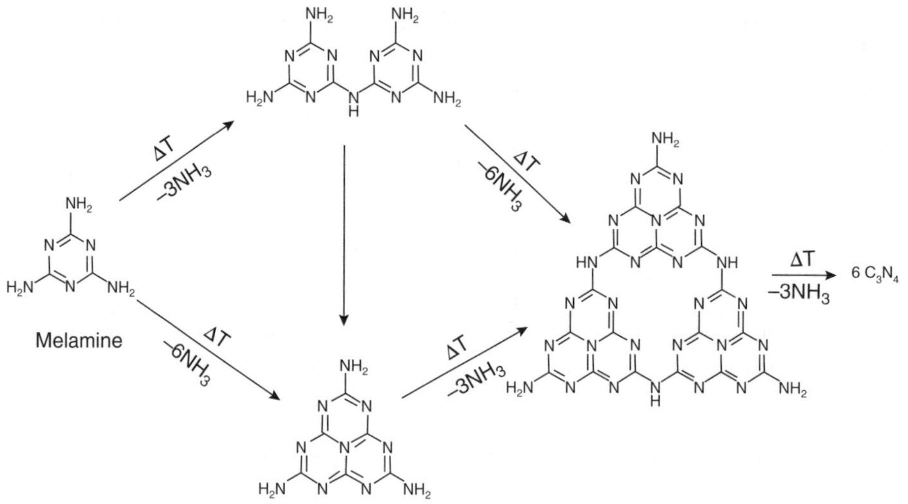

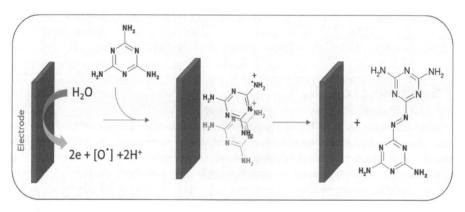

Scheme 9.2 Reaction path for the formation of graphitic C_3N_4 starting from melamine [14].

Figure 9.2 Representation for electrochemical preparation of g-C_3N_4 from melamine. (*See color plate section for the color representation of this figure.*)

or other dimer radicals. The g-C_3N_4 is finally produced by further oxidation of the oligomers and coupling of melamine radicals or other oligomer radicals [23].

The structural properties and morphology of g-C_3N_4 can be studied via X-ray diffraction (XRD), scanning electron microscope (SEM), Fourier transform infrared spectroscopy (FTIR), transmission electron microscopy (TEM), and diffuse reflectance spectroscopy (DRS). According to XRD patterns (Figure 9.3a), the pristine g-C_3N_4 has two strong diffraction peaks at about 13.1° and 27.3° characteristic to the inter-layer structural packing and inter planar stacking, respectively. FTIR spectrum of pristine g-C_3N_4 in Figure 9.3b depicts, the bands in the range of 1230–1650 cm^{-1} which are related to the stretching vibrations of C–N and C=N in heterocycles. Moreover, the band at 806 cm^{-1} is characteristic to the heptazine breathing mode. The broad absorption band at 3000–3300 cm^{-1} corresponds to the terminal NH_2 or NH groups at the defect position of g-C_3N_4 aromatic rings. The SEM image of the fresh sample depicts

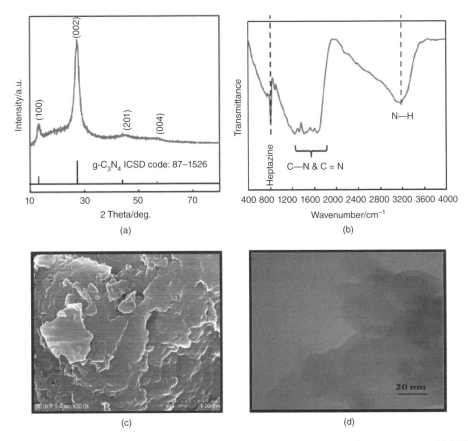

Figure 9.3 (a) XRD pattern of the g-C₃N₄ along with the standard pattern. (b) FTIR spectrum (c) SEM (d) TEM images of g-C₃N₄. (*See color plate section for the color representation of this figure.*)

relatively regular planner morphologies (Figure 9.3c). Furthermore, TEM technique also confirmed the sheet-like morphology and the image is shown in Figure 9.3d [14].

9.3 Electrocatalytic Behavior of g-C₃N₄

The excellent catalytic activity of g-C₃N₄ makes them promising noble metal-free catalyst [9, 23]. However, the low electrical conductivity of g-C₃N₄ makes it of limited use for electrocatalytic processes [24]. Several strategies, including P-doped and protonation, have been therefore proposed to enhance its electrical conductivity [25]. To date, carbon incorporated g-C₃N₄ composite such as g-C₃N₄-carbon black mixtures [26] and graphene composites have been regarded as the most promising materials as they show higher electrical conductivity [27]. The rather low specific surface area of bulk g-C₃N₄ is another issue to be overcome for electrocatalytic applications. The use of porous silica templates [28] for g-C₃N₄ or the use of polar solvents such as water and ethanol can increase the specific surface area of g-C₃N₄ [29]. A hollow mesoporous carbon nitride

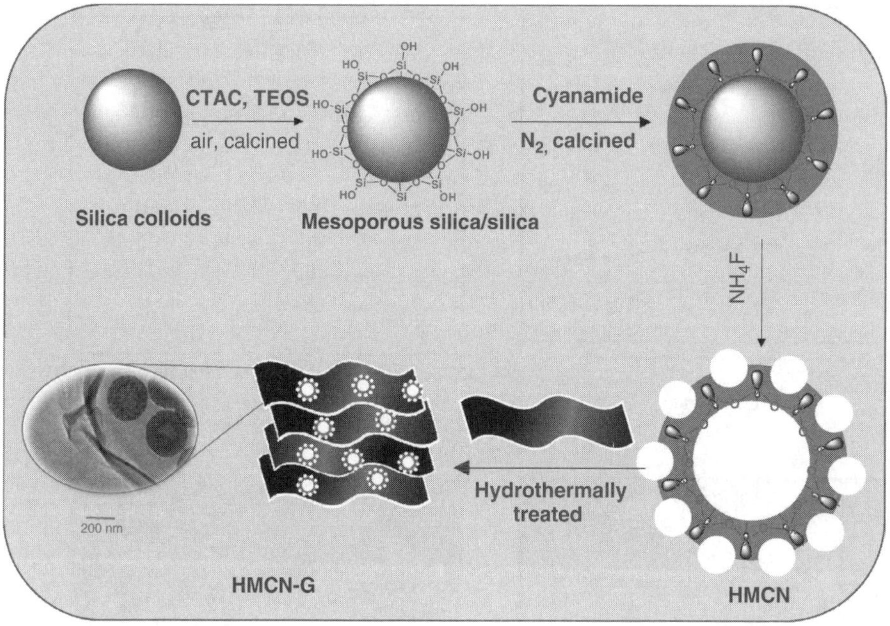

Figure 9.4 Synthetic route to HMCN-G composite.

nanosphere (HMCN) prepared via etching from hollow mesoporous silica as a template has been reported by Quin *et al.* (Figure 9.4). The final product has uniform spherical particles with a diameter of ~300 nm and a high specific surface area up to $439 \, m^2 g^{-1}$. Hollow mesoporous carbon nitride nanosphere/three-dimensional graphene composite (HMCN-G) is subsequently fabricated through a hydrothermal treatment of HMCN with graphene oxide. As an electrocatalyst for oxygen reduction reaction (ORR), the HMCN-G shows significantly enhanced electrocatalytic activity compared to bulk graphitic carbon nitride (g-C_3N_4) and HMCN in terms of the electron-transfer number, current density and onset potential. The electrocatalytic performance of the HMCN-G composite has been greatly improved in several aspects including; increased density of catalytically active sites, enhanced accessibility to electrolyte by the hollow and mesoporous architecture of HMCN, and high conductivity. Furthermore, HMCN-G exhibits superior methanol tolerance compared to Pt/C catalyst. Therefore, it is a promising candidate as a metal-free electrocatalyst for polymer electrolyte membrane fuel cell [25].

Dispersed solutions of ultrathin g-C_3N_4 nanosheets and graphene oxide can be mixed easily under ultrasonication [30].The final three-dimensional porous architecture shows outstanding features including; high surface area, multilevel porous structure, good electrical conductivity, efficient electron transport network, and fast charge transfer kinetics. It facilitates the diffusion of O_2, electrolyte, and electrons in the porous frameworks during ORR. In the other hand, efficient electron tunneling through g-C_3N_4 barrier is in charge of rich electrode–electrolyte–gas three-phase boundaries in ultrathin g-C_3N_4 nanosheets. Accordingly, electron diffusion distance from reduced graphene oxide to O_2 is reduced [31].

g-C_3N_4 has also been introduced as an advanced supporting material for Pt nanoparticles (NPs) due to its excellent stability and abundant Lewis acid sites for anchoring metal NPs. However, its electrochemical applications are still impeded by its non-conductive nature and low surface areas. Herein, a π-π stacking method is presented to prepare graphene/ultrathin g-C$_3$N$_4$ nanosheets composite as PtRu catalyst support. The 2D layered structure can compensate the g-C$_3$N$_4$ draw backs to a great extent. The significantly enhanced performance of the novel PtRu catalyst are attributed to: the homogeneous dispersion of PtRu NPs on g-C$_3$N$_4$ nanosheets due to its abundant Lewis acid sites for anchoring PtRu NPs; the magnificent mechanical stability in acidic and oxidative environments; the increased electron conductivity of support by forming a layered structure and the strong interaction between metal NPs and g-C$_3$N$_4$ NS [32].

9.4 Electroanalysis with g-C$_3$N$_4$ Nanostructures

The material g-C$_3$N$_4$ has also been used as an electrode material for electroanalytical purposes (Table 9.2). Depending on the electrochemical detetion method used they can be divided into:

1) Electrochemiluminescent sensors
2) Photo-electrochemical sensors
3) Voltammetric sensors.

9.4.1 Electrochemiluminescent Sensors

Electrochemiluminescence (ECL) is chemiluminescence triggered upon an electrical bias. Owing to its versatility, simple instrumentation, good stability against photobleaching, and low background signals, ECL detection has received considerable attention. ECL sensors are designed by a series of the common ECL emitters including Ru complexes [33], luminol [34], metal oxide semiconductors [35] and quantum dots [36]. Recently, g-C$_3$N$_4$ with a stacked 2D structure has shown to be added as efficient ECL emitter. Compared to the traditional luminophores, g-C$_3$N$_4$ presents many advantages, such as easy preparation, metal-free, nontoxic, low cost, as well as excellent biocompatibility.

Anodic ECL of g-C$_3$N$_4$ nanosheets was first observed by Liu *et. al.* The ECL signal was 40 times stronger than the bulk g-C$_3$N$_4$ ECL in the presence of triethylamine (Et$_3$N) used as co-reactant due to the large surface-to-volume ratio. At pH 7.0, the g-C$_3$N$_4$ nanosheets modified electrode prepared with $0.75\,mg\,mL^{-1}$ g-C$_3$N$_4$ nanosheets in 0.025% chitosan solution presents good stable and reproducible results in the presence of 30 mM Et$_3$N. The ECL mechanism of g-C$_3$N$_4$ /Et$_3$N system was attributed to the excited g-C$_3$N$_4$, formed via electro-oxidized of C$_3$N$_4$ in the presence of Et$_3$N. A sensitive and specific dopamine sensor was designed with a detection limit of 96 pM [18].

g-C$_3$N$_4$ nanosheets, as a cathodic ECL emitters in the presence of dissolved oxygen, were reported to produce H$_2$O$_2$ on the electrode surface [37] The aforesaid report applied the emission ability for an ECL sensitive DNA biosensor through the g-C$_3$N$_4$ affinity to single-stranded DNA (ssDNA). The presence of hemin-labeled ssDNA on g-C$_3$N$_4$ nanosheets leads to dissolved oxygen consumption, as the co-reactant producer via hemin-mediated electrocatalytic reduction. Accordingly the ECL emission of g-C$_3$N$_4$ nanosheets is quenched. Once the double stranded target DNA interacts wih

Table 9.2 g-C$_3$N$_4$ nanostructure used in electroanalysis.

Nanostructure	Synthesis method	Detection method	Precursor material	LOD (Mol L^{-1})	DLR (Mol L^{-1})	Analyte	Reference
Graphitic phase carbon nitride (g- C$_3$N$_4$) doped graphene oxide	pyrolysis	DVP	Dicyandiamide	AA: 1.4×10^{-4}–5.0×10^{-4} DA:1.0×10^{-6}–2.0×10^{-5} UA:1.0×10^{-5}–2.0×10^{-4}	AA:1.172×10^{-6} DA:0.096×10^{-6} UA:0.228×10^{-6}	Ascorbic acid Dopamine Uric acid	[29]
g-C$_3$N$_4$ nanosheet/chitosan	hydrothermal	DPV	Melamine	1.0×10^{-6} to 8.0×10^{-5} and 1.0×10^{-7} to 5.0×10^{-6}	1.0×10^{-8}	Hg^{2+}	[14]
Mesoporous carbon nitride material (MCN)	Template synthesis	Chronoamperometry	Ethylene diamine and carbon tetrachloride	Ph:5.00×10^{-8}–9.50×10^{-6} CC:5.00×10^{-8}–1.25×10^{-5}	Ph: 1.50×10^{-8} C:1.024×10^{-8}	Phenol-Catechol	[52]
Ordered mesoporous carbon nitride (OMCN)	Template synthesis	CV	melamine	H$_2$O$_2$:4.0×10^{-5}–1.2×10^{-2} NB:0.5×10^{-6}–1.0×10^{-3} NADH:2.0×10^{-6}– 2.2×10^{-3}	H$_2$O$_2$:1.52×10^{-6} NB:0.18×10^{-6} NADH:0.82×10^{-6}	H$_2$O$_2$ Nitrobenzene NADH	[51]
Pd Wormlike Nanochains/Graphitic Carbon Nitride Nanocomposites	hydrothermal	CV	Melamine	1.0×10^{-9}–1.5×10^{-5} 3.8×10^{-9}–2.1×10^{-5}	—	Ethyl paraoxon. Huperzine-A	[19]
Graphitic carbon nitride polymers (g-CN)	hydrothermal	DPV	Melamine	5.0×10^{-6}–0.6×10^{-3}	1.0×10^{-9}	Dopamine	[53]
Three-dimensional (3D) g-C$_3$N$_4$ nanosheets	pyrolysis	CV EIS	Dicyandiamide	CC:1.0×10^{-6}–2.0×10^{-4} HQ:1.0×10^{-6}–2.5×10^{-4}	9.0×10^{-8}	Catechol hydroquinone	[50]

g-C$_3$N$_4$	One-step thermal-induced self-condensation of melamine	ECL	Melamine	0.2×10^{-6}–4.5×10^{-5}	1.4×10^{-7}	Rutin	[54]
Graphitic-phase carbon nitride	Polymerization of melamin	ECL	Melamine	1.0×10^{-14}–1.0×10^{-8}	2.0×10^{-15}	DNA	[34]
g-C$_3$N$_4$	Polymerization of melamine	ECL	Melamine	1.0×10^{-9}–1.0×10^{-7}	9.6×10^{-11}	Dopamine	[18]
g-C$_3$N$_4$		ECL	Melamine	1.87×10^{-10}–7.48×10^{-9}	3.74×10^{-8}	Nuclear MatrixProtein 22	[39]
Two-dimensional ultrathin g-C$_3$N$_4$ nanosheets	Directly pyrolysis of melamine	ECL	Melamine	1.2×10^{-7}–9.6×10^{-7}	2.4×10^{-11}	DNA	[37]
g-C$_3$N$_4$ nanohybrids	Polymerization of carbamide molecules	ECL	Carbamide	1.45×10^{-11}–7.2×10^{-8}	7.2×10^{-12}	Alpha fetoprotein	[55]
g-C$_3$N$_4$ nanosheet		ECL	Melamine	1.0×10^{-15}–1.0×10^{-9}	0.5×10^{-15}	MicroRNA	[56]
g-C$_3$N$_4$ nanosheet	pyrolysis	ECL	Dicyandiamide	—	2.5×10^{-7}, 1×10^{-7}, 2×10^{-8}	Cu^{2+}, Ni^{2+}, Cd^{2+}	[57]

the platform, the hemin-labeled ssDNA detaches and the ECL emission signal recovers. The sensitive sensing strategy provides a new pattern for the ultrasensitive detection method designs. The Figure 9.5 demonstrates the overall interactions of the biosensor.

Chen et al. used molecular imprinted polypyrrole (MIP) modified 2D ultrathin g-C_3N_4 nanosheets as a cathodic ECL emitter with $S_2O_8^{2-}$ as co-reactant for the detection of perfluorooctanoic acid. The prepared MIP functionalized g-C_3N_4 nanosheets exhibited a stable and significantly amplified ECL signal. Perfluorooctanoic acid is oxidized by the electro-generated strong oxidants of SO_4^{2-} more efficiently than well-established liquid chromatography-tandem mass spectrometry (LC-MS/MS) [38].

In another research, a label-free ECL immunosensor based on g-C_3N_4 and gold nanoparticles (Au NPs) was developed by *Han et al.* for the detection of Nuclear Matrix Protein 22 (NMP 22). Firstly, g-C_3N_4 combined with Au NPs, which promoted electron transfer and enhanced the ECL intensity of g-C_3N_4 to a great extent. Secondly, anti-NMP 22 was immobilized on the electrode. The connection between Au NPs and antibody affected the Au NPs conductivity, leading to a ECL intensity decrease. Then, bovine serum albumin (BSA) solution was dropped on the surface to block nonspecific binding sites. Finally, after the specific immune-reaction between NMP 22 and anti-NMP 22, the ECL intensity suppressed greatly, due to the protein blocking effect on the electrode surface toward luminescent reagents and electrons diffusion to the electrode surface. The proposed ECL immune-sensor provides a rapid, simple, and sensitive immunoassay strategy for protein detection, with clinical application potentials [39].

9.4.2 Photo-electrochemical Detection Schemes

Photo-electrochemical (PEC) detection based on photon-to-electricity conversion has attracted considerable interest due to its high sensitivity as well as simple and cheap instrumentation. The selection of an appropriate photo-electric semiconductor material and a highly sensitive and selective recognition system set up play important roles in PEC sensor performance. A series of semiconductor materials to form the photo-active layers are as follows; TiO_2 [40], Cd series quantum dots [41], ZnO [42], gold nanoclusters [43], WO3 [44], and so on. The photoelectrochemical activity of bare g-C_3N_4 is limited due to the high recombination rate of photo-generated electron hole pairs. Coupling g-C_3N_4 with the other materials is an effective way to overcome the above problem by improving the photo-electron chemical property of the g-C_3N_4.

A novel photo-electrochemical strategy for the detection of Cu^{2+} with AgX/g-C_3N_4 (X = Br, I) hybrid materials has been designed [45].When the hybrid material system was irradiated with visible light, the g-C_3N_4 and AgX nanoparticles in the composite materials both absorb photons, as well as excite electron and hole pairs (Figure 9.6a). Based on the band gap positions, the conduction band (CB) and valence band (VB) edge potentials of g-C_3N_4 were −1.12 eV and +1.57 eV, respectively The CB and VB edge potentials of AgBr were at 0 eV and +2.6 eV, respectively (conduction band of AgI: −0.15 eV, valence band: 2.95 eV). The CB and VB potentials of g-C_3N_4 were more negative than those of AgBr and AgI. Therefore, the photo-generated electrons in the CB of g-C_3N_4 layer structures could be easily transferred to the surface of the AgBr or AgI nanoparticles, and the holes generated in the VB of AgBr or AgI nanoparticles could migrate to the surface of g-C_3N_4, which promoted the effective separation of photoexcited electron–hole pairs and decreased the probability of electron and hole recombination. Then, the electron

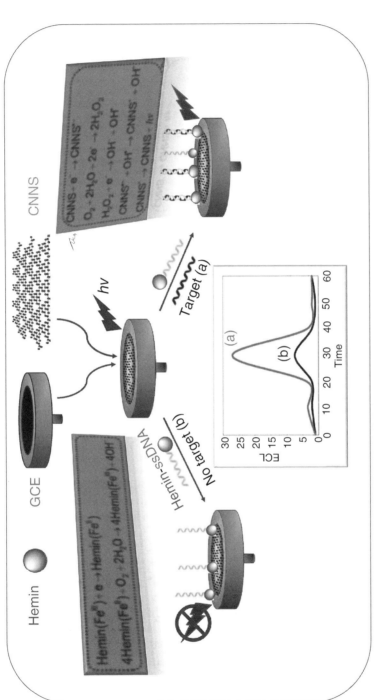

Figure 9.5 Schematic illustration of Carbon nitride nanosheets based ECL sensing platform for DNA detection.

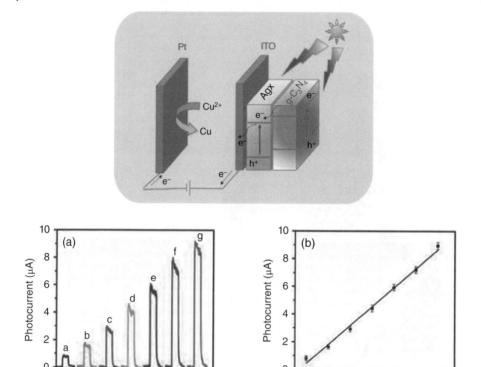

Figure 9.6 Schematic illustration of exciton trapping mechanism and photoelectrochemistry for sensing of Cu^{2+}. (a) Effect of different concentrations of Cu^{2+} on the photocurrent intensity of the ITO/ $(AgBr/g-C_3N_4)$ electrode (b) The linear calibration curves of the ITO/ $(AgBr/g-C_3N_4)$ electrode. *Source*: Xu 2013 [45]. Reproduced with permission of Elsevier.

transferred to the ITO electrode and generated photocurrent because the energy level of the conduction band of ITO was low. The effect of different copper concentrations on the photo-current intensity of the ITO/$(AgBr/g-C_3N_4)$ electrode and its calibration curve exhibited in Figure 9.6b and 6c. Li *et al.* showed the interest of polythiophene modified $g-C_3N_4$ nanosheets PT/$g-C_3N_4$ [46].

A highly efficient PEC biosensor based on $g-C_3N_4$ nanosheets modified with CdTe quantum dots (QDs), was designed by Hao *et al.* Using dissolved oxygen as an electron acceptor, the hybrid photocathode showed a sensitive photocurrent response at −0.2 V bias potential under 405 nm illumination. The modified photocathode results in 100% photocurrent increase compared to CdTe QDs modified electrode, owing to the heterojunction formation through the two semiconductors contact. The sensitive PEC sensor for Cu^{2+} shows a good linear range from 20 nm to 100 μM and a detection limit of 3.3 nM, and was successfully applied in the detection of Cu^{2+} in human hair samples [47].

Fang *et al.* proposed an effective PEC sensor for the determination of trace amounts of chromium in water samples under visible-light irradiation. A unique nanostructured graphitic carbon nitride with incorporated formate anions (F-g-C_3N_4) was integrated with a Cr(VI) ion-imprinted polymer (IIP) as a photoactive electrode (denoted as

IIP@F-g-C₃N₄). The F-g-C₃N₄ materials exhibits an enhanced charge separation with substantially improved PEC responses versus g-C₃N₄. The low-cost and sensitive sensor has been successfully applied for speciation determination of chromium in environmental water samples [48].

Kang *et al.* developed lately a sensitive detection for ascorbic acid using g-C₃N₄/TiO₂ nanotubes hybrid film via a double-channel photo-electrochemical detection set up. The response is considered as the measuring and reference PEC cells signal difference. This approach by its differentiating nature, takes the advantage of removing the excitation source drift, leading to better signal-to-noise ratio and hence higher measurement sensitivity, compared to the established single-channel PEC mode. Additionally, the differential measurement scheme reduces the matrix interference. The as-prepared PEC sensor shows signal-enhancement response to ascorbic acid ranging from 1 nm to 10 μM with a detection limit of 0.3 nM. Tracking the activity of alkaline phosphatase in serum samples resulted in a dynamic range of 0.3 mU L^{-1}–1 U L^{-1} with a detection limit of 0.1 mU/L [49].

9.4.3 Voltammetric Determinations

Voltammetric techniques such as differential pulse voltammetry (DPV), square wave voltammetry (SWV) and stripping methods have been used for trace analysis of many important biological, pharmaceutical and environmental compounds. A g-C₃N₄ nanosheets-graphene oxide (CNNS-GO) composite was synthesized by Zhang *et al.* and the electrochemical performance of the composite investigated. Due to the synergistic effects of layer-by-layer structures, arising from π-π stacking or charge-transfer interactions, g-C₃N₄ nanosheets-graphene oxide composite has a great performance in terms of conductivity, electrocatalytic and selective oxidation. A possible mechanism is shown in Figure 9.7. GO acts as an electron trap and transporters, containing unoxidized aromatic

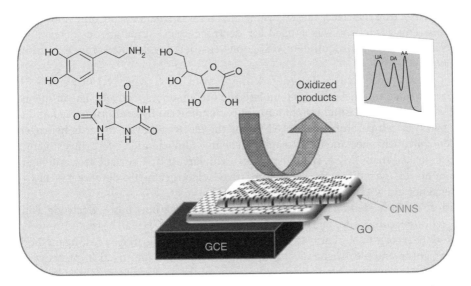

Figure 9.7 Schematic drawing of electrochemical oxidize AA, DA and UA on CNNS -GO / glassy carbon electrode (GCE).

rings with rich delocalized π electrons. GO with oxygen-containing groups interacts with ascorbic acid (AA), uric acid (UA) and dopamine (DA) via hydrogen bonding. This interaction catalyzes their oxidation. During the over-oxidization process, the oxygen-containing groups such as C-O, C = O and OH-C = O are generated on the surface of CNNS-GO, providing a selective interface via hydrogen bonding with the proton-donating group of AA, DA and UA. Finally, g-C_3N_4 nanosheets with a graphite-like structure have strong covalent bonds between carbon and nitride atoms. Therefore, nitrogen atoms incorporated in the carbon architecture enhance electrical properties. GO acting as electronic conductive channels in CNNS -GO can efficiently transmits electrons. Due to the synergistic effects of GO and CNNS, the modified electrode displayed charming selectivity and catalytic activity toward AA, DA and UA determination [28].

A new electrochemical biosensor for organophosphorus pesticides (OPs) and huperzine-A (hupA) detection based on Pd wormlike nanochains/graphitic carbon nitride (Pd WLNCs/g-C_3N_4) nanocomposites and acetylcholinesterase (AChE) was developed by Wang *et al.* Enzymes were effectively immobilized on the Pd WLNCs/g-C_3N_4 nanocomposites. Under the optimum condition, the dynamic range for the determination of OPs and hupA were 1.00 nm to 14.96 mM and 3.89 m to 20.80 mM, respectively. Not only, the biosensor owned good reproducibility and stability, but also it has potentials for practical analysis, e.g. a promising method for pesticide analysis [19].

In another report, a three-dimensional g-C_3N_4 nanosheets-carbon nanotube (CNNS-CNT) composite was synthesized via hydrothermal reaction of 2D CNNS and 1D CNT-COOH by π-π stacking and electrostatic interactions. The CNNS-CNT composite offers excellent conductivity compared with the individual components e.g. CNNS and CNT-COOH. The R_{ET} value of GCE was about 950 Ω. When modified with CNNS, R_{ET} value increased to 104 Ω, due to the poor electrical conductivity of the semiconductor CNNS. After electrode modification with CNT and CNT-COOH, the R_{ET} values decreased to 160 and 135 Ω, respectively. Obviously, when the CNNS-CNT was modified on the electrode surface, the R_{ET} value (125 Ω) was smaller than other modified electrodes. This composite was applied for electrochemical simultaneous determination of catechol and hydroquinone with good sensitivity, wide linear range and low detection limit [50].

Zhang *et al.* introduced a 2D ordered mesoporous C_3N_4 (OMCN) using SBA-15 mesoporous silica and melamine as template and precursor respectively. The influence of BET surface area and different amounts of N-bonding configurations formed at different pyrolysis temperatures of OMCN-x for the electrocatalysis towards hydrogen peroxide, nitrobenzene, and nicotinamide adenine dinucleotide were investigated. Results indicated that OMCN treated at 800° C with largest BET surface area and highest amounts of pyrindinic N showed improved electrocatalytic activity for H_2O_2, nitrobenzene, and NADH in neutral solution [51].

Amiri *et al.* have demonstrated the modification of carbon paste electrode with g-C_3N_4/chitosan composite. The positively charged chitosan (CH) can be settled between exfoliated sheets of g-C_3N_4. The g-C_3N_4/CH electrodes were prepared by casting method. The mechanical and electrochemical properties of electrode improved in presence of CH because it behaves as a binder and separating nanosheets. Experimental results show the superb adsorptive properties of g-C_3N_4/chitosan composite through Hg(II). To prove the interaction between g-C$_3$N$_4$ and Hg^{2+}, DRS and ATR spectroscopy have been employed. Comparison between DRS spectra of g-C$_3$N$_4$ and g-C$_3$N$_4$ with adsorbed Hg^{2+} exhibits a new broad peak appeared near 500 nm in the presence of Hg^{2+} which can be related to charge transfer from C$_3$N$_4$ to Hg^{2+}. ATR spectra for g-C$_3$N$_4$ and g-C$_3$N$_4$ with adsorbed Hg^{2+} have been shown two new peaks appear in 633 cm^{-1} and 478 cm^{-1} which is assigned to Hg-N bonds which can confirm the interaction between Hg^{2+} and g-C$_3$N$_4$. Differential pulse voltammetry was applied for quantitative determinations [14].

9.5 Introduction to SiC

Silicon carbide (SiC) is one of the most important wide bandgap semiconductors and has shown unique properties such as thermal shock resistance, low thermal expansion, good thermal conductivity (3.6– 4.9 W cm^{-1} K^{-1}), a high breakdown electric field of typically larger than 2 MV cm^{-1}, and chemical inertness [58, 59]. SiC is a covalently bonded IV-IV compound, where each C (or Si) atom is surrounded by four Si (or C) atoms (Figure 9.8) in tetrahedral sp^3-hybridized bonds, causing the unique thermal and chemical stability of the material. The SiC crystallographic structures exhibit a close-packed stacking of double-layers of Si and C atoms. SiC can be formed in monocrystalline, polycrystalline and amorphous solid forms. Different stacking sequences of C-Si double-layers can produce different crystallographic forms, called polytypes leads to a tunable band gaps of 2.4–3.2 eV depending on its crystal structure [60].

Compared with bulk SiC, its nano-materials, with their large ratio of surface-to-volume and possible quantum effects show exceptional mechanical, electrical, luminescent, thermal, electrochemical and wetting properties to develop novel nano-devices [61]. The first preparation of SiC nano-materials has been reported by Zhou *et al.* in 1994 by reaction carbon nanoclusters and SiO to fabricate SiC nanorods [62].

Recently, a significant interest on producing SiC nanostructures with various morphologies has been developed due to their size-dependent mechanical, optical and electrical properties [61]. There are various kinds of SiC nanostructure such as nanowires [63] (Figure 9.9a), hollow nanospheres [64] (Figure 9.9b), nanotubes [65] (Figure 9.9c), nanocrystals [66] (Figure 9.9d), nanoflakes [67], nanowalls [68] and nanorods [69].

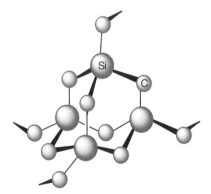

Figure 9.8 SiC crystals are formed via bi-layers of C and Si.

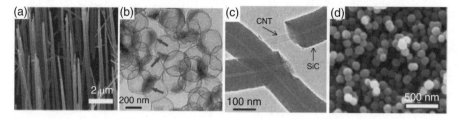

Figure 9.9 (a) SiC nanowire, (b) SiC hallow nanosphere, (c) SiC nanotubes, (d) SiC nanocrystal

Besides the above SiC nanostructures, core-shell nanospheres that are composed of a SiC core (inner material) and a shell (out layer material) have also been developed. The fabrication of core-shell nanostructures is leading to advantages from the intrinsic properties of both components. Surface functionalization is an effective strategy to exhibit multifunctionality of nano-materials and improve their performance. For instance, the bio-functionalized SiC nanowires with DNA surface coverage has been introduced for DNA sensor [70]. Figure 9.10 demonstrate the procedure to bio-functionalizing SiC surface to produce and DNA sensor.

However, SiC nanomaterials have low electrical conductivity, surface-functionalizing SiC nanostructures with high-conductive materials like CNTs can be solving this problem and causes superhydrophobicity [71]. It has been found that SiC nanowires show the intrinsic n-type semiconductor behavior due to high-density donor states under the construction band edge if they are not intentionally doped [72]. The electrical properties of SiC nanowires can be changed substantially via surface functionization. SiC crystal defects revealed better electronic properties because of its unique crystal structure, the enhancement of average oxygen concentration and the accumulation of oxygen [73].

SiC nanostructures were also used to fabricate superhydrophobic surface because of their inherent surface roughness and excellent stability under harsh conditions. Chen *et al.* reported the SiC nanowires with tunable hydrophobicity/hydrophilicity. They demonstrated SiC nanowires switching from hydrophobic to hydrophilic due to

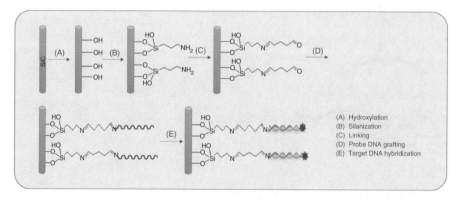

Figure 9.10 Steps for functionalizing SiC nanowires with DNA.

the surface-tethered hydrophilic layer as well as increasing interspace between nanowires [74].

The distinct mechanical properties of SiC nanostructures are important for their applications in electronic and optical devices. The Young modulus of SiC is higher than that of Si, and its high breakdown field ~2 MV cm^{-1} [75], is two times than that of Si. The bulk SiC materials show, however, low luminescent efficiency due to their indirect band gap, which limites their applications in optoelectronic devices. However, the luminescent intensity can be significantly increased with decreasing of crystallite sizes of SiC to several nanometers. This is due to the improved radiative recombination rates and suppressed non-radiative recombination rates in confined clusters. With reducing the sizes of SiC crystals, their bandgap increase and the emission is blue shifted [76].

Scientists have made numerous efforts to investigate the unique properties of SiC nanostructures for field emitters [77], field effect transistors [78], catalyst [79], gas sensing [80], hydrogen production and storage [81], supercapacitors [82], and bio-imaging probes [83]. Owing to their exceptional properties, SiC nano-materials have demonstrated as good candidates as reinforcements in different composites such as polymer-matrix composites [84], ceramic-matrix composites [82] and carbon composites [80] to improve the mechanical, thermal and barrier properties of the pure matrix. However, SiC materials have been proposed to be of interest due to their hardness, chemical inertness, electrical conductivity, pore volume, surface area and roughness and high thermal stability in an oxidative environment.

9.6 Synthesis of SiC Nanostructures

Nanosized SiC powders are commercially available and can be performed by various methods with control over size, composition and crystallinity. Umar *et al.* synthesized thin films of TiCx/SiC/α-C:H on Si substrates using a complex mix of high energy density plasmas [85]. A DC plasma reactor for SiC nanoparticle production has been developed by Yu *et al.* SiC nanopowders were synthesized using this system and the synthesized primary particles have nearly spherical structures, mostly β-SiC phase with a particle size of 10–30 nm [86]. Ni/SiC nanocomposite coatings were obtained by electrochemical co-deposition of SiC nanoparticles with nickel, from an additive-free Watts type bath [87]. Bastwros *et al.* has reported ultrasonic spray deposition of SiC nanoparticles for laminate metal composite fabrication [88]. Continuous synthesis of SiC nanoparticles by RF thermal plasma method has been demonstrated by Karoly *et al.* The resulted SiC nanoparticles were crystallized mainly in β phase with trace amount of α [89]. SiC nanostructures can also be produced by other techniques including the pulsed laser [90], the mechanical ball milling [91] microwave [92], chemical vapor deposition (CVD) [93] and so on. Parkash summarized synthesis methods for SiC nanostructure in a useful review [74].

Nanoscale SiC networks were prepared by using carbothermal reaction between silica xerogels and carbon nanoparticles in an argon atmosphere on the silicon substrate. Nickel has been used as catalyst for growth of SiC nanostructures. Figure 9.11a shows a typical synthesis procedure. SEM image of as-prepared product are shown uniform nanostructures with network-like morphology was densely stacked all over the

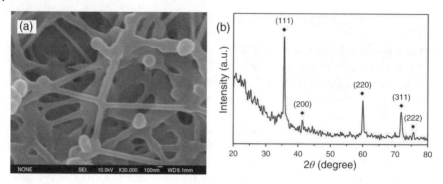

Figure 9.11 (a) SEM image of as-prepared SiC nanowire networks. (b) XRD pattern of SiC crystals. *Source*: Qi 2014 [94]. Reproduced with permission of Elsevier.

substrate. It is demonstrated typical junctions connected through two branches or more during growth. Figure 9.11b is the XRD pattern of the as-prepared product. The diffraction peaks correspond to the (111), (200), (220), (311) and (222) planes of 3C-SiC (JCPDS, no. 29-1129) [94].

9.7 Electrochemical Behavior of SiC

SiC was investigated more than 70 years ago as an electrode material by Hume and Kolthoff [95]. They reported that a SiC electrode can be used as an oxidation-reduction indicator electrode in potentiometric titrations of potassium iodide with permanganate and with ceric sulfate of ferrous iron with permanganate, of titanous chloride with ferric chloride, and of hydrochloric acid with sodium hydroxide. The electrodes were made with single crystal SiC and their behavior was studied by measuring their potential against a calomel electrode. The authors concluded that the SiC electrode behaved similarly to an oxidation-reduction indicator material such as Au and Pt.

SiC as a promising electrode material, has attracted thus considerable industrial interest in electronics and sensors due to its wide band gap and high electron mobility [96]. Because of its high specific surface area, excellent electrochemical performance, and compatibility with various electrolytes, SiC electrodes with different morphology and structure exhibit high surface efficiency, rate capability and cycle lifetime [61]. Moreover, the SiC surface functionalization process predicts the possibility to be a state-of-the-art electrode for biosensing applications owing to its high biocompatibility, high chemical stability, non-toxicity and the robust semiconducting properties [61]. The electronic properties of SiC have made it a suitable material for sensing devices, especially those that are based on a surface impedance change or are electrochemistry related. Nanocrystalline cubic β-SiC exhibits much higher electron mobility than single crystalline SiC because its average oxygen concentration is enhanced and the oxygen is accumulated at crystal defects, resulting in a much better electronic conductivity [74]. For electrochemical applications, the dopants, doping level, surface termination and choice of SiC polytype play an important role, since the background current and voltammetric reactivity of SiC electrodes would be affected. When SiC is appropriately doped, the conductivity of this material dramatically increases and exhibits electrical characteristics similar to carbon materials [97]. But in contrast to carbon, the close-packed hexagonal or cubic-SiC structure should afford a well-defined surface for electron transfer [98]. Meier *et al.* [94] used a CVD method using tetramethylsilicon and a resistively heated carbon fiber to fabricate a concentric SiC conductor for voltammetric measurements. They found that the SiC electrode in 0.1 M H_2SO_4 showed a wide potential window, free from interference from +1.4 V to −1.2 V vs Ag/AgCl electrode. These findings led to the construction of different SiC electrode applications.

Khare *et al.* [99] have studied electrochemical impedance spectroscopy (ESI) for characterization of glassy carbon electrode (GCE) and SiC nanoparticles/glassy carbon electrode (SiC NPs-GCE). The Nyquist plots for quinalphos were obtained for both electrodes. It was observed that the diameter of the semicircle for SiC NPs-GCE (Figure 9.12a) was smaller than that of bare GCE. This is an indication that SiC NPs-GCE has less charge transfer resistance than that of bare GCE which is also an evidence for surface modification. The charge transfer resistance (R_{ct}) for SiC NPs- GCE and GCE were found to be 635.24 Ω and 1631.51 Ω, respectively. Therefore, it could be inferred that charge transfer resistance of the electrode surface decreased and charge transfer rate increased by employing SiC NPs- GCE, thus facilitating the QNP towards the electrode surface.

Sarno *et al.* [100] prepared a supercapacitor electrodes made of exhausted activated carbon- derived SiC nanoparticles coated by graphene by low pressure CVD method. The results show a very high capacitance up to 114.7 F g^{-1} for SiC alone and three times higher in the presence of graphene with an excellent cycle stability.

Figure 9.12b shows the cyclic voltammograms of the graphite paper supported SiC nanowire film at various scan rates [101]. The SiC nanowire film exhibits stable electrochemical performance at the potential window ranging from 0 to 0.5 V vs Ag/AgCl in 0.1 M H_2SO_4. At 10 mV s^{-1}, the CV is almost symmetrical, suggesting an ideal capacitive behavior [102]. The obvious increase of current and persistence of the symmetric CVs shape along with scan rates reveal good rate capability and low contact resistance for the SiC nanowire film electrode [103]. The excellent electrochemical performance of SiC is attributed to the unique 3D graphite paper-supported nanowire film electrode. Nanowires are intertwined together by van der Waals interaction to form

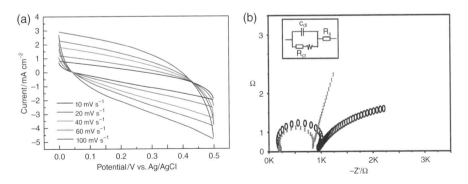

Figure 9.12 (a) CVs of the graphite paper-supported SiC nanowire at the scanning rates 10, 20, 40, 60 and 100 mV^{-1}. (b) Nyquist plots for EIS measurements (3.38×10^{-5} M) QNP at GCE (black), SiCNP-GCE (grey) Inset, equivalent circuit used for data fitting. *Source*: Chen 2014 [101]. Reproduced with permission of Elsevier.

interconnected porous structure, favoring easier pathway of the electrolyte and ions to the active materials by providing large reaction surface and inner space, which decreases charge transfer resistance and ion diffusion resistance, and is favorable for the charge redistribution process. The large specific surface area of nanowires increases the electrochemical reaction rate at the interfaces, facilitating the electrolytic ion transport [104]. In addition, the interactions between electrolyte and SiC surface increase because the SiO_2 layer improves hydrophilicity of nanowires, which provides SiC film better accessibility for electrolytic ions. These mean that the nanowire film electrode has fast reaction kinetics [105].

9.8 SiC Nanostructures in Electroanalysis

In addition to its outstanding electronic and mechanical properties, SiC nanostructures are suitable for electrochemical and photo-electrochemical sensor applications (Table 9.3). SiC NPS are however only recently recognized as candidate for biosensor platforms. A superoxide dismutase (SOD) biosensor based on SiC nanoparticles has been demonstrated by Rafiee-Pour *et al.* [106]. The analytical performance of the biosensor was based on direct voltammetry and amperometry of immobilized SOD onto the surface of a GCE electrode modified with silicon carbide nanoparticles. The recorded cyclic voltammogram in Figure 9.13a indicating no redox peaks was observed for GCE modified with SiC NPs at potential range 0.65 to –0.65 V. However, for GCE modified with SiC-SOD a pair of reduction-oxidation peaks with formal potential –0.03 V is clearly observed. Thus, SiC NPs must have a great effect on the kinetics of electrode reaction and provide a suitable environment for the SOD to transfer electrons with underlying GCE. Figure 9.13b shows the Nyquist plot of GCE , GCE/SiC NPs and GCE modified with SiC-SOD in 0.1 M KCl solution containing 1 mM $[Fe(CN)6]^{3-/4-}$. As sindicated in Figure 9.13b, the bare GCE shows an almost straight line that is characteristic of diffusion limited electrochemical process. Furthermore, R_{et} is about $100\,\Omega$ for GCE/SiCNPs. After the immobilization of SOD onto GCE/SiC NPs the value of R_{et} is significantly increased to about $1700\,\Omega$ (curve c). The results suggest that SOD immobilized onto electrode surface may inhibit the electrochemical contact between the electrode transfer indicator increased value of R_{et} obtained for GCE/SiC NPs-SOD indicates hindrance to the electron transfer, confirming the successful adsorption of SOD onto the SiC film. The authors reported that due to high electron transfer of immobilized SOD and its excellent biocatalytic activity toward superoxide dismutation, the proposed biosensor can be used for micromolar detection of superoxide.

Salimi *et al.* [107] used SiC NPs to modify a GCE for insulin dosing. The electrode exhibited excellent electrocatalytic activity for insulin oxidation and minimizes surface fouling effects of insulin and their oxidation products. The same research group also reported an amperometric nitrite sensor based on a nanocomposite containing an amine-terminated ionic liquid (1-(3-Aminopropyl)-3-methylimidazolium bromide) and SiC NPs (NH_2-IL/SiC NPs) [108].The peak potential shifted to lower potential and the peak current increased significantly at the surface of NH_2-IL/SiC NPs /GCE.

Yang et al. prepared a titanium dioxide–silicon carbide nanohybrid (TiO_2–SiC) with enhanced electrochemical performance through a facile generic *in situ* growth strategy [109]. Monodispersed ultrafine palladium nanoparticles (Pd NPs) with a uniform

Table 9.3 SiC nanostructure in electroanalysis.

Nanostructure	Detection method	LOD (Mol.L^{-1})	DLR (Mol.L^{-1})	Analyte	Reference
Silicon Carbide Nanoparticles	AdS DPV	6.69×10^{-9}–1.34×10^{-6}	1.34×10^{-9}	Quinalphos	[100]
Silicon carbide nanoparticles	DPV	Up to 6.0×10^{-10}	3.3×10^{-12}	Insulin	[108]
Silicon carbide nanoparticles	CV	0.05×10^{-6}–0.35×10^{-6}	0.02×10^{-6}	Nitrite	[109]
A titanium dioxide–silicon carbide nanohybrid (TiO$_2$–SiC)	DPV	HQ:0.01×10^{-6}–5.0×10^{-6}, 5.0×10^{-6}–2.0×10^{-4} BPA: 0.01×10^{-6}–5.0×10^{-6}, 5.0×10^{-6}–2.0×10^{-4}	HQ: 5.5×10^{-9} BPA:4.3×10^{-9}	Hydroquinone Bisphenol	[110]
Au@SiC nanohybrids film	DPV	0.1×10^{-7}–1.0×10^{-4}	2.5×10^{-9}	Tadalafil	[112]
Silicon carbide nanoparticles	DPV	G:0.1×10^{-6}–1.2×10^{-5} A: 0.1×10^{-6}–1.2×10^{-5} T:1.2×10^{-6}–1.36×10^{-4} C:1.2×10^{-6}–1.36×10^{-4}	G:0.015×10^{-6} A:0.015×10^{-6} T:0.14×10^{-6} C:0.14×10^{-6}	Guanine Adenine Thymine Cytosine	[113]
Silicon carbide nanoparticles and graphene nanosheets hybrid	CCFFTCV	0.5–120×10^{-8}	5.2×10^{-9}	Candesartan cilexetil	[116]

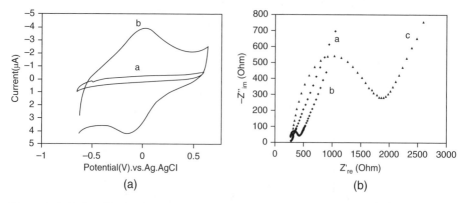

Figure 9.13 (a) CVs of GCE/SiCNPs (a) and GCE/SOD in pH 7 (b) at scan rate 200 mV s⁻¹. (b) impedance for GCE (a), GCE/SiCNPs (b), and GCE/SiCNPs -SOD (c) in 0.1 M KCl solution containing 1 mM [Fe(CN)6]³⁻/⁴⁻. *Source*: Rafiee-Pour 2010 [106]. Reproduced with permission of John Wiley and Sons.

size of ~2.3 nm were deposited on the TiO_2–SiC surface using a chemical reduction method (See Figure 9.14). Pd-loaded TiO_2–SiC nanohybrid (Pd@TiO_2–SiC) glassy carbon electrode has been applied for simultaneous electrochemical determination of hydroquinone (HQ) and bisphenol (BPA) with significant improvement in peak current. This improvement indicates that SiC significantly promotes electron transport and communication between the solution and electrode. The excellent adsorptive capacity of SiC can also lead to high current responses. The oxidation currents of HQ and BPA on the Pd NPs-modified GCE improved by approximately 30% compared with that on the bare GCE due to the excellent catalytic activities of the Pd NPs. On the TiO_2–SiC/GCE, the HQ and BPA oxidation currents significantly increased compared with those on the bare and SiC-modified GCEs, which indicates that the high surface

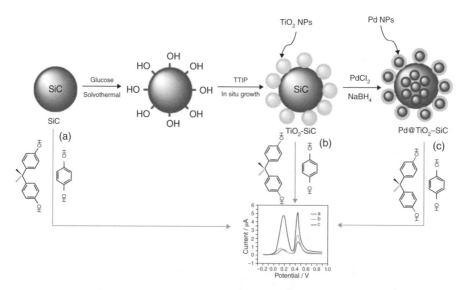

Figure 9.14 The illustration of the Pd@TiO2-SiC nanohybrides for sensing HQ and BPA.

area and adsorptive capacity of the TiO$_2$–SiC composite increase the effective electrode area and improve catalytic activity for HQ and BPA oxidation. In the case of Pd@TiO$_2$/GCE and Pd@SiC/GCE, the HQ and BPA oxidation currents remarkably increased compared with those on the SiC, Pd, and TiO$_2$–SiC-modified GCEs, which could be ascribed to the synergistic effects between Pd NPs and TiO$_2$ or SiC. Oxidation currents increased dramatically when the Pd@TiO$_2$–SiC nanohybrid was immobilized onto the GCE surface. This result may be attributed to the synergistic effect between the three components. And the remarkable conductivities and electrocatalytic activities of ultrafine-monodispersed Pd NPs may be the main contribution that can amplify the electrochemical signals.

Yang *et al.* [110] presented the surface modification of nanocrystalline SiC with diazonium salts via electrochemical methods for later DNA bonding with a nitrophenyl film to the modified electrodes. They showed that the modified electrode demonstrated a wider potential window and lower background current than GC electrodes. The authors also presented successful DNA immobilization on the modified SiC surfaces using Cy5 labeled cDNA with increased red fluorescence on the nitrophenyl SiC surfaces compared to the bare SiC which shows biocompatible property of SiC. Likewise, the voltammogram of hybridized DNA indicated the presence of target DNA. The same research group also developed dual β-cyclodextrin functionalized Au@SiC nanohybrids for the electrochemical determination of tadalafil in the presence of acetonitrile [111]. Uniform and monodispersed ~5.0 nm Au NPs were anchored on the SiC-NH$_2$ surface via a chemical reduction process by using poly-ethyleneglycol and sodium citrate as dispersant and stabilizing agent. The tadalafil electrochemical signal was dramatically amplified by introducing 40% of acetonitrile in buffer medium and further enhanced by the host–guest molecular recognition capacity of β-cyclodextrin.

A glassy carbon electrode modified with SiC NPs has been presented for simultaneous determination of DNA bases by Ghavami *et al.* [112]. For unmodified GCE, the voltammetric response for oxidation of guanine and adenine is negligible but SiC NPs-GCE shows well-defined oxidation peaks for G and A. The detection limits of the SiC NPs-GCE electrodes toward guanine, adenine, thymine and cytosine determinations were 0.015, 0.015, 0.14, and 0.14 µM, respectively.

Godignon's group [113] developed an impedimetric microsystem consisting of four Pt electrodes on an isolated semi-insulating SiC substrate. A Pt serpentine conductor acted as the temperature sensor with the aim of distinguishing impedance changes due to either tissue or thermal effects. In their work, the authors also identified the advantages and potential of using SiC for DNA polymerase chain reaction (PCR) electrophoresis chips because of the high electric field strength and resistivity of semi-insulating SiC, in addition to its high thermal conductivity.

In the similar way Fradetal *et al.* [114] have functionalized two kinds of SiC nanopillar arrays (i) top-down SiC nanopillars with a wide pitch of 5 mm and (ii) dense array (pitch: 200 nm) of core@shell Si@SiC Nanowires by carburization of silicon nanowires. Depending on both the pillar morphology and the pitch, different results in terms of DNA surface coverage were obtained. Particularly, in the case of the wide pitch array, it has been shown that the DNA molecules are located all along the nanopillars. It was concluded that to achieve a DNA sensor based on a nanowire-field effect transistor, the functionalization must be conducted on a single SiC nanowire or nanopillar that constitutes the channel of the field effect transistor and further experimentally verified.

The localization of the functionalization in a small area around the nanostructures guarantees high performances to the sensor.

The determination of Candesartan cilexetil using SiC NPs /graphene nanosheets hybrid mixed with ionic liquid (1-Butyl-3-methylimidazolium hexafluorophosphate ([bmim] [PF_6]) on a glassy carbon electrode has been determined [115]. Wu *et al.* [116] developed a method to resolve the overlapping voltammetric responses of ascorbic acid, dopamine and uric acid on a SiC-coated glassy carbon electrode, and the selective determination of DA in the presence of AA and UA with a sensitivity of $16.9\,A\,M^{-1}\,cm^{-2}$ and a detection limit of $0.05\,\mu M$. Wright *et al.* [117], Yakimova *et al.* [118] and Oliveros *et al.* [61] have reviewed the different kind of SiC materials for sensor applications. These results show that SiC can be an interesting transducer material for applications in sensor applications.

9.9   Conclusion

In this chapter, the properties, synthesis methods and applications of g-C_3N_4 and SiC nanostructures and their use for electroanalytical applications have been reviewed. Several articles have shown that the use of g-C_3N_4 nanostructures in electroanalysis can improve the sensing properties in several ways. The easy in the synthesis of g-C_3N_4 nanostructures and the presence of hydrogen and lone pair in the outer electron shell of nitrogen makes g-C_3N_4 rich with surface properties that can be used to preconcentrate analytes to the surface of the electrode. The exceptional photoluminescence and photo-electrochemical properties of g-C_3N_4, can be further used in electrochemiluminesence and photoelectrochemical assays. Next to g-C_3N_4 nanostructures, SiC nanostructures have shown their potential for electoanlaysis. SiC nanoparticles have been used in elec-trocatalyst to enhancement peak currents which results in increased sensitivity and lower detection limits. Owing to high biocompatibility, high adsorption ability and little harm to the biological activity of biomolecules, SiC nanoparticles have been choose for immobilizing biological compounds and fabricating new biosensors.

Acknowledgements

The author thanks Dr. A. Bezaatpour, Dr. R. Saberi, Dr. L. Fotouhi, F. Rezapour and Y. Sefid-Sefidehkhan for all their kind help in completing this chapter.

References

1 Liebig, J. Uber Einige Stickstoff: Verbindungen. Annalen der Pharmacie, 1834, 10, 1–47.
2 Semencha, A.V.; Blinov, L.N. Theoretical prerequisites, problems, and practical approaches to the preparation of carbon nitride: A review. Glass Physical Chemistry 2010, 36, 199–208.
3 Franklin, E. C. The Ammono carbonic acids. Journal of American Chemical Society, 1922, 44, 486–509.
4 Teter, D.M.; Hemley, R.J. Low. Compressibility carbon nitrides. Science,1996, 271, 53–55.

5 Lowther, J.E. Relative stability of some possible phases of graphitic carbon nitride. Physical Review B: Condensed Matter, 1999, 59, 11683–11686.

6 Sehnert, J.; Baerwinkel, K.; Senker, J. Ab initio calculation of solid-state NMR spectra for different triazine and heptazine based structure proposals of g-C_3N_4. Journal of Physical Chemistry B, 2007, 111, 10671–10680.

7 Wang X.; Maeda K.; Thomas A.; *et al.* A metal-free polymeric photocatalyst for hydrogen production from water under visible light. Nature Materials, 2009, 8, 76–80.

8 Loh, K.P.; Bao, Q.; Eda G.; Chhowalla, M. Graphene oxide as a chemically tunable platform for optical applications. Nature Chemistry, 2010, 2, 1015–1024.

9 Zhu, J.; Xiao, P.; Li, H.; Carabineiro, S.A. Graphitic carbon nitride: Synthesis, properties, and applications in catalysis. ACS Applied Materials and Interfaces, 2014, 6, 16449–16465.

10 Mortazavi, B.; Cuniberti, G.; Rabczuk, T. Mechanical properties and thermal conductivity of graphitic carbon nitride: A molecular dynamics study. Computational Materials Science, 2015, 99, 285–289.

11 Gong Y.; Wang J.; Wei Z.; *et al.* Combination of carbon nitride and carbon nanotubes: synergistic catalysts for energy conversion. ChemSusChem, 2014, 7, 2303–2309.

12 Xu J.; Shalom M.; Piersimoni F.; *et al.* Color-tunable photoluminescence and NIR electroluminescence in carbon nitride thin films and light-emitting diodes. Advanced Optical Materials, 2015, 3, 91–917.

13 Li X.; Ye J. Photocatalytic degradation of rhodamine B over $Pb_3Nb_4O_{13}$/fumed SiO_2 composite under visible light irradiation. Journal of Physical Chemistry C, 2007, 111, 13109–13116.

14 Ye, S.; Wanga, R.; Wu, M.Z.; Yuan, Y.P. A review on g-C_3N_4 for photocatalytic water splitting and CO_2 reduction. Applied Surface Science, 2015, 358, 15–27.

15 Gong, Y.; Li, M.; Wang, Y. Carbon nitride in energy conversion and storage: Recent advances and future prospects. ChemSusChem, 2015, 8, 931–946.

16 Meng, Z.; Xie, Y.; Cai, T.; *et al.* graphene-like g-c3n4 nanosheets/sulfur as cathode for lithium–sulfur battery. Electrochimica Acta 2016, 210, 829–836.

17 Liu, Y.; Wang, Q.; Lei, J.; *et al.* Anodic electrochemiluminescence of graphitic-phase C_3N_4 nanosheets for sensitive biosensing. Talanta, 2014, 122, 130–134.

18 Wang, B.; Ye, C.; Zhong, X.; *et al.* Electrochemical biosensor for organophosphate pesticides and huperzine-A detection based on Pd wormlike nanochains/graphitic carbon nitride nanocomposites and acetylcholinesterase. Electroanalysis, 2016, 28, 304–311.

19 Baig, R.B.; Verma, S.; Varma, R.S.; Nadagouda, M.N. Magnetic Fe@g-C_3N_4: A photoactive catalyst for the hydrogenation of alkenes and alkynes. ACS Sustainable Chemical Engineering 2016, 4, 1661–1664.

20 Zhu, J.; Wei, Y.; Chen, W.; *et al.* Polymeric carbon nitride as a metal-free catalyst for NO decomposition. Chemical Communications, 2010, 46, 6965–6967.

21 Rong, M.; Lin, L.; Song, X.; *et al.* fluorescence sensing of chromium (vi) and ascorbic acid using graphitic carbon nitride nanosheets as a fluorescent 'switch.' Biosensors and Bioelectronics, 2015, 68, 210–217.

22 Thomas A.; Fischer A.; Goettmann F.; *et al.* Graphitic carbon nitride materials: Variation of structure and morphology and their use as metal-free catalysts. Journal of Material Chemistry, 2008, 18, 4893–4908.

23 Lu Q.; Deng J.; Hou Y.; *et al.* One-step electrochemical synthesis of ultrathin graphitic carbon nitride nanosheets and their application to the detection of uric acid. Chemical Communications, 2015, 51, 12251—12253

24 Quin, Y.; Li, J.; Yuan, J.; *et al.* Hollow mesoporous carbon nitride nanosphere/three-dimensional graphene composite as high efficient electrocatalyst for oxygen reduction reaction. Journal of Power Sources, 2014, 272, 696–702.

25 Zhang Y.; Mori T.; Ye J. Phosphorus-doped carbon nitride solid: Enhanced electrical conductivity and photocurrent generation. Journal of the American Chemical Society, 2010, 132, 6294–6295

26 Wang, G.; Zhang, J.; Hou, S. g-C$_3$N$_4$/conductive carbon black composite as Pt-free counter electrode in dye-sensitized solar cells. Materials Research Bulletin, 2016, 76, 454–458.

27 Zhang, H.; Huang, Q.; Huang, Y.; *et al.* graphitic carbon nitride nanosheets doped graphene oxide for electrochemical simultaneous determination of ascorbic acid, dopamine and uric acid. Electrochimica Acta, 2014, 142, 125–131.

28 Xu, J.; Wu, H.; Wang, X.; *et al.* A new and environmentally benign precursor for the synthesis of mesoporous g-C$_3$N$_4$ with tunable surface area. Physical Chemistry Chemical Physics, 2013, 15, 4510–4517.

29 Zhang, X.; Xie, X.; Wang, H.; *et al.* Enhanced photoresponsive ultrathin graphitic-phase C$_3$N$_4$ nanosheets for bioimaging. Journal of American Chemical Society, 2013, 135, 18–21

30 Tian, J.; Ning, R.; Liu, Q.; *et al.* Three-dimensional porous supramolecular architecture from ultrathin g-C$_3$N$_4$ nanosheets and reduced graphene oxide: Solution self-assembly construction and application as a highly efficient metal-free electrocatalyst for oxygen reduction reaction. ACS Applied Materials and Interfaces, 2014, 6, 1011–1017.

31 Tian, J.; Ning, R.; Liu, Q.; *et al.* Three-dimensional porous supramolecular architecture from ultrathin g-C$_3$N$_4$ nanosheets and reduced graphene oxide: Solution self-assembly construction and application as a highly efficient metal-free electrocatalyst for oxygen reduction reaction. ACS Applied Materials and Interfaces, 2014, 6, 1011–1017.

32 Li, C.Z.; Wang, Z.B.; Sui, X.L.; *et al.* Ultrathin graphitic carbon nitride nanosheets and graphene composite material as high-performance PtRu catalyst support for methanol electrooxidation. Carbon, 2015, 93, 105–115.

33 Yue, X.; Zhu, Z.; Zhang, M.; Ye, Z. Reaction-based turn-on electrochemiluminescent sensor with a Ruthenium(II) complex for selective detection of extracellular hydrogen sulfide in rat brain. Analytical Chemistry 2015, 87, 1839–1845.

34 Chu, H.H.; Yan. J. L.; Tu, Y.F. Study on a luminol-based electrochemiluminescent sensor for label-free DNA sensing. Sensors, 2010, 10, 9481–9492.

35 Liu, X.; Wang, N.; Zhao W, Jiang H. Electrochemiluminescent pH sensor measured by the emission potential of TiO$_2$ nanocrystals and its biosensing application. Luminescence, 2015, 30, 98–101.

36 Liu, S.; Zhang, X.; Yu, Y.; Zou, G. A monochromatic electrochemiluminescence sensing strategy for dopamine with dual-stabilizers-capped CdSe quantum dots as emitters. Analytical Chemistry, 2014, 86, 2784–2788.

37 Feng, Y.; Wang, Q.; Lei, J.; Ju, H. Electrochemiluminescent DNA sensing using carbon nitride nanosheets as emitter for loading of hemin labeled single-stranded DNA. Biosensors and Bioelectronics, 2015, 73, 7–12.

38 Chen, S.; Li, A.; Zhang, L.; Gong, J. Molecularly imprinted ultrathin graphitic carbon nitride nanosheets based electrochemiluminescence sensing probe for sensitive detection of perfluorooctanoic acid, Analytica Chimica Acta, 2015, 896, 68–77.

39 Han, T.; Li, X.; Li, Y.; *et al.* Gold nanoparticles enhanced electrochemiluminescence of graphite-like carbon nitride for the detection of nuclear matrix protein 22. Sensors and Actuators B, 2014, 205, 176–183.

40 Shia, H.; Zhao, G.; Liu, M.; Zhu, Z. A novel photoelectrochemical sensor based on molecularly imprinted polymer modified TiO_2 nanotubes and its highly selective detection of 2,4-dichlorophenoxyacetic acid. Electrochemistry Communications, 2011, 13, 1404–1407.

41 Wang, X.; Yan, T.; Li, Y.; *et al.* A competitive photoelectrochemical immunosensor based on a CdS-induced signal amplification strategy for the ultrasensitive detection of dexamethasone. Science Reports, 2015, 5, 17945.

42 Zhang, B.; Lu, L.; Hu, Q.; *et al.* ZnO nanoflower-based photoelectrochemical DNAzyme sensor for the detection of Pb^{2+}. Biosensors and Bioelectronics, 2014, 56, 243–249.

43 Zhang, J.; Tu, L.; Zhao, S.; *et al.* Fluorescent gold nanoclusters based photoelectrochemical sensors for detection of H_2O_2 and glucose. Biosensors and Bioelectronics, 2015, 15, 296–302.

44 Zhang, X.; Li, L.; Peng, X.; *et al.* Non-enzymatic hydrogen peroxide photoelectrochemical sensor based on WO_3 decorated core–shell TiC/C nanofibers electrode. Electrochimica Acta, 2013, 108, 491–496.

45 Xu L.; Xia J.; Xu H.; *et al.* AgX/graphite-like C_3N_4 (X= Br, I) hybrid materials for photoelectrochemical determination of copper(II) ion. Analyst, 2013, 138, 6721–6726.

46 Li, R.; Liu, Y.; Li, X.; *et al.* A novel multi-amplification photoelectrochemical immunoassay based on copper(II) enhanced poly-thiophene sensitized graphitic carbon nitride nanosheet. Biosensors and Bioelectronics, 2014, 62, 315–319.

47 Hao, Q.; Lei, J.; Wang, Q.; *et al.* Carbon nitride nanosheets sensitized quantum dots as photocathode for photoelectrochemical biosensing. Journal of Electroanalytical Chemistry, 2015, 759, 8–13.

48 Fang, T.; Yang, X.; Zhang, L.; Gong, J. Ultrasensitive photoelectrochemical determination of chromium(VI) in water samples by ion-imprinted/format anion-incorporated graphitic carbon nitride nanostructured hybrid. Journal of Hazardous Materials, 2016, 312, 106–113.

49 Kang, Q.; Wang, X.; Ma, X.; *et al.* Sensitive detection of ascorbic acid and alkaline phosphatase activity by double-channel photoelectrochemical detection design based on g-C_3N_4/TiO_2 nanotubes hybrid film. Sensors and Actuators B, 2016, 230, 231–241.

50 Zhang, H.; Huang, Y.; Hu, S.; *et al.* Self-assembly of graphitic carbon nitride nanosheets-carbon nanotube composite for electrochemical simultaneous determination of catechol and hydroquinone. Electrochimica Acta, 2015, 176, 28–35.

51 Zhang, Y.; Bo, X.; Nsabimana, A.; *et al.* Fabrication of 2D ordered mesoporous carbon nitride and its use as electrochemical sensing platform for H_2O_2, nitrobenzene, and NADH detection. Biosensors and Bioelectronics, 2014, 53, 250–256.

52 Zhou Y.; Tang L.; Zeng G.; *et al.* Mesoporous carbon nitride based biosensor for highly sensitive and selective analysis of phenol and catechol in compost bioremediation. Biosensors and Bioelectronics, 2014, 15, 519–525.

53 Jianga T.; Jianga G.; Huanga Q.; Zhou H. High-sensitive detection of dopamine using graphitic carbon nitride by electrochemical method. Materials Research Bulletin, 2016, 74, 271–277.

54 Cheng C.; Huang Y.; Wang J.; *et al.* Anodic electrogenerated chemiluminescence behavior of graphite-like carbon nitride and its sensing for rutin. Analytical Chemistry, 2013, 85, 2601–2605.

55 Zheng X.; Hua X.; Qiao X.; *et al.* Simple and signal-off electrochemiluminescence immunosensor for alpha fetoprotein based on gold nanoparticle-modified graphite-like carbon nitride nanosheet nanohybrids. RSC Advances, 2016, 6, 21308–21316.

56 Feng Q.; Shen Y.; Li M.; *et al.* Dual-wavelength electrochemiluminescence ratiometry based on resonance energy transfer between Au nanoparticles functionalized g-C_3N_4 nanosheet and Ru(bpy)$_3^{2+}$ for microRNA detection. Analytical Chemistry, 2016, 88, 937–944.

57 Zhou Z.; Shang Q.; Shen Y.; *et al.* Chemically modulated carbon nitride nanosheets for highly selective electrochemiluminescent detection of multiple metal-ions. Analytical Chemistry, 2016, 88, 6004–6010.

58 Presser R.; Nickel K.G. Silica on silicon carbide. Critical Reviews of Solid State, 2008, 33, 1–99.

59 Levinshtein ME, Rumyantsev SL, Shur M.S, (eds.) Properties of Advanced Semiconductor Materials GaN, AlN, SiC, BN, SiC, SiGe. John Wiley and Sons, Inc. 2001, p. 93–148.

60 Oliveros A.; Guiseppi-Elie A.; Saddow S.E. Silicon carbide: A versatile material for biosensor applications. Biomedical Microdevices, 2013, 15, 353–368.

61 Wu R.; Zhou K.; Yue C.; *et al.* Recent progress in synthesis, properties and potential applications of SiC nanomaterials. Progress in Materials Science, 2015, 72, 1–60.

62 Zhou W.M.; Yang B.; Yang Z.X.; *et al.* Large-scale synthesis and characterization of SiC nanowires by high-frequency induction heating. Applied Surface Science, 2006, 252, 5143–5148.

63 Wang L.; Li C.; Yang Y.; *et al.* Large-scale growth of well-aligned SiC tower-like nanowire arrays and their field emission properties. ACS Applied Materials and Interfaces, 2015, 7, 526–533.

64 Li H.; Yu H.; Zhang X.; *et al.* Bowl-like 3C-SiC nanoshells encapsulated in hollow graphitic carbon spheres for high-rate lithium-ion batteries. Chemistry of Materials, 2016, 28, 1179–1186.

65 Gu Z.; Yang Y.; Li K.; *et al.* Aligned carbon nanotube-reinforced silicon carbide composites produced by chemical vapor infiltration. Carbon, 2011, 49, 2475–2482.

66 D'angelo M.; Deokar G.; Steydli S.; *et al.* In-situ formation of SiC nanocrystals by high temperature annealing of SiO_2/Si under CO: A photoemission study. Surface Science, 2012 , 606, 697–701.

67 Dragomir M.; Valant M.; Fanetti M.; Mozharivskyj Y. A facile chemical method for the synthesis of 3C–SiC nanoflakes. RSC Advances, 2016, 6, 21795–21801.

68 Hu M.; Kuo C.; Wu C.; *et al.* The production of SiC nanowalls sheathed with a few layers of strained graphene and their use in heterogeneous catalysis and sensing applications. Carbon, 2011, 49, 4911–4919.

69 Ruemmeli M.; Borowiak-Palen E.; Gemming T.; *et al.* Modification of SiC based nanorods via a hydrogenated annealing process. Synthetic Metals, 2005, 153, 349–352.

70 Fradetal L.; Stambouli V.; Bano E.; *et al.* Bio-functionalization of silicon carbide nanostructures for SiC nanowire-based sensors realization. Journal of Nanoscience and Nanotechnology, 2014, 14, 3391–3397.

71 Román-Manso B.; Vega-Díaz S.M.; Morelos-Gómez A.; *et al.* Aligned carbon nanotube/silicon carbide hybrid materials with high electrical conductivity, superhydrophobicity and superoleophilicity. Carbon, 2014, 80, 120–126.

72 Seong H.K.; Choi H.J.; Lee S.K.; *et al.* Optical and electrical transport properties in silicon carbide nanowires. Applied Physical Letters, 2004, 85, 1256–1258.

73 Prakash J.; Venugopalan R.; Tripathi B.M.; *et al.* Chemistry of one dimensional silicon carbide materials: Principle, production, application and future prospects. Progress in Solid State Chemistry 2015, 43, 98–122

74 Chen J.; Zhai F.; Liu M.; *et al.*SiC nanowires with tunable hydrophobicity/hydrophilicity and their application as nanofluids. Langmuir, 2016, 32, 5909–5916.

75 Harris G.L. Properties of Silicon Carbide. INSPEC, Institution of Electrical Engineers, United Kingdom, 1999.

76 Fan J.Y.; Wu X.L.; Chu P.K. Low-dimensional SiC nanostructures: Fabrication, luminescence, and electrical properties. Progress in Material Science, 2006, 51, 983–1031.

77 Chen S.; Ying P.; Wang L.; *et al.* Highly flexible and robust N-doped SiC nanoneedle field emitters. NPG Asia Materials, 2015, 7, e157, 1–7.

78 Ollivier, M.; Latu-Romain, L.; Salem, B.; *et al.* Integration of SiC-1D nanostructures into nano-field effect transistors. Materials Science in Semiconductor Processing, 2015, 29, 218–222.

79 You D.J.; Jin X.; Kim J.H.; *et al.* Development of stable electrochemical catalysts using ordered mesoporous carbon/silicon carbide nanocomposites. International Journal of Hydrogen Energy, 2015, 40, 12352–12361.

80 Chen, J.; Zhang, J.; Wang, M.; Li, Y. High-temperature hydrogen sensor based on platinum nanoparticle-decorated SiC nanowire device. Sensors and Actuators, B: Chemical, 2014, 201, 402–406.

81 Mishra G.; Parida K. M.; Singh S. K. Facile fabrication of S-TiO$_2$/β-SiC nanocomposite photocatalyst for hydrogen evolution under visible light. ACS Sustainable Chemical Engineering, 2015, 3, 245–253.

82 Zhao, Y.; Kang, W.; Li, L.; *et al.* Solution blown silicon carbide porous nanofiber membrane as electrode materials for supercapacitors. Electrochimica Acta, 2016, 207, 257–265.

83 Beke, D.; Jánosi, T.Z.; Somogyi, B.; *et al.* Identification of luminescence centers in molecular-sized silicon carbide nanocrystals. Journal of Physical Chemistry C, 2016, 120, 685–691.

84 Lu, C.; Yuan, Q.; Simons, R.; *et al.* Influence of SiC and VGCF nano-fillers on crystallization behaviour of PPS composites. Journal of Nanoscience and Nanotechnology, 2016, 16, 8366–8373.

85 Umar Z.A.; Rawat R.S.; Tan K.S.; *et al.* Hard TiCx/SiC/a-C:H nanocomposite thin films using pulsed high energy density plasma focus device. Nuclear Instruments and Methods in Physics Research B, 2013, 301, 53–61

86 Yu I.K.; Rhee J.H.; Cho S.; Yoon H.K. Design and installation of DC plasma reactor for SiC nanoparticle production. Journal of Nuclear Materials, 2009, 386–388, 631–633.

87 Özkan S.; Hapçı G.; Orhan G.; Kazmanlı K. Electrodeposited Ni/SiC nanocomposite coatings and evaluation of wear and corrosion properties. Surface and Coatings Technology, 2013, 232, 734–741.

88 Kim G. Y. Ultrasonic spray deposition of SiC nanoparticles for laminate metal composite fabrication. Powder Technology, 2016, 288, 279–285.

89 Karoly Z.; Mohai I,; Klebret S.; *et al.* Synthesis of SiC powder by RF plasma techniques. Powder Technology, 2011, 214, 300–305

90 Zhang H.X.; Feng P.X.; Makarova V.; *et al.* Synthesis of nanostructured SiC using the pulsed laser deposition technique. Materials Research Bulletin, 2009, 44, 184–188.

91 Kang P.; Zhang B.; Wua G.; *et al.* Synthesis of b-SiC nanowires by ball milled nanoparticles of silicon and carbon. Journal of Alloys and Compounds, 2014, 604, 304–308.

92 Wei G.; Qin W.; Zheng K.; *et al.* Synthesis and properties of SiC/SiO$_2$ nanochain heterojunctions by microwave method. Crystal Growth and Design, 2009, 9,1431–1435.

93 Meier F.; Giolando D.M.; Kirchhoff J.R. Silicon carbide: A new electrode material for voltammetric measurements. Chemical Communications, 1996, 22, 2553–2355

94 Qi X.; Liang J.; Yu C.; *et al.* Facile synthesis of interconnected SiC nanowire networks on silicon substrate. Materials Letters, 2014, 116, 68–70.

95 Hume D.N.; Kolthoff I.M. The silicon carbide electrode. Journal of American Chemical Society, 1941, 63, 2805–2806.

96 Yang N.; Zhuang H.; Hoffmann R.; *et al.* Electrochemistry of nanocrystalline 3C silicon carbide films. Chemistry: A European Journal, 2012, 18, 6514–6519.

97 Chu V.; Conde J.P.; Jarego J.; *et al.* Transport and photoluminescence of hydrogenated amorphous, silicon-carbon alloys. Journal of Applied Physics, 1995,78, 3164–3173.

98 Pierson H. (ed.) Handbook of Chemical Vapor Deposition, 2nd edition. Noyes Publication, Norwich, NY, 1999, pp. 499.

99 Khare N.G.; Dar R.A.; Srivastava A.K. Adsorptive stripping voltammetry for trace determination of quinalphos employing silicon carbide nanoparticles modified glassy carbon electrode. Electroanalysis 2015, 27, 503–509.

100 Sarno M.; Galvagno S.; Piscitelli R.; *et al.* Supercapacitor electrodes made of exhausted activated carbon- derived sic nanoparticles coated by grapheme. Industrial Engineering and Chemical Research, 2016, 55, 6025–6035.

101 Chen J.; Zhang J.; Wanga M.; *et al.* SiC nanowire film grown on the surface of graphite paper and its electrochemical performance. Journal of Alloys and Compounds, 2014, 605, 168–172.

102 Yuan Y.F.; Pei Y.B.; Guo S.Y.; *et al.* Sparse MnO$_2$ nanowires clusters for high-performance supercapacitors. Material Letters, 2012, 73, 194–197.

103 Yan Z.; Ma L.; Zhu Y.; *et al.* Three dimensional metal graphene nanotube multifunctional hybrid materials. ACS Nano 2013, 7, 58–64.

104 Bao L.H.; Li X.D. Towards textile energy storage from cotton t-shirts. Advanced Materials, 2012, 24, 3246–3252.

105 Yuan Y.F.; Xia X.H.; Wu J.B.; *et al.* Hierarchically porous Co$_3$O$_4$ film with mesoporous walls prepared via liquid crystalline template for supercapacitor application. Electrochemistry Communications, 2011, 13, 1123–1126.

106 Rafiee-Pour H.; Noorbakhsh A.; Salimi A.; Ghourchian H. Sensitive superoxide biosensor based on silicon carbide nanoparticles. Electroanalysis, 2010, 22, 1599–1606.

107 Salimi A.; Mohamadi L.; Hallaj R.; Soltanian S. Electrooxidation of insulin at silicon carbide nanoparticles modified glassy carbon electrode. Electrochemistry Communications, 2009, 11, 1116–1119.

108 Salimi A.; Kurd M.; Teymouriana H.; Hallaj R. Highly sensitive electrocatalytic detection of nitrite based on SiC nanoparticles/amine terminated ionic liquid

modified glassy carbon electrode integrated with flow injection analysis. Sensors and Actuators B, 2014, 205, 136–142.

109 Yang L.; Zhao H.; Fan S.; *et al.* A highly sensitive electrochemical sensor for simultaneous determination of hydroquinone and bisphenol A based on the ultrafine Pd nanoparticle@TiO$_2$ functionalized SiC. Analytica Chimica Acta, 2014, 852, 28–36.

110 Yang N.; Zhuang H.; Hoffmann R.; *et al.* Nanocrystalline 3C-SiC electrode for biosensing applications. Analytical Chemistry, 2011, 83, 5827–5830.

111 Yang L.; Zhao H.; Li C.; *et al.* Dual β-cyclodextrin functionalized Au@SiC nanohybrids for the electrochemical determination of tadalafil in presence of acetonitrile. Biosensors and Bioelectronics, 2015, 64, 126–130.

112 Ghavami R.; Salimi A.; Navaee A. SiC nanoparticles-modified glassy carbon electrodes for simultaneous determination of purine and pyrimidine DNA bases. Biosensors and Bioelectronics, 2011, 26, 3864–3869.

113 Godignon P. SiC materials and technologies for sensors development. Materials Science Forum, 2005, 483, 1009–1014.

114 Fradetal L.; Stambouli V.; Bano E.; *et al.* Bio-functionalization of silicon carbide nanostructures for SiC nanowire based sensors realization. Journal of Nanoscience Nanotechnology, 2014, 14, 3391–3397.

115 Norouzi P.; Pirali-Hamedani M.; Ganjali M.R. Candesartan cilexetil determination by electrode modified with hybrid film of ionic liquid- graphene nanosheets-silicon carbide nanoparticle using continuous coulometric fft cyclic voltammetry. International Journal of Electrochemical Science, 2013, 8, 2023–2033.

116 Wu W.C.; Chang H.W.; Tsai Y.C. Electrocatalytic detection of dopamine in the presence of ascorbic acid and uric acid at silicon carbide coated electrodes. Chemical Communications, 2011, 47, 6458–6460.

117 Wright N.G.; Horsfall A.B.; Vassilevski K. Prospects for SiC electronics and sensors. Materials Today, 2008, 11, 16–21.

118 Yakimova R.; Petoral R.M.; Yazdi G.R.; *et al.* Surface functionalization and biomedical applications based on SiC. Journal of Physics D: Applied Physics, 2007, 40, 6435–6442.

Index

a

acetaminophene   94, 112, 151, 166
affinity   44, 106, 140, 145, 156, 161, 175, 186, 188, 233
amino acid   87, 91, 92, 98, 156, 157, 175, 189
amorphous   3, 4, 8–10, 21–23, 59, 77, 85, 188, 199, 208, 241, 256
amperometric   14, 16, 20, 23, 25, 32, 51, 52, 68, 69, 77, 79, 82, 87, 113, 150, 154, 155, 160, 169, 171, 185, 188, 194, 196, 211, 221, 226, 246
angiogenin   158, 159, 171
antibacterial   154
antibody   44–46, 86, 107, 110, 129, 147, 154, 158, 159, 170, 171, 183, 184, 236
anti-fouling   67, 80
antigen   44–46, 86, 103, 105, 115, 129, 158, 159, 161, 170, 171
aptamer   103, 109, 114, 115, 147, 149, 151, 154, 157, 158, 160, 161, 168, 171, 180, 181, 183, 194
array   2, 8, 21, 22, 33, 34, 36, 44, 50, 52, 53, 56, 59, 61–64, 76, 79, 82, 173, 179, 201, 219, 220, 249, 254
ascorbic acid   13, 33, 49, 55, 57, 58, 63, 64, 67, 69, 72, 74–78, 81, 90, 112, 126, 128, 131, 135, 136, 144, 145, 147, 149–151, 164, 166, 168, 169, 180, 188, 196, 234, 239, 240, 250–253, 257
ATP   80, 107, 128

b

biomarkers   146, 157

biosensors
biosensors   1–3, 21, 27, 28, 32, 33, 35–44, 47–49, 51–53, 74, 76, 77, 79, 80, 82, 97, 111–114, 116, 132, 135, 136, 156, 164–177, 179–181, 183, 185, 187–196, 209, 250–253, 257
bisphenol A   99, 113, 147–149, 169, 257
boron-doped diamond (BDD)   1, 20, 24, 63, 68, 76, 79–81

c

capacitance   4, 46, 60, 124, 211–214, 245
capillary electrophoresis (CE)   1, 51
carbon dots   85, 86, 94, 97, 110–113, 115
carbon nanofibers (CNFs)   1, 27, 29, 30, 31, 33–35, 37, 39, 41, 42, 47–53, 57, 60
carbon nanotubes (CNTs)   1, 20, 23, 24, 27, 53, 56, 74–78, 80, 82, 85, 89, 100, 103, 114, 115, 133, 162–164, 166, 173, 177, 180, 188, 191–193, 195, 196, 201, 221, 223, 251
carbon nitride   8, 22, 23, 227, 231, 232, 234, 235, 237, 238, 240, 250–254
catalyst   20, 28–30, 32, 47–51, 59, 61, 72, 75, 82, 86–89, 91, 92, 98, 99, 108, 111, 135, 160, 166, 167, 173, 174, 178, 197, 204, 205, 209, 214, 215, 224, 225, 231–233, 243, 250–252, 255
cells   1, 20, 22, 27, 41, 47, 62–64, 72, 76, 83, 87–90, 103, 105, 110, 114–116, 124, 125, 141, 154–156, 159, 165, 168, 173, 191, 195, 225, 239, 252

Nanocarbons for Electroanalysis, First Edition.
Edited by Sabine Szunerits, Rabah Boukherroub, Alison Downard and Jun-Jie Zhu.
© 2017 John Wiley & Sons Ltd. Published 2017 by John Wiley & Sons Ltd.

chitosan   46, 57, 75, 82, 92, 93, 97–99, 107,
    112, 113, 131, 136, 148–152, 158, 161,
    164, 170, 185, 186, 188, 233, 234, 240
cholesterol   146, 147, 153
chronoamperometry   61, 234
conducting polymers   2, 90
conductivity   2, 3, 8, 11, 21, 27–30, 32, 33,
    41, 43, 45, 47, 52, 57–59, 61, 64, 67,
    72, 78, 86, 87, 89–92, 95, 97, 101,
    103, 105, 107, 121, 122, 127, 128,
    136, 139, 140, 159, 180, 183, 197,
    201, 212, 216, 217, 228, 231–233,
    236, 239–243, 249, 251, 255
contact angle   6, 11, 12, 125, 217
covalent   46, 57, 95, 98, 106, 107, 143, 157,
    177, 180, 185, 189, 224, 240, 241
cyclic voltammetry   32, 59, 74, 80, 87, 99,
    146, 183, 212, 225, 257
cyclodextrin   57, 75, 147, 158, 160, 161,
    168, 171, 249, 257
cysteine   81, 98, 102, 104, 113, 114, 142,
    151, 158, 167

d
delivery   48, 163, 173, 194
diagnostic   132, 140, 149, 151, 153, 157,
    163, 173, 179, 191
diamond like carbon(DLC)   3, 23, 59, 67,
    77, 82, 217, 219
diazonium   3, 21, 143, 164, 249
Diels–Alder   175, 176
differential pulse voltammetry (DPV)   32,
    44, 58, 97, 99, 101, 151, 156, 211,
    239, 241
DNA   1, 7, 15, 20, 22, 24, 75, 82, 86, 93, 95,
    100, 101, 105–107, 115, 116, 143,
    146, 156–158, 160, 163–165, 168,
    170–172, 174, 177–183, 191, 193,
    194, 209, 210, 219, 224, 233,
    235–237, 242, 249, 252, 253, 257
drug   10, 21, 40, 48, 55, 88, 99, 111, 147,
    163, 164, 173, 183, 194

e
E. coli   159, 163, 182, 185, 194
electrochemical impedance spectroscopy
    (EIs)   32, 33, 46, 97, 129, 245

electrochemiluminescence (ECL)   46, 86,
    95, 101, 104, 114–116, 154, 160,
    161, 171, 181, 195, 233, 251–254
electrostatic   46, 65, 67, 71, 86, 87, 91, 93,
    95, 142–144, 146, 170, 178, 189, 195,
    226, 240
enzyme   2, 20, 28, 40–44, 51, 55, 60,
    86–95, 97, 99, 101, 111, 112, 134,
    147, 152–154, 158, 160, 170, 181,
    183, 185–188, 191, 194–196, 240
estradiol   147, 153, 154, 167

f
fluorescence   90, 96, 106, 115, 139, 227,
    249, 251
fluorescence (or Forster) resonance energy
    transfer (FRET)   106
fouling   5, 6, 56, 61, 65, 67–73, 78, 80, 81,
    226, 246
functionalization   67, 86, 95, 165, 166,
    173–177, 191, 192, 209, 210, 217,
    224, 242, 244, 249, 254, 257

g
glassy carbon (GC)   1, 8, 14, 22, 24, 49–51,
    57, 65, 68, 75, 76, 79–82, 87, 91, 93,
    95, 99, 111–113, 131, 133, 136, 146,
    155, 159, 160, 169, 170, 179, 183, 186,
    195, 239, 245, 248–250, 256, 257
glucose   14–17, 33, 34, 36–43, 48, 52, 87,
    90, 91, 97–99, 111, 170, 251

h
health   36, 91, 92, 97, 111, 113, 127, 149,
    151, 154, 156, 191
heavy metal ions   1, 9, 39, 92, 112, 146, 156
hexacyanoferrate   182, 186
highly oriented pyrolytic graphite
    (HOPG)   1, 68, 81
high performance liquid chromatography
    (HPLC)   1, 20
histidine   44, 57, 74, 157, 158, 170
human serum   49, 72, 89, 155, 157, 159, 171
hydrogen peroxide (H_2O_2)   10, 11, 13, 14,
    24, 25, 33, 34, 36, 42, 50, 52, 80, 87,
    89, 95, 98, 101, 102, 107, 111, 113,
    135, 148, 149, 155, 159, 168–170,
    240, 253

hydrophilic   6, 8, 13, 23, 67–70, 91, 97,
    153, 197, 217, 223, 242, 243, 246, 255
hydrophobic   11, 67–70, 97, 124, 125, 134,
    156, 178–180, 201

i

immunosensors   15, 28, 44–46, 52, 86, 95,
    102, 103, 114, 146, 170, 172, 183,
    194, 195
insulin   40, 51, 158, 171, 246, 247, 256
interdigitated   2, 21

l

laccase   42, 43, 51, 52, 99, 182, 187, 188
lactate   32, 42–44, 49, 52, 158, 160, 171
limit of detection (LOD)   17, 35, 57,
    125–127, 132, 149, 156, 159, 162, 211
linear sweep voltammetry (LSV)   32, 99
lithographic technique   2, 5
localized surface plasmon resonance
    (LSPR)   140, 163

m

mannitol   15, 17–19
mesoporous   33, 34, 40, 47, 49, 51, 57, 58,
    76, 103, 114, 156, 168, 209, 224, 231,
    232, 234, 240, 252, 253, 255, 256
metal nanoparticles (NPs)   5, 13, 28, 30,
    32–34, 36, 56, 67, 89, 164, 199, 204
microelectrode   20, 21, 23, 37
molybdenum disulphide (MoS$_2$)   99
M. tuberculosis   159
multi-walled carbon nanotubes
    (MWCN)   75, 188

n

NADH   126, 135, 148, 154, 155, 160, 168,
    234, 240, 253
nanoelectrode   36, 44, 53, 56, 57, 62–64,
    79, 165, 179, 194
nanowires   20, 95, 96, 108, 110, 112, 117,
    173, 197–199, 201, 209–216, 219,
    220, 223, 224, 226, 241–243, 245,
    246, 249, 254–256
nitric oxide   72, 149, 170
nitride   8, 22, 23, 227, 231, 232, 234, 235,
    237, 238, 240, 250–254
non-covalent   57, 180, 185

o

oligonucleotides   7, 8
oxygen reduction reaction (ORR)   8, 10,
    89, 111, 135, 165, 232, 252

p

plasma   3, 5, 8, 11, 22, 28, 59, 69, 128, 175,
    179, 192, 194, 197–203, 208, 209,
    219, 220, 223, 243, 255
platinum   1, 20, 24, 25, 32, 34, 44, 50, 52,
    55, 59, 76, 89, 119, 135, 167, 181,
    185, 220, 225, 226, 255
photo-electrochemical   233, 236, 239,
    246
photoresist   2, 3, 21
polyaniline   30, 34, 36, 42, 52, 57, 75, 96,
    112, 129, 132, 151, 165, 167, 172
poly(diallyldimethylammonium chloride
    (PDDA)   142–147, 149, 151, 152,
    155, 158, 159, 164
polyelectrolyte   146, 165, 168, 171
poly(3, 4-ethylenedioxythiophene
    (PEDOT)   57, 58, 67, 69, 74, 80,
    158, 159
poly-ethylene glycol (PEG)   45, 109, 143
polyimide   3, 79
poly(*p*-Phenylene Vinylene) (PPV)   2, 3
polypyrrole   30, 36, 42, 57, 72, 74, 82, 98,
    99, 113, 148, 169, 179, 196, 203, 213,
    236
polyvinyl alcohol (PVA)   61, 68, 69, 99
prostate specific antigen   103, 115, 159
Prussian blue   41, 42, 52, 98, 113, 159,
    180
pyrolysis   2, 3, 5, 14, 15, 31, 49, 59, 234,
    235, 240

r

ractopamine   147, 151, 167
radio frequency (RF)   3, 8, 14, 198, 219
Raman spectroscopy   8, 128, 163, 164, 206,
    221
real-time   3, 58, 62, 72, 73, 79, 82, 83
reduced graphene oxide (rGO)   21, 72, 76,
    136, 137, 161, 163–171, 232, 252
redox reaction   55, 62, 65–67, 73, 87, 90,
    92, 94, 124, 145

resolution   36, 55, 57–59, 62–67, 72, 78,
       126, 146, 199, 208
resonance energy transfer (RET)   106, 114,
       115, 116, 254

s
screen-printed electrodes   76
signal-to-noise ratio   63, 239
silica   33, 76, 103, 114, 116, 156, 168, 201,
       202, 231, 232, 240, 243, 254
silicon carbide (SiC)   221, 241, 246, 247,
       254–256
single-walled carbon nanotubes (SWCN)   8,
       80, 188, 193
supercapacitors   27, 124, 212, 222, 225,
       243, 255, 256
surface chemistry   69, 81

t
template   28, 33, 47, 72, 88, 94, 99, 111,
       130, 162, 167, 197, 200–204, 213,
       217, 218, 222, 223, 231, 232, 234,
       240, 256
temporal   55, 57, 58, 62, 63, 65–67, 69, 72, 78
thionine   126, 148, 152, 153, 158, 167

thrombin   101, 158, 160, 161, 171,
       181–183
toluidine blue (TB)   142, 143, 149, 156,
       158, 164, 168
transmission electron microscopy
       (TEM)   199, 208, 219, 230
trinitrotoluene   90, 99, 113, 154, 168

u
uric acid   33, 49, 55, 57–59, 69, 70, 72,
       75–77, 82, 90, 92, 94, 112, 125–127,
       129, 131, 135, 136, 146, 147,
       149–151, 166, 168, 211, 234, 240,
       250–252, 257
urine   72, 94, 128, 129, 151, 168, 257

v
virus   100
visible light   108–110, 116, 117, 236, 238,
       251, 255
vitamin E   13
voltammetric   20, 21, 23, 32, 76, 81, 83, 96,
       112, 124, 127, 129, 132, 149, 167,
       179, 180, 188, 211, 233, 239, 244,
       245, 249, 250, 256